D0919214

THIRST

THIRST

WATER AND POWER
in the ANCIENT WORLD

STEVEN MITHEN

with SUE MITHEN

HARVARD UNIVERSITY PRESS

Cambridge, Massachusetts

2012

First published in Great Britain in 2012 by
Weidenfeld & Nicolson

First Harvard University Press edition, 2012

Typeset by Input Data Services Ltd, Bridgwater, Somerset

Library of Congress Cataloging-in-Publication Data

Mithen, Steven J.
Thirst : water and power in the ancient world / Steven Mithen.
p.   cm.
Includes bibliographical references and index.
ISBN 978-0-674-06693-9 (cloth : alk. paper)
1. Water supply—History.   2. Water use—History.   3. Water consumption—
History.   4. Civilization, Ancient.   I. Title.
TD345.M644 2012
333.91009—dc23   2012027504

For Heather Mithen, with love and thanks

CONTENTS

LIST OF FIGURES

LIST OF PHOTOGRAPHS

ACKNOWLEDGEMENTS

My interest in water arose in Wadi Fyanan, southern Jordan, during the excavation of a Neolithic settlement with Bill Finlayson, Sam Smith, Emma Jenkins and Darko Maričević. We had many conversations about water: the likelihood of a flash flood in the wadi, whether we had enough water for our flotation tanks, how the Neolithic inhabitants had managed their water supply, how the local Bedouin could survive with so little water to drink. Those conversations sparked a more general interest in water, and so my first thanks must go to Bill, Sam, Emma and Darko.

My interest was further developed in the 'Water, Life and Civilisation' project that I directed at the University of Reading between 2005 and 2010, generously funded by the Leverhulme Trust. That project was concerned with the climate, environment and cultural history of the Jordan Valley, bringing together some of the university's archaeologists, meteorologists, hydrologists, geoscientists and geographers to work on a single integrated study. I found this project was so rewarding that I was compelled to look at the bigger picture, the role of water in the entirety of the ancient world – or at least as much of it as I could cover. With the Leverhulme funds exhausted, I was forced to plough a lone furrow – or perhaps dig my own ditch would be a better metaphor. So I am most grateful to my colleagues on the Water, Life and Civilisation project for drawing me further into the world of water and its relationship to past climate and society. In particular, I would like to thank Emily Black, Stuart Black, David Brayshaw, Bill Finlayson, Rebecca Foote, Brian Hoskins, Emma Jenkins, Stephen Nortcliff, Rob Potter, Claire Rambeau, Bruce Sellwood, Julia Slingo, Sam Smith, Andy Wade and Paul Whitehead. Of these, I would like to especially thank Rob Potter whose studies of the inequalities of water distribution in modern-day Amman were particularly insightful and inspiring.

Another set of 'thank you's must go to those who share the other half of my academic life, that in management at the University of Reading. This book was written while I was a Pro Vice Chancellor (International) at the University. The opportunity to find the mental space to occasionally think about things other than student fees, recruitment statistics, business plans and so forth reflects the enormously supportive environment created by my colleagues on the 'third floor' of Whiteknights House. I would therefore like to thank Tricia Allen, Dianne Berry, Gavin Brooks, Tony Downes, Cindy Isherwood, Sue Jones, Gordon Marshall, Rob Robson, Wanda Tejada and Christine Williams. Of these, Sue Jones, my PA, requires a special 'thank you' for the tasks she undertook to support this book: searching for and downloading academic papers, ordering books, checking manuscripts, arranging my overseas travel and managing my diary to find me precious gaps and crevices in which to read some archaeology as well as the next set of committee papers. Much of the writing for this book was done at airports and on planes while I was travelling on university business which also enabled me to visit a few archaeological sites that would have otherwise been out of reach. So I must also thank those at the university who appointed me as PVC International – although my current writing desk at Ashgabat airport, Turkmenistan, while waiting at three in the morning on 23 April 2012 for a flight home, leaves much to be desired.

Other colleagues in the university have been supportive in the preparation of this book – sometimes unknowingly. Yinshan Tang has been a constant source of advice about China and took me to the Forbidden City in Beijing. Dr Alex Gong kindly arranged and accompanied my visits to Dujiangyan and the Three Gorges Dam. Beth Reed read and provided insightful comments on several draft chapters. She kindly accompanied me on archaeological visits in Sudan and inspired my interest in the ancient kingdoms of Africa – although little of that has managed to enter these pages. Beyond the university I am grateful to those archaeologists and academics who provided me with copies of published and unpublished papers, as well as images of their sites, and answered queries. In particular, I would like to thank Patrice Bonnafoux, Michael Boyd, Brendan Buckley, John Oleson, Yosef Garfinkel, Sumio Fujii and Tony Wilkinson. Also, special thanks to

Arnoldo from Cusco, Peru, without whom Sue and I may never have reached Pisac and Ollantaytambo.

I am grateful to Edwin Hawkes, my agent, for getting this book off the ground and providing invaluable advice throughout its preparation. For that I am also grateful to Lucy Martin for preparing the line drawings and to the editorial staff at Orion for finessing my manuscript. I am grateful to an anonymous reviewer – a specialist on the Americas, I believe – who provided some useful comments in the final stages.

Finally, I wish to thank my family. Sue, my wife, spent part of her 2011/12 sabbatical from school teaching in helping with the research. Without her help I would never have mastered the chronology of the Maya, understood Ctesibius' water clock or found my way around Machu Picchu – both academically and literally. The notes she prepared on water management by the Incas were so detailed and elegantly written that I had little to do other than include them directly as Chapter 11. I thank Sue for being my constant companion when visiting archaeological sites – whether in person or just in my thoughts – and for sharing my intellectual journey into the ancient world of water and power, although I gave her little choice. I am grateful to Hannah and Nick for their interest in my work and undertaking the admittedly not too onerous task of helping Sue explore the aqueducts and Baths of Caracalla in Rome. But it is to Heather, my youngest daughter, to whom I give my greatest thanks. With Hannah and Nick living away, she was left with not just one parent but two who were obsessed about Minoan toilets one day and Chinese canals the next. Fortunately we had Heather to pull the plug and make us laugh at ourselves. Without her I would have drowned and not written a word. So with my sincere thanks, I dedicate this book to Heather.

THIRST

For knowledge of the past and lessons for the future

There is a narrow sidewalk next to US Route 93 at the boundary between Nevada and Arizona. I am not usually inclined to walk alongside four-lane highways, especially those used by 20,000 vehicles daily and in the 43°C heat of a July afternoon. But this took me to the centre of a bridge, 270 metres above the Colorado River and to a perfect view of the Hoover Dam.[1] Constructed almost seventy years ago, this remains a dramatic icon of the human endeavour to control the most precious resource on planet earth: water.

The Hoover Dam was constructed between 1931 and 1936 to protect settlements from floods, to provide irrigation water and to generate hydroelectricity. With its grey concave concrete wall abutting both sides of Boulder Canyon, trapping the deep blue river waters behind to create Lake Mead, it represented a triumph of modern art as much as engineering: an eloquent statement of man's ability to transform the natural world (Photograph 1).

President Franklin Roosevelt had anticipated my emotions precisely in his speech on the morning of 30 September 1935 to dedicate the dam. Standing on his podium in front of 10,000 people he declared that 'This morning I came, I saw and I was conquered, as everyone would be who sees for the first time this great feat of mankind.'[2] The silent gaze of the few others who stood alongside me on that July afternoon in 2011 indicated that they too had been 'conquered'.

The visitor centre at the Hoover Dam explains how it transformed the American West by making possible the growth of Los Angeles, Phoenix, Denver, Salt Lake City and San Diego.[3] By so doing, it also transformed the United States. One might make similar claims for how the Aswan Dam, completed in 1970, has transformed Egypt and how the Three Gorges Dam, completed in 2009, might do the same for China.[4]

Such dams are merely the most striking statements about a fundamental truth of the modern world: it and we are absolutely dependent upon a managed water supply and hence upon hydraulic engineering. This book asks a simple question: was such dependency also the case for the ancient world, for the civilisations of Mesopotamia, Greece and Rome, for the ancient Maya, Incas and all those other long-lost cultures?

This is not a question merely about the presence or otherwise of dams, aqueducts and reservoirs, a work of tick-box archaeology. *Thirst* is about the driving forces behind the rise and fall of civilisations, the quest for power by ancient kings and the long-term relationship between people, culture and nature. *Thirst* is also about whether the past can provide lessons for the present.

Quest for the past

There are few topics more intriguing than the rise and fall of ancient civilisations. Our fascination with the ruins of temples and palaces found in the depths of rainforests, the middle of arid deserts and on isolated mountain peaks continues with the same excitement as when they were first discovered by explorers in the 19th and early 20th centuries. The artistic and scientific achievements of the Ancient Sumerians, Egyptians, Mayas and others continue to astonish, as does the complexity of their ideology and politics. As we extend our knowledge via archaeological excavation and the application of new scientific methods, we discover that there is yet more to learn. But however much data we gather, there will never be definitive accounts of ancient civilisations because what we perceive as important is influenced by our own contemporary concerns.

One such concern is water. The significance of the managed water supply has risen in prominence during recent years. Monumental dams are both tourist attractions and centres of education; at least one million people visit the Hoover Dam each year and are treated to an exhibition of its construction. Moreover, our day-to-day lives have become immersed in public debates about bottled water, building on flood plains, water meters, new reservoirs, leaking pipes and the profits of water companies. Our water awareness has grown because droughts and floods appear to be increasing in their frequency and severity, both in the UK and throughout the world.

Is this a new era for human society? Or has water management, and the impact of its absence or failure, been of prime significance throughout human history? Current archaeological studies of the ancient world emphasise the critical role of food surpluses to support craftsmen, bureaucrats and the political elite, the role of trade to provide staples and exotic goods, kings that claimed divine authority to rule and, of course, a not inconsiderable use of violence.[5] What about the role of water management and hydraulic engineering? If these played a role in the rise and fall of ancient civilisations, should we not take more notice when deciding how to manage our water supply today?

Twenty-first-century water management: failure and success

Despite the Hoover, Aswan and other monumental dams, the world is facing a water crisis. The need for management of the water supply has never been greater and is getting more serious by the day. The statistics of the thirsty, unwashed and inundated are horrendous: one billion people – a seventh of the global population – do not have access to safe drinking water. Two billion people have inadequate sanitation. By 2025 more than half the world's nations will face shortages of freshwater; this is predicted to have risen to 75 per cent of the global population by 2050.[6]

On 12 November 2011 (the day that I am writing), the BBC has announced catastrophic flooding in Thailand. Entire towns have been submerged with at least 100,000 people displaced, a quarter of the rice crop destroyed and $47.5bn of damage to the economy. Thailand's dams, canals and flood control basins were simply inadequate to cope with the intensity of the monsoon. On the previous day, the news reported a drought in the south-east of the United States of America, from Florida to Texas. This was its worst for 60 years, destroying up to a third of the wheat crop and further depleting the aquifers. Fortunately the reservoirs had been filled with rain in 2010, enabling the domestic water supply and some for irrigation to continue – for the time being at least.[7] Throughout July and August of 2011 I watched

3

reports of the drought in the Horn of Africa threatening the lives of 13 million people – there was simply no water to manage. In the autumn of 2010 it had been the turn of Pakistan, with floods arising from intense monsoon affecting 20 million people.[8] Expenditure of $900m by the Federal Flood Commission of Pakistan for flood control projects had been entirely ineffective.[9] All I could do was to make my charitable donations and wonder which type of water crisis is the worst: to have too much or too little?

Doom and gloom are all too easy. There is, of course, another side of the water management story. If one billion people lack access to safe drinking water a remarkable six billion people do have such access. Personally, I have an unlimited supply, as did my parents and their parents, as I expect my children and hopefully their children will enjoy. Our town planners and politicians celebrate their control over water by display: cities throughout the world, even those suffering severe water shortages, have ostentatious fountains that spill and waste water on the ground.

Our political leaders build far more than fountains. The monumental dams are matched by colossal canals: in Jordan a 200-kilometre canal is planned to transport water from the Red Sea to the Dead Sea to halt the dwindling of the latter's levels;[10] in China the South–North Water Transport Project plans to divert more than 40 billion cubic metres of water;[11] the Libyan desert has a partially completed 3,500-kilometre man-made river of water pipes large enough to drive trucks through, originally planned by Colonel Gaddafi to bring fresh water from boreholes in the desert to the Mediterranean coast.[12]

Unanswerable questions

To what extent will such projects solve, or at least ameliorate, an existing or anticipated water crisis? To what extent are such monumental projects driven by the most rational evaluation of how to manage the water supply for the common good? Is it even possible to make such evaluations when the future is unknown?

Even with meticulous historical studies such as that about the Hoover Dam,[13] we are unable to answer such questions satisfactorily. The success or otherwise of projects that took decades to complete must be measured in centuries or millennia and hence insufficient time has elapsed to pass judgement. Although the Hoover Dam had been built in 1936, its ability to control flooding was first put to a serious test in 1983, when exceptional rainfall made the Colorado a rampaging river. The dam was found wanting: hundreds of houses, farms and tourist resorts were destroyed or swept away. A second test, one of drought, began in 1999 and is on-going. By 2002 the river had dropped to its lowest level since 1906.[14] Despite the Hoover Dam and a multitude of others built in its wake, the south-west United States suffers from severe water stress, the optimism of unlimited water having generated unsustainable urban growth. The approach to water management has now shifted away from massive public works to conservation and recycling.

With regard to the motivations for building monumental dams and canals, we are perhaps too embedded within our present day to effectively evaluate the case. Were these undertaken for the common good or for the self-aggrandisement of those with power or who desired to attain it? For this we also need the hindsight of history. That of the Hoover Dam is as much about the political infighting and commercial competition, the quest to build personal reputations and gain financial success, as it is about the mechanics of tunnelling through rock to create spillways and pouring concrete to create the dam. But our understanding of the 1930s is itself undergoing constant change and the Hoover Dam's history will need rewriting. Herbert Hoover's massive public spending on the dam, $49m, was a response to the Great Depression, a means to create jobs and attempt to kick-start the economy. As the world today teeters on another depression and water stress in the south-east United States becomes more severe, our evaluation of Herbert Hoover is bound to change as these current conditions unfold.

Hydraulic engineering and human history

There are three potential lessons from the past. The first is simply for us to learn about the role that hydraulic engineering has played in the course of human history, enabling us to put the present day into a long-term perspective. We are familiar with the supposed turning points of history, ranging from the so-called Neolithic revolution of 10,000 years ago that gave rise to the first settled farming communities, through the industrial revolution of the 18th and 19th centuries, to the digital revolution of the 21st century. Was there ever a water revolution? What role did hydraulic engineering play in the emergence of the ancient civilisations and in the manner in which they functioned?

We tend to see ambitious water management projects today as something new. Whether we are proud, dismayed or horrified by them, we are staggered by the amount of money, time and materials they involve. Consider China's Three Gorges Dam. It was first mooted in 1919, but construction work began in 1992 and it took 17 years for the dam to become operational; the project cost the equivalent of £17.3bn and required 27 billion tons of concrete.[15] But is such scale of work really new? Did our forebears within the ancient world engage in projects of an equivalent scale? Maybe they were even more ambitious. Might even Gaddafi's desire for a man-made river in the Libyan Desert – started in 1983, with an on-going cost of at least US$27bn in 2007,[16] and an uncertain future following Gaddafi's demise – appear to be nothing more than a trivial undertaking by a modest man?

While long-term history covering centuries and millennia is essential for understanding both the recent past and the present, in other respects it is limited. When exploring ancient civilisations we rarely gain insights into individual lives and experience. When we do, they tend to concern the kings and other members of the elite, those identified by rare inscriptions. We lack accounts of the workers who built ancient dams and aqueducts, those who drew upon their water for their day-to-day lives and those who suffered most from floods and droughts.

Again we might look to the building of the Hoover Dam, this time to appreciate the human emotion and drama that hydraulic engineering entails. As well as the machinations of the politicians and deliberations of the engineers, we know about the day-to-day conditions for the thousands of workers who, in the middle of the depression, flocked to the area to work on the dam: the heat they suffered during the summer of 1931 when the *average* temperature was 48.8°C; the 112 deaths from drowning, rockslides, truck collisions, heat exhaustion and carbon monoxide poisoning in the tunnels; the workers' poverty, protests and ultimate pride in building a monumental dam. Within *Thirst* we will encounter constructions within the ancient world which were at least the equivalent of the Hoover Dam and assess their role in long-term history. But we will lack any knowledge of those who toiled to build them.

Water and power

A second lesson from the past is that about the relationship between water and power. We are increasingly aware of how the control of the water supply is a means of securing and maintaining power in the modern world. Such power might be political, and hence the investment by leaders, elected and otherwise, in the construction of dams and canals either from a genuine concern to provide a water supply or as a demonstration of their ability to do so. Power from the control of water can also be commercial in nature, and hence the intense competition today between the multi-national water companies for our cash.

We have become familiar with the term 'water wars' to characterise a wide range of conflicts over access to water, whether between nations or between multi-national companies and local populations whose water sources they exploit. During the 20th and 21st centuries, nations have rarely, if ever, gone to war over water.[17] But disputes about access to aquifers and the impact of dams have been major sources of tension between nation states, notably in the Middle East – between Israel, Palestine, Syria and Jordan – and in the Nile Basin – between Egypt, Ethiopia and Sudan. The resolution of such disputes and water shortages has played a fundamental role in the content of treaties and international trade agreements and hence in mediating power relations between nation states.[18]

What about in the past? Has controlling water always been a pathway to power, or perhaps even the predominant route? If so, what form has this taken? Was water the cause of conflict in the past between competing city-states or was it a medium for cooperation and mutual growth?

Archaeologists have previously addressed such questions. Karl Wittfogel's 1957 volume entitled *Oriental Despotism: A Comparative Study of Total Power* argued that ancient civilisations were dependent upon large-scale irrigation works.[19] He proposed the 'hydraulic hypothesis' that such works required forced labour and a large bureaucracy, both of which were features of 'despotic rule'. More recently, Vernon Scarborough, Professor of Anthropology at the University of

Cincinnati, entitled his 2003 cross-cultural analysis of ancient water system *The Flow of Power*.[20] The relationship between water management and power requires further exploration by drawing on newly available archaeological evidence and reflecting on how this might influence our understanding of the present day.

How can we survive?

The global water crisis appears to get worse from year to year. Whether this is really the case or a misapprehension, arising from near-instantaneous global media coverage of floods and droughts conspiring with short memories, is unclear. But with the world's population having breached seven billion, continuing mega-urbanisation, climate change altering the global distribution of rainfall, and the increasing frequency of extreme events, the future looks bleak. Is it inevitable that ever greater numbers of people will be affected by drought and flood? Will the number of people living in unsanitary conditions continue to increase? Will environmental degradation continue unabated? Will we reach a threshold of water demand that even our most sophisticated systems of hydraulic engineering cannot satisfy?

While the past cannot answer such questions it can provide a long-term perspective on the success or otherwise of hydraulic engineering within the ancient world. Which, if any, of the ancient systems proved most successful? Is there anything we should copy or avoid? Why did the ancient hydraulic systems ultimately fail – if indeed they did? Will a greater understanding of the past provide us with a sense of optimism for the future, perhaps by enhancing our faith in finding technological and engineering solutions to the water crisis? Or will the past demonstrate that these will never suffice because, as appears to be the case with the Hoover Dam, they cannot keep pace with ever-changing environmental, demographic, social and political conditions?

A journey through the ancient world

Rather than undertaking a single in-depth study of one selected ancient civilisation, I have chosen to make a global survey of the ancient world, selecting nine case studies ranging from the Sumerian

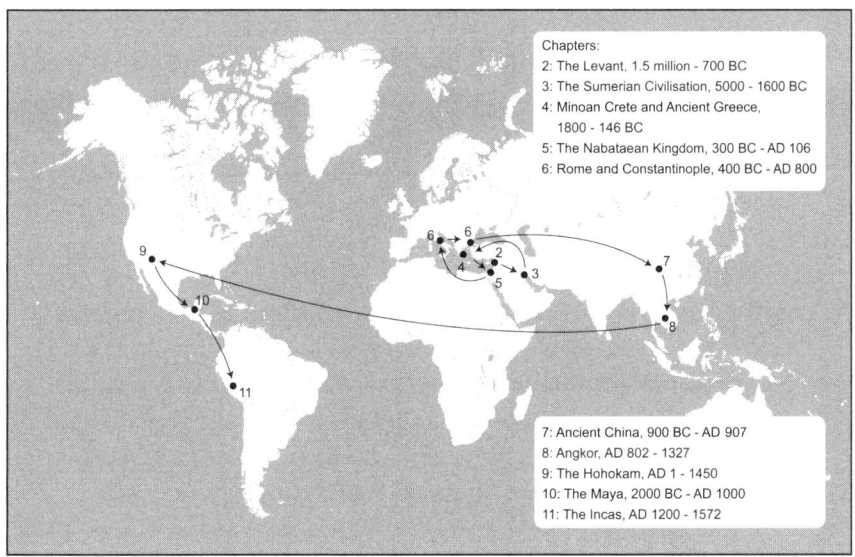

Chapters:
2: The Levant, 1.5 million - 700 BC
3: The Sumerian Civilisation, 5000 - 1600 BC
4: Minoan Crete and Ancient Greece,
   1800 - 146 BC
5: The Nabataean Kingdom, 300 BC - AD 106
6: Rome and Constantinople, 400 BC - AD 800

7: Ancient China, 900 BC - AD 907
8: Angkor, AD 802 - 1327
9: The Hohokam, AD 1 - 1450
10: The Maya, 2000 BC - AD 1000
11: The Incas, AD 1200 - 1572

1.1 Societies of the ancient world visited within this book

civilisation of 3000 BC in Mesopotamia to the Inca Empire centred in Peru and which fell to the Spanish less than 500 years ago (Figure 1.1).

This book has been, to use a cliché, a journey for me both intellectually and quite literally. It might seem odd not to have included Ancient Egypt because of its dependence on the waters of the Nile. But that, I supposed, was already widely known and further information is readily accessible in the swathe of books about Ancient Egypt.[21] Moreover, the natural hydrology of the River Nile made water management in Ancient Egypt less of a challenge than elsewhere in the ancient world, such as in Mesopotamia, where people had the greater task of manipulating the natural world for their own needs.[22] Rather than covering Egypt, I have chosen to explore less widely known societies such as the Khmer civilisation in Cambodia and the Hohokam in the south-west United States, both of which had a remarkable history of water management.

A truly comprehensive study would have included not only Ancient Egypt, but also the hydraulic-enginering achievements of the Assyrians, the Indus civilisation, the Aztecs, the Yoruba–Benin culture of West Africa, the Sinhua Civilisation (ancient Ceylon), nineteenth-century Bali and many more cultures of the ancient world.[23] It would also have covered all techniques of hydraulic engineering: I am conscious that several are missing from my case studies, notably the qanat system of subterranean conduits that were widespread in the Near East, and which spread to Europe and the New World. Also the noria, a chain of pots attached to a large wheel turned by tethered oxen as used to draw water from the Nile in the sixth century BC.[24] A comprehensive study would have required a multi-volume textbook and any insights into the past would have been crushed by its weight. While my intention has been relatively modest, I have nevertheless selected my nine case studies to provide a representative sample of the types of water management and hydraulic-engineering systems that were undertaken in the ancient world. They include lowland and highland civilisations, those in deserts and rainforests, with Stone Age and Iron Age technology, and with and without writing.

I start not with a civilisation but with the earliest prehistoric communities in the Near East, specifically in the Levant – the area of modern-day Syria, Lebanon, Israel, Palestine and Jordan. This is where the earliest farming societies arose sometime after 9500 BC and then the first urban communities around 3000 BC. In that chapter we will track the earliest forms of hydraulic engineering and their relationships, if any, to the origins of farming and urbanism that laid the foundations for civilisation throughout the Old World.

The earliest ancient civilisation arose in Mesopotamia, present-day Iraq, which forms the subject of my third chapter. From there my global tour will take you to Ancient Greece (Chapter 4), the Nabataean Kingdom of southern Jordan and Petra, its capital (Chapter 5), and then the Roman World, focusing on its two capitals, Rome and Constantinople (Chapter 6). We then jump to Asia with case studies of Ancient China (Chapter 7) and the first of two rainforest civilisations, Ancient Angkor, located in modern Cambodia (Chapter 8). We next go west, with three studies in the Americas – the Hohokam of the south-west United States of America (Chapter 9), the Maya of Mesoamerica, our second rainforest civilisation (Chapter 10), and finally the Inca Empire, spending some time at Machu Picchu, the iconic site for all ancient civilisations (Chapter 11). In the last chapter we will return to the 21st century and see what lessons we may have learned, both about the past and for the future of our planet.

So let me start by taking you with me from a busy highway in the United States to one in the Middle East, to an inconspicuous lay-by that I like to think of as the centre of the world.

2

THE WATER REVOLUTION

The origins of water management in the Levant,
1.5 million years ago to 700 BC

The sparkling blue waters of the Dead Sea come suddenly into view as one drives south and downhill from Amman, heading towards the Red Sea port of Aqaba. As the expanse of water appears, I often pull into a lay-by, step out into stifling heat, pause, and reflect about the very special place on planet earth at which I have arrived.

A few kilometres further south, one reaches the earth's lowest point, 423 metres (1,388 feet) below sea level in the middle of the Dead Sea Rift, one of the geological wonders of the world. The road from Amman will pass below steep towering cliffs of water-lain sediments which record the dramatic rise and fall of the Dead Sea and its fore-runner, Lake Lisan, as the earth's climate swung in and out of ice ages during the last million years.

To the west of my parking place, across the Jordan Valley and into the disputed West Bank, is an archaeological wonder: the ruins of ancient Jericho. Claimed by some to be where farming and town life were invented 10,000 years ago, its surrounding modern town is visible from the lay-by on a clear day. A mere 20 kilometres beyond sits Jerusalem, an architectural, religious and political wonder: holy city for Judaism, Christianity and Islam, whose unresolved status is central to the persisting Israeli–Palestinian conflict.

Jerusalem cannot be seen from that lay-by, but further south along the Dead Sea road there are signs eastwards to Kerak, where a monumental crusader castle reminds one of the great battles to control that city. One also passes close to the supposed baptism site of Jesus Christ and the Bronze Age sites of Bab edh-Dhra and Numeira, claimed by some to be the biblical settlements of Sodom and Gomorrah.

I could list many sites more in the vicinity of that lay-by: sites that help document the Roman and Byzantine empires, the Islamic caliphates, Ottoman rule, British colonialism, and the establishment of the present-day Hashemite Kingdom of Jordan.

My viewing place over the Dead Sea is often filthy with litter, smelly and noisy from the enormous trucks that thunder down the hill from Amman. But such potential distractions go unnoticed as I become overwhelmed by a profound sense of being at the centre of the world, at the crux of both geological and human history. In that lay-by I stand in the middle of the Levant, the tumultuous meeting place for Africa, Asia and Europe.

Sometimes known as the western arm of the Fertile Crescent, the Levant is the region covered today by southern Turkey, Syria, Israel, Lebanon, Palestinian Territories, Jordan and Cyprus. The Fertile Crescent is where both farming and civilisation first arose.

If the management of water was as critical to the ancient world as it is to the 21st century, where else would one possibly go to search for its origin, to find the sinking of the first well and building of the first dam? Might water management have begun within the earliest Neolithic farming communities such as Jericho at 8000 BC? Maybe it had appeared long before, developed by hunter-gatherers 20,000 years ago or even earlier, a necessary precursor to farming? Or perhaps water management was a relatively recent invention, the catalyst that transformed Neolithic villages into Bronze Age towns at 3500 BC? Alternatively, it may have been a consequence of urbanism itself, enabling towns to become cities and communities to become civilisations.

In this chapter I will take you on an archaeological trail to find the answer (Figure 2.1). Or, at least, an answer for the Levant.

Even though the Levant is where the earliest farming and urban communities in the world appeared, there is no necessity for it to also have provided the earliest water management. The origin and history of this needs to be individually traced for each region of the world and may show quite different relationships with economic, social and cultural developments to those we will discover in the Levant. In the New World, for instance, water management is known to date as far back as 10,000 years ago, following the discovery of a well at the site of San Marcos Necoxtla, Mexico,[1] and I would not be surprised to find similar dates in, say, China, India and Australia. But this book can provide only one account and my aim is to resolve when and where water management began in the Levant prior to exploring its role in the early civilisations of the Old World.

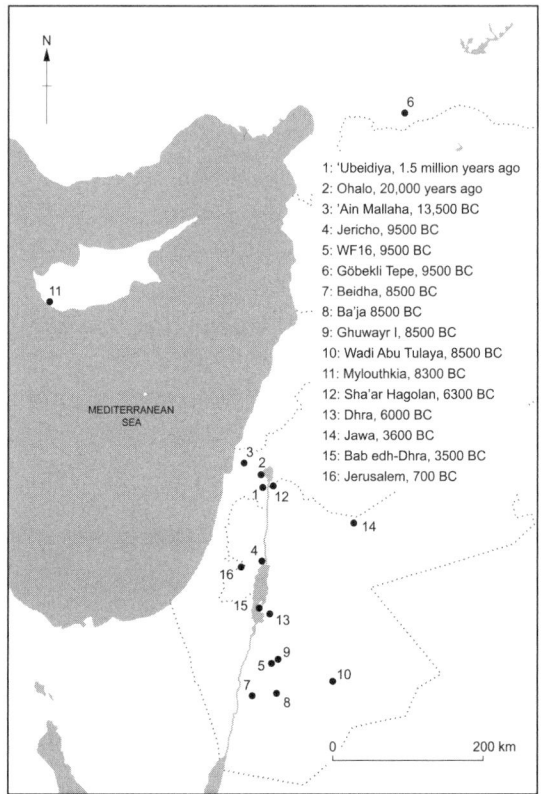

2.1 *Map of the Levant showing archaeological sites and localities referred to in the text*

The archaeological evidence from the Levant is often fragmentary and difficult to interpret. We are, after all, dealing with prehistory, the time before writing began, and hence lack any documentary evidence to help with our interpretations. A great deal still remains to be discovered, the existing archaeological evidence providing a still incomplete story of how water management evolved. It is just possible, however, that the evidence we do possess will illuminate a few key moments in Levantine history: the sinking of the first well, the building of the first dam, the construction of the first cistern and aqueduct. These moments are at least equivalent in their importance to the emergence of civilisation as were the planting of the first seeds and the smelting of the first copper.

The origin of farming was once described as the Neolithic Revolution and that of towns as the Urban Revolution. I'm writing about a third revolution of equal significance: the Water Revolution.

A beginning at 'Ubeidiya

Where does one start when writing the history of water management? People, *Homo sapiens* and our past ancestors and relatives, must always have been managing water in some manner as far back as six million years, the date at which we shared a common ancestor with the chimpanzee. Chimpanzees are known to crumple up leaves to use as sponges for transporting water from hollow tree trunks to their mouths;[2] we must assume that our ancestors were not only doing the same but carrying water short distances, whether cupped in their hands, within folded leaves or in skin containers. While we lack any direct archaeological evidence for such water carrying, it is implied by the remnants of camping or activity sites found in locations distant from water sources. Such evidence is itself problematic, however, because sufficiently detailed environmental reconstructions to pinpoint the specific location of a river course or the nearest pool of standing water are difficult for the earliest periods of prehistory.

Early prehistoric archaeological sites – those of the Lower and Middle Palaeolithic – are found throughout the Levant from at least 1.5 million years ago. It was around that date, perhaps as much as 0.5 million years earlier, that *Homo ergaster* arrived in the Levant, dispersing via the Nile Valley from East Africa, where this species had evolved. Further types of humans followed: *Homo heidelbergensis* after 0.5 million years ago and *Homo neanderthalensis* after 250,000 years ago.[3] All were responding to increased levels of rainfall caused by a climate change that created rich riverine woodland for hunting and gathering and plentiful sources of water for drinking. None of the arrivals necessarily remained for prolonged periods because the return of arid conditions drove them away or into localised extinction. The next up-turn in climate brought new arrivals from Africa, perhaps the same or perhaps a new species of human.

There are thousands of archaeological sites throughout the Levant testifying to the presence of *Homo ergaster*, *heidelbergensis* and *neanderthalensis*. The vast majority are no more than scatters of stone artefacts: sharp nodules and flakes of stone used for cutting into animal carcasses, chopping plants and digging for roots. Any remnants of tools made from wood or bone have long since rotted away, this also being the fate of their own bones and those of any animals they hunted or scavenged. Fortunately, there are rare exceptions arising from freak conditions of geology, aridity or waterlogging which have led to better preservation.

The most notable is the site of 'Ubeidiya in Israel, located a few kilometres south of Lake Tiberias, otherwise known as the Sea of Galilee.[4] This site was repeatedly visited by early humans between 1.5 and 0.85 million years ago. Along with many thousands of stone artefacts, the bones of deer and an extinct form of hippopotamus are preserved on what are believed to be actual living floors – dense concentrations of tools and bone fragments where butchering and eating has taken place. I found it a most remarkable site to visit during the course of the 1999 excavations by Professor John Shea of New York University. He had an extraordinary excavation challenge because geological activity had tilted those ancient living floors to 70 degrees from the horizontal. So rather than digging down into the ground, he was digging sideways into a near vertical rock-face and mapping its mosaic of archaeological finds.[5]

Like many, probably the majority, of early stone-age sites, 'Ubeidiya appears to have been located immediately adjacent to an ancient lake, the predecessor of Lake Tiberias. This in itself suggests that it would have been a temporary location: with prowling hyenas and lions, lakeside locations are hardly suitable for our early ancestors to have resided overnight, especially without the control of fire. Neither at 'Ubeidiya nor at any other of these early sites is there any evidence that lake or river waters were being managed in any manner, such as being channelled towards a campsite or dammed to create a standing pool for drinking or washing.

The dilemma we face is that such work would not necessarily leave any archaeological trace. I have been in the desert with Bedouin and watched them construct small dams from pebbles and sand to store water following a rain storm, quite content for their work to be soon washed away, another being built when required. Early humans might have done likewise. The sophistication of their tool making certainly indicates they had the capacity to do so. But we have no evidence, and their transient hunter-gatherer lifestyle – always being on the move – makes me doubt whether they had the need.

The first modern humans and a campsite at Ohalo

For the next site on our origin trail for water management we need to travel no more than 20 kilometres northwards from 'Ubeidiya to the western lake shore of Lake Tiberias, where the site of Ohalo is located. Although short in distance, this is a vast journey in time, from 1.5 million to a mere 20,000 years ago. Ohalo is the best-preserved site in the Levant for hunter-gatherers at the peak of the last ice age, when much of the northern hemisphere was covered in glaciers (Photograph 2).[6] This made the Levant relatively cold and dry when compared to earlier and later periods, a landscape of extensive grasslands and open woodland. By 20,000 years ago, the Neanderthals and their predecessors were all extinct: Ohalo, and much of the rest of the world, was occupied by *Homo sapiens*.

Modern humans, *Homo sapiens*, arrived in the Levant on a permanent basis around 45,000 years ago although some early forms may have had a temporary presence as early as 120,000 years ago. The later permanent arrivals had also dispersed from Africa, but appear to have taken a round-about route in getting to the Levant. They initially dispersed eastwards from the Horn of Africa into the Arabian Peninsula and only into the Levant via Central Asia.[7]

Modern humans, with the added advantage of a symbolic capacity and spoken language, soon pushed any remaining Levantine Neanderthals into extinction.[8] Their archaeological sites are more abundant but rarely better preserved, with a continuing absence of anything other than stone artefacts except in a few instances. New types of tools were made, suggesting both more effective hunting technology and the encoding of information about individual and social identity into the design of arrowheads. Otherwise, the *Homo sapiens'* lifestyle had great similarities with that of the early humans who had butchered hippopotamus bones at 'Ubeidiya: hunters and gatherers who were always on the move, living in groups of between 25 and 50 persons which were allied by kinship ties, with periodic large-scale aggregations during which 'rites of passage', marriage ceremonies, and the exchange of information and materials would occur.

Ohalo is a quite exceptional site because a great number of plant remains and bones have been preserved, along with features including hearths, hut floors and a burial. For this we must be grateful to both climate change and water. The 19,400-year-old campsite was on the shore of Lake Tiberias when its waters were at a particularly low level because of the ice age conditions. The site appears to have been abandoned after a fire had burned down its brushwood huts. During the next few hundred, perhaps thousands, of years the climate ameliorated and rainfall increased. As a consequence the lake waters rose and gently flooded the campsite debris and charred remains of the huts. The water now protected the site from further decay.

A drought in 1989 caused a nine-metre drop in the lake waters, exposing the archaeological site. Professor Dani Nadel of Haifa University began a meticulous excavation that soon astonished archaeologists throughout the world by showing the diversity of plants and animals that had been exploited in the Levant close to the height of the last ice age. The remnants of six dwellings were excavated, with dense accumulations of seeds from wild grasses, fruits and nuts, all coming from around 100 different species, concentrated around wooden grinding slabs. Gazelle had been hunted and butchered at the site, while fish had been caught in the lake, probably with nets made from twisted plant fibres. After two years of excavation, the drought came to an end and the site was again inundated by the lake water. Dani Nadel had to wait until 1999 for another drought year to cause the lake waters to recede and enable his excavation to be resumed. Fortunately, I was travelling in Israel at that time and able to visit this remarkably well-preserved site to gain my best ever glimpse into the Stone Age and ice age past.

Ohalo provides us with the best opportunity imaginable to find evidence for water management by ice age hunter-gatherers. Ironically, although the presence of water so nearby allowed such outstanding preservation, it also meant that there was no necessity for the original inhabitants to undertake any water management at Ohalo, although they might have chosen to do so. One might imagine modifications being made to the lake shore, perhaps digging channels to bring water to the immediate vicinity of the huts or to ease access for fishing.

No such evidence exists at Ohalo for any form of water management, nor at any other hunter-gatherer site in the Levant. The most likely explanation is that there was simply no need: hunter-gatherer groups were relatively small and mobile; they were always able to locate themselves in the vicinity of a water source. If that dried up, they could easily relocate.

'Settling down'

The hunter-gatherers who lived in the Levant 5,000 years after those at Ohalo may have been an exception to the 'small and mobile' groups that were still the norm elsewhere. Between 14,500 and 13,000 years ago Levantine hunter-gatherers may have 'settled down' and begun living in permanent settlements with up to 100 inhabitants. The opportunity to do so arose because of significantly warmer and wetter conditions during this period, technically known as the 'late glacial inter-stadial'. This has been identified in the various measures of global temperature that can be extracted from the Greenland ice cores and from within sedimentary evidence from the Levantine region itself.[9] Its impact was to make plant foods and wild game especially abundant, alleviating the need to continually find new locations with fresh supplies.

Some of the archaeological sites of this period have stone-walled dwellings of a much larger size than those at Ohalo. They also have cemeteries, which archaeologists often take as a sign that territories are being marked by the presence of ancestors, suggesting that hunter-gatherers had become less mobile. Large stone mortars and grinding stones testify to the intensive gathering and processing of wild plants. A new range of art objects and body decorations suggest changes in social organisation with the emergence of either wealthy and/or high-status individuals – unlikely in constantly moving hunter-gatherers who cannot accumulate possessions.

Archaeological sites with these attributes are referred to as belonging to the Early Natufian culture. The economy remained entirely dependent on wild plants and animals, and the extent to which these communities were sedentary remains a matter of substantial debate among archaeologists. But what no one disputes is that the Early Natufian hunter-gatherers remained dependent upon an entirely natural water supply. One can find no trace of wells, cisterns, terrace walls or channels to divert the flow of rivers or capture rainfall among the new range of architecture and artefacts.[10]

The Early Natufian site of 'Ain Mallaha dating to 14,500 years ago is a typical example.[11] When visiting this site today, regrettably close to a sewage works, one can see the remains of many stone-built dwellings and massive stone mortars. These provide the impression of a thriving settlement, indicating a level of investment in architecture that hardly equates to a short-term campsite. Several beautifully carved animal figurines have come from 'Ain Mallaha and a few of its inhabitants were buried with impressive shell-bead head-dresses. But there is no evidence for water management: no wells, cisterns, dams, aqueducts. Again there was no need: 'Ain Mallaha was dependent upon a local spring, after which the site is named – ''Ain' being Arabic for 'spring'. This spring not only provided water to drink but also ponds with abundant fish that attracted migrating birds – three valuable resources in one.

This more complex and possibly sedentary hunter-gatherer lifestyle of the Early Natufian did not last long, a mere two millennia at most: by 13,000 years ago colder and drier conditions had returned. This is also recorded in the Greenland ice cores and is thought to have been a consequence of the great North American ice sheets collapsing into the Atlantic Ocean. The return to ice age conditions was felt more strongly in northern latitudes than in the Levant.[12] But even here the colder and drier conditions reduced the abundance of wild plant foods and game, forcing hunter-gatherer groups to become smaller and more mobile once again. 'Ain Mallaha was not entirely abandoned, but became a seasonally visited hunter-gatherer campsite rather than a (semi-?) permanent village. The dead were brought from afar to be interred within its cemetery.

At Jericho – without pottery

After a millennium and a half of cold and dry conditions, the climate turned again, this time for good – at least so far. At around 11,500 years ago there was a dramatic increase in global warming that brought the ice age to a final end. So began the post-glacial Holocene period, with its relatively warm, wet and stable climate that we currently enjoy and which was critical for the emergence of the first farming communities.

The Levantine hunter-gatherer response was similar to that at 14,500 years ago. Larger and more permanent settlements reappear, together with an intensive exploitation of the newly abundant plant foods, especially wild cereals. They included the ancestors of our modern-day barley and wheat.

Kathleen Kenyon (1906–78), one of the greatest British archaeologists of the 20th century, discovered the first of these sites in the 1950s when excavating Tell es-Sultan at Jericho.[13] The Tell is a huge mound of collapsed rectangular mud-brick buildings that had principally accumulated during the Bronze and Iron Age periods (Photograph 3). Kenyon sunk deep soundings to explore what was at its base and discovered a number of circular mud-brick structures, quite different to any buildings of the later periods in the mound. On the western side, they had been enclosed by a massive stone wall and a large circular tower – a scale of architecture which was then, and remains, completely unprecedented prior to the Bronze Age.

When Kenyon was excavating, the Neolithic had already been identified in Europe as the period of the first farming settlements. Kenyon found many similarities between the structures and artefacts at the base of Tell es-Sultan and those from the earliest European Neolithic, and she assumed that she was dealing with a farming economy. There was one key difference, however: an absence of pottery. During the 1930s, Gordon Childe (1892–1957), the greatest archaeologist of the pre-war period and a major influence on Kenyon's thinking, had made pottery one of the key defining features of the European Neolithic.[14] But there was no trace of pottery in the basal levels of Tell es-Sultan – it had not yet been invented. As a consequence, Kenyon described her discovery as belonging to the 'Pre-Pottery Neolithic'. That term has not only stuck, but has been divided into two: the Pre-Pottery Neolithic A and B, referred to in shorthand as PPNA and PPNB.[15]

The PPNA is now known to have lasted for just over 1,000 years, from 11,500 to 10,200 years ago, and is the key period of transition from hunter-gatherer to farming lifestyles. During this period hunter-gatherers began to cultivate wild cereals and legumes – watering, weeding, transplanting seedlings and removing pests. This gradually led to the evolution of domestic varieties that were as dependent upon those cultivating them as the 'cultivators' became dependent on the much larger yields that were now being produced. As such, the invention of farming was as much by accident as by design.[16] Once plants had become domesticated, and people had become tied down to living in permanent villages to tend their surrounding fields, the domestication of sheep, goats and eventually cattle soon followed.

The ensuing PPNB period, lasting from about 10,200 to 8,300 years ago, is one of established farming villages with both domesticated crops and animals. It is marked by dramatic changes in architecture, technology and the size of settlements, some now reaching populations of 2,000 or more. This is followed by the 'Pottery Neolithic' period, which is partly self-explanatory because pottery is invented, but it also marks a significant change to the farming economy with the probable appearance of transhumant herding of sheep and goats. Then, at 5,600 years ago, came the Bronze Age with the first urban communities.

Here we need to make a switch from referring to dates as 'years ago' to years BC (Before Christ). Archaeologists who study the Palaeolithic period prefer to use 'years ago', but Neolithic and Bronze Age specialists prefer to use either BC or BCE (Before the Common Era). The difference is about 2,000 years and this level of approximation is sufficient for our needs in this book.[17] So according to this dating scheme, the PPNA begins at 9500 BC rather than 11,500 years ago, the PPNB lasts between 8200 and 6300 BC, and the Bronze Age starts at 3600 BC.

Was water management necessary for the origin of farming?

With that awkward switch of dating terminology out of the way, we can return to our search for the origins of water management. Is there any evidence for this at Jericho or elsewhere within the PPNA? We might expect there to be some: crops need to be watered, whether they are still wild and being cultivated or whether they are domesticated and being farmed. It is not unreasonable to suppose that the earliest PPNA Neolithic farmers might have needed to excavate wells to access additional water sources, and to build aqueducts to transport water to their fields, irrigation channels to take water to their plants, and cisterns to store water in case of drought at critical times of germination and grain maturation.

One can 'suppose', but there is a complete lack of evidence for water management of any type. While Kenyon found evidence for monumental architecture in the form of walls and towers and complex mortuary rituals involving the removal of skulls from internments, there was nothing to suggest an approach to water that differed from that adopted by the people of 'Ain Mallaha at 14,500 years ago (12,500 BC) and Ohalo at 19,400 years ago – and even 'Ubeidiya at one million years ago.

Jericho also had a spring, 'Ain es-Sultan. This appears to have met all of the inhabitants' needs, while crops were grown on alluvial soils which were well-watered by local streams.[18] So either 'location, location, location' had once again alleviated the need for water management, or the absence of water management practices had constrained the whereabouts of the Jericho settlement, requiring it to be placed adjacent to a spring.

One must, as always in archaeology, be cautious. By having to work at the bottom of deep, narrow trenches cutting through a massive overburden of collapsed Bronze and Iron Age mud-brick buildings, Kenyon was only able to expose 10 per cent or less of the PPNA settlement. The area remaining covered by the tell might have the evidence we are missing of wells, aqueducts and cisterns.

It might, but I doubt it. Since Kenyon excavated at Jericho, numerous PPNA settlements have been excavated without providing any evidence of water management. Not far from Jericho, on the West Bank, the sites of Netiv Hagdud and Gilgal received extensive excavation during the 1970s and 1980s.[19] Neither had a substantial overburden of later occupation debris and hence large areas of the settlements could be exposed, especially at Netiv Hagdud. Just like at Jericho, the PPNA inhabitants had relied on an unadulterated, natural water supply. Netiv Hagdud used water from the spring of 'Ain Duyuk and exploited wetlands of the nearby delta of the River Jordan as it emptied into the Dead Sea.[20]

More recently, three PPNA sites have been excavated further south and on the east side of the Jordan Valley: Zahrat edh-Dhra', Dhra' and my own rather oddly named 'WF16' (the 16th site found during an archaeological survey of Wadi Faynan).[21] These are especially

interesting because, unlike the West Bank sites, these were not positioned on the well-watered alluvial soils of the Jordan Valley itself, but
on terraces in side wadis that led to the Jordanian plateau. All three
have complex architecture, mortars and grinding stones suggestive
of permanent communities engaged in plant cultivation if not farming itself. But, once again, they lack any signs of water management.
Despite this, a brief visit to WF16 is called for to explore how the
PPNA met their water needs in what is now a severely arid landscape.

The dynamics and symbolism of water flow in Wadi Faynan

My colleague Bill Finlayson and I discovered WF16 in 1996 as a scatter of stone artefacts on a small knoll just above the floor of Wadi
Faynan. Our excavations between 1997 and 2003, and then between
2008 and 2010, have exposed an extensive area of densely clustered
mud-walled semi-subterranean dwellings, workshops and storage
areas (Photograph 4). These are adjacent to a massive walled structure
that appears to have been for communal activity, possibly of a ritual
nature involving the grinding of seeds in light of mortars embedded
into its floor.

Wadi Faynan is extremely arid. The only rain falls in the winter
months and rarely exceeds 10 centimetres per annum, the threshold
for farming without irrigation. The landscape around WF16 is almost
entirely barren of vegetation; the wadi floor is totally dry except for
short periods after the winter rains that often cause violent flash
floods.

Close to WF16 there is a substantial Roman settlement, one known
as Phaino in the classical texts, along with its cemetery, where slaves
who had been forced to mine copper ore in the surrounding mountains were buried. Phaino had been entirely reliant on sophisticated
water management involving an aqueduct bringing water from a
spring several kilometres away, a large reservoir and a complex of
field walls to guide run-off water on to the fields and protect them
from flash floods. The present-day Bedouin of Wadi Faynan are also
dependent upon that spring, now using black plastic pipes rather than
a stone-built channel to bring water to their tomato and melon fields.

While the PPNA population of WF16 may have been considerably lower than that of the Roman settlement and of today's, their

water needs must have been substantial. They not only required water for drinking and for watering wild/domesticated crops, but also for building. To build the dense agglomeration of mud-walled and mud-floored buildings would have required many thousands of litres of water. We reconstructed just one small building and found that we needed 3,000 litres to make sufficient mud for its walls and floors. But there is no evidence for a PPNA aqueduct from the spring or any other means of controlling water.

The extent of rainfall in the PPNA does not appear to have been significantly different from that of today; it also appears to have been similarly distributed throughout the year with long dry summers and winter rain coming in intermittent storms.[22] Nevertheless, the dynamics of the PPNA water flow in the wadi may have been quite different, alleviating the need to interfere in its distribution in the manner of the Romans and present-day Bedouin. Evidence from our excavation suggests that the surrounding landscape had carried a substantially greater amount of vegetation than it does today – grasses, herbs and trees. These were lost through a combination of overgrazing by domesticated goats from the PPNB onwards and by deliberate vegetation clearance from the Bronze Age onwards to provide fuel for copper smelting.

Without the vegetation cover, winter rainfall flows readily across the sun-baked ground surface to accumulate in the wadi and create flash floods. When vegetation is present, however, the rain can infiltrate the ground and resupply the ground water reserves that in turn feed the local springs. As such, there would have been a constantly flowing stream along the wadi during the PPNA rather than destructive flash floods, providing a more accessible supply of water than that enjoyed by the Roman and modern populations of Wadi Faynan.

Although 'water stress' may have been absent in the PPNA, I nevertheless sense that water may have been on the minds of those who lived within the mud-walled dwellings and especially those who undertook ritual performances within their large communal structure. We have excavated numerous 'art' objects from WF16. While these include a few human and animal figurines, the majority of finds are abstract geometric designs. There is one recurrent and particularly striking motif of a wavy line, found incised on stone slabs and into the wall of the communal structure – a finger having traced this out in the wet mud plaster. Whenever I see this motif, I cannot help but think that it represents water, the shape reflecting both the bends of Wadi Faynan and ripples on the surface of water. I have no evidence that this is the case, but sense that even if water was not being physically managed, it was gaining a greater significance in human consciousness.

I gained the same impression when I visited the truly remarkable PPNA site of Göbekli Tepe at the far northern end of the Levant in southern Turkey.[23] When discovered in 2001, this astonished the archaeological world by being quite unlike any PPNA site ever seen before. It was perched on the top of a limestone hill and had a number of stone-built enclosures in which massive pillars had been erected, with the images of wild and dangerous animals carved on to them – foxes, wild boar, snakes, raptors. This was a hill-top Neolithic sanctuary, most likely a gathering place for people coming from throughout the Levant and perhaps further afield for shared ritual. Their water needs must have been substantial, but there is no evidence to indicate how these were met, and certainly no traces of water management. The site appears as a chimera of Stonehenge and Lascaux cave with heightened emotional content coming from a spectacular highland setting.

The majority of commentators about Göbekli Tepe remark on the theme of dangerous animals within its art. They note how curious this concern with the wild appears at a time when nature – wild cereals, sheep and goats – was being domesticated. On my visit to Göbekli Tepe in 2003, however, I was struck by a standing stone that had a quite different image: ducks or some other waterfowl standing on a wavy criss-cross pattern that must surely be a representation of water. Here, too, water was on the Neolithic mind.

Farming villages appear

One might excuse the absence of water management in the PPNA by arguing that the settlements remained small and their occupants more like hunter-gatherers than farmers. But when we find that water management remains almost entirely absent throughout the PPNB, there being only two known exceptions, both located at the extreme margins of the Levant, either the archaeological record or people's past behaviour – or both – becomes baffling.

The PPNB is a period (8200–6500 BC) of established farming settlements, villages for want of a better term. Unlike the small settlements of the PPNA with circular, subterranean dwellings, the PPNB has 'proper' architecture with substantial rectangular houses, workshops, storerooms and courtyards. These sites are found throughout the Levant, appearing to develop first in the north, where the site of Jerf el Ahmar in Syria has an architectural sequence showing a transition from round to rectangular structures.[24] PPNB buildings often have thick plaster floors and two storeys, showing a completely new level of time, effort and materials being invested in architecture compared to anything that had come before.

These people were fully fledged farmers, with fields of wheat and barley and herding goats, which may still not have been fully-domesticated. Hunting was continued, now using a new range of large arrowheads. This may have been for now prestigious 'wild' meat – much like stag hunting by English Victorians. Accompanying these developments was an equivalent increase in artistic and religious activity. A skull cult existed throughout the region involving the moulding of plaster around skulls exhumed from burials to re-create faces, using cowrie shells for eyes. These were either displayed or used in ritual, and were then buried once again. One site – 'Ain Ghazal, located close to Amman – has produced a set of half-sized clay figures modelled on straw frames that some claim to be a representation of the PPNB gods.[25]

A great many PPNB settlements are known, showing how agriculture led to substantial population growth. Some of them reached a considerable size. A number of these along the edge of the Jordanian plateau, including 'Ain Ghazal and Basta, are described by archaeologists as 'mega-sites', believed to have had populations of up to 2,000 people. And yet there remains a complete absence of wells, cisterns, aqueducts and dams within such sites. I find this remarkable: by the start of the PPNB at 8200 BC people had learned how to manipulate stone into axes, knives and arrowheads, turn mud into walls and floors, control wild animals and transform plants into their domesticated forms. And yet, within these farming settlements, the approach to water appears to have remained unchanged from that of the simplest of hunter-gatherers. There are, however, two exceptions – the earliest well and the earliest dam known in the Levant. These are found far away from the concentration of PPNB settlement at the geographical margins of the Levant. Before going there, we must briefly visit three other sites to explore the PPNB relationship with water in a little more detail.

Visiting Beidha, Ba'ja and Ghuwayr 1

First we must go the southern reaches of the Levant and visit Beidha. This PPNB village was discovered and excavated by Dianne Kirkbride (1915–97) during the 1960s, providing one of the largest areas exposed of any such village (Photograph 5).[26] During eight field-seasons she excavated 65 buildings, from which the growth of the settlement from a small hunter-gatherer campsite to a substantial farming village can be traced. Beidha is especially notable for having a number of two-storey 'corridor' buildings, the lower floors providing small work-shops and storerooms while the upper floors were open plan with finely made plaster floors, used for eating, entertaining and sleeping. The settlement also had large non-domestic buildings, possibly used as meeting places for the community.

Beidha is located in the beautiful sandstone country close to Petra, capital of the Nabataean Kingdom between 800 BC and AD 106. We will be visiting Petra later in this book (Chapter 5) and learn how the Nabataeans excelled at managing the meagre water supply in this arid region. Beidha is literally surrounded by Nabataean rock aqueducts and cisterns that had captured and stored the sparse rainfall. In con-trast, the Neolithic occupants of Beidha, despite being so adept at complex architecture, a variety of crafts and farming, chose to rely on local springs alone – and suffered the consequences.

Dr Claire Rambeau, while working as my colleague at the University of Reading, located the sedimentary-rock deposits left by one of these springs and used a variety of scientific techniques to reconstruct the periods when the water had been flowing and to estimate local tem-peratures.[27] She found a strong correlation between the flow of the water and occupation of the site: when it first ceased at around 13,000 years ago, during the cold and dry period following the late glacial inter-stadial, the Natufian hunter-gatherers abandoned their camp-site at Beidha; flowing water returned by 9500 BC and so did people, now living as Neolithic farmers. But the flow of water stopped again at around 6500 BC and people once again abandoned the settlement.

Why, one might legitimately ask, did not the Beidha people start to manipulate the water supply as was done so effectively during the later Nabataean period?

Not far from Beidha there is another equally intriguing PPNB settlement with a curious story about water management – or the lack of it. This is Ba'ja, located in one of the most inaccessible places imaginable: it is 'hidden in the Petra mountains' to quote Professor Hans Gebel, who excavated the site between 1997 and 2007.[28] To reach Ba'ja one has to walk for several kilometres along the narrow gorge of a steeply winding canyon, referred to as a siq, often having to climb over sheer rock faces as the wadi ascends in height. The gorge's walls have been polished smooth by millennia of flowing water to levels way above one's own height: it is easy, and rather terrifying, to imagine torrents of gushing and swirling water flowing down the wadi after a winter rainstorm. After several hours of walking, climbing to a height of more than 1,000 metres, one reaches Ba'ja – a cluster of stone-walled rectangular houses perched on a terrace above the wadi floor which Hans Gebel believes had once housed a 600-person community.

The walk to Ba'ja is thirsty work and must have also been so during the Neolithic. But the settlement has neither a spring nor a water course of any type to quench one's thirst. To find water at Ba'ja one has to continue climbing for another hour at least, ascending a further 300 metres to reach what is still only a seasonal spring.

Hans Gebel has pondered why Ba'ja should have been built in such an inaccessible and water-less location. There are no convincing explanations relating to cultivating crops or tending to animals – both would have been extremely challenging at Ba'ja. One possibility is that its people were outcasts from the PPNB 'mega-sites' that developed at this time, forcing them to live in hiding. Another is that luxury goods such as sandstone rings were manufactured at Ba'ja and hence the site needed to be in a location that was easily protected. A third, according to Gebel, is that the setting of the site was 'ritually attractive because it was only accessible through the vagina-type of channel of the siq'.

These are all equally implausible to Gebel and they are so to me. Gebel's most convincing solution is that Ba'ja was located to take advantage of natural accumulations of water within the siq. He argues that the sharp, right-angle turns which often follow each other in quick succession would have acted to slow down the flow of water within the siq. He assumes that water flowed throughout the year, aided by an extensive catchment and a landscape that still carried considerable vegetation. The straight and relatively horizontal stretches of the gorge that often follow the sharp turns would then have acted as natural cisterns, perhaps with the aid of stone-built dams positioned just before the next descent of the gorge. This is both an appealing and plausible theory, explaining how such a community could have survived within what today seems such an inhospitable location. It is, of course, quite un-testable: any dams that might have once been built within the gorge have long since been washed away by the flash floods of recent times.

Returning now to Wadi Faynan we have to walk a mere 500 metres from the PPNA site of WF16 to find Ghuwayr 1, a medium-sized PPNB site a little further up the wadi, heading east towards the Jordanian escarpment and where the name changes from Wadi Faynan to Wadi Ghuwayr.[29] Occupation at Ghuwayr 1 begins immediately after the occupation of WF16 comes to an end, around 8200 BC. My guess is that the PPNA population simply relocated to access larger expanses of terrace for their fields of barley and build a new kind of settlement – one suitable for farmers. The new village had two-storey rectangular buildings, storerooms and so forth, but continued to lack any sign of water management, although it was closer to the spring that later fed the Roman aqueduct and still feeds the modern Bedouin pipes.

Just like Beidha, Ghuwayr 1 became abandoned at around 6500 BC. The likely reason is that excessive grazing by goats and collecting of wood for burning and building destabilised the soils, causing erosion, and reduced the ability of rain to infiltrate the soil. As a result the dynamics of water flow within Wadi Faynan changed, ushering in the flash floods that continue today. The settlement of Ghuwayr 1 had to be abandoned.

Many other PPNB sites were abandoned around the same time, notably the large 'mega-sites', probably for the same reason. One that continued was Jericho. Kenyon had interpreted the wall surrounding the PPNA village as a means of defence against human enemies, perhaps those wishing to gain access to the spring. This has always seemed questionable in light of the lack of evidence for tribal warfare in the Neolithic period. In 1986 Professor Ofer Bar-Yosef, the doyen of Near Eastern Prehistory, offered an alternative interpretation: the wall was a means of defence not against people but against mud flows.[30] This seems most likely to me: just like in Wadi Faynan, Neolithic vegetation clearance, caused by browsing goats, and gathering of wood for fuel and for building material, destabilised the soils on the surrounding slopes leaving the inhabitants and their houses threatened by water and mud. Unlike at Ghuwayr 1, however, settlement continued – Jericho was protected from mud flows by its wall and had the advantage of a prodigious spring. One might describe that wall itself as a means of water management, constraining the flow of water contained within the mudslides. If so, then this would be the earliest water management structure in the Levant, and potentially the world.

The first dam: in the Jafr Basin

There are two exceptions to the absence of water management in the PPNB, the earliest known dam and the earliest known well. To find the dam we must visit the extremely arid Jafr Basin in the far south of Jordan at the border with Saudi Arabia. Professor Sumio Fujii from Kanazawa University, Japan, has been working within that basin since 1997 and has discovered numerous prehistoric settlements.[31] The basin is a flat expanse of bedrock and sand in which it is easy to lose all sense of direction and of time: I know because I did just that on my visit to the Jafr Basin, becoming bewildered below a dazzling burning sun in a landscape that lacked any features as far as I could see in all directions.

Although it is likely to have received a slightly higher amount of rainfall during the Neolithic period than today, the Jafr Basin is unlikely to have ever provided the lush oasis-like environment that Jericho and a few other PPNA and PPNB villages enjoyed. Indeed, that might be precisely why this is where the earliest known evidence for flood control in the world has been discovered.

In 2001 Sumio Fujii discovered a small PPNB settlement in Wadi Abu Tulayha; his 2005 excavations showed that this had been used by people who cultivated cereals and pulses and herded sheep, probably occupying what he described as an 'outpost' for a few months each year as part of a transhumant round. The final buildings within the complex have been dated to around 7500 BC. Close to the settlement and located in the base of the wadi there is a large, amorphous structure that Sumio interpreted as a cistern for capturing the wadi flow in order to provide drinking water for people and their animals. His interpretation is based on the location and form of the structure, its depth – more than two metres – and the fact that it is full of silt rather than occupation debris. It is a persuasive interpretation, as is Sumio's argument that it dates to the earlier phase of the occupation that remains undated.

Sumio calculates that this cistern could have held up to 60 cubic metres of water, sufficient for a few dozen people with their livestock staying at the outpost for about a month. Another structure was found about 100 metres away: a V-shaped wall across the wadi that Sumio describes as a barrage – in other words a dam. Circumstantial evidence suggests that this is chronologically later than the cistern. Sumio suggests that the difficulty in emptying the cistern of silt every spring had led to its gradual disuse and the adoption of the barrage as an alternative water run-off management system (Figure 2.2; Photograph 6). Rather than creating a reservoir, the barrage would have enhanced infiltration into the ground and facilitated the accumulation of sediment suitable for cultivation. Evidence for such cultivation comes not only from the cereals and pulses that Sumio excavated, but also from quern stones and pounders.

It seems unlikely that in PPNB times the annual precipitation by itself would have been sufficient to have allowed cultivation. So Sumio argues that this barrage wall enabled a few hectares of fields along the winding course of the wadi to have been developed. About eight kilometres away a similar barrage wall was found across the floor of another wadi (Wadi Ruweishid). This lacks any associated settlement but distinctive artefacts within the walls and the absence of any other settlement in the immediate vicinity suggests that this is also PPNB in date, making a basin-irrigated crop field or pasture possible within an extremely arid region.

The first wells: on Cyprus and under the sea

It may not be surprising to find the earliest known dam in the extremely arid environment of the Jafr Basin where water would have been so precious. But it may be surprising to find the earliest known wells come from a relatively well-watered landscape. Rather than in the far south, we must now go to the far west of the Levant, to the island of Cyprus.

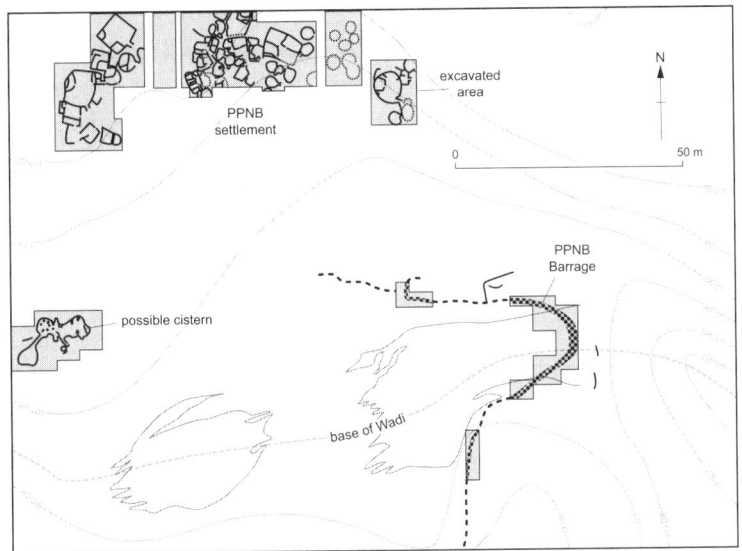

*2.2 The Neolithic barrage at Wadi Abu Tulayha, Jafr Basin
(after Fujii 2007a)*

Although the dating remains contentious, it seems likely that Cyprus was first colonised by farming communities of the PPNA culture, probably sailing from the coast of Syria around 9000 BC. The earliest settlements on Cyprus have PPNB characteristics, including those in the region of Mylouthkia on the west coast of Cyprus where several Neolithic wells have been excavated.[32] They are two metres in diameter and had been sunk at least eight metres through sediment to reach underground water courses that ran along channels in the bedrock heading towards the sea. The wells lacked any internal structures or linings other than small niches within the walls, interpreted as hand- and footholds to allow access during construction and for cleaning. When abandoned, the wells were filled with domestic rubbish which dates from around 8300 BC, indicating that the wells had been built at or just before this date.

The wells have several intriguing features. One is that there would have been no evidence on the ground to indicate the location of the

33

water course eight metres below. So how did the Neolithic people know where to dig the wells? Another is why wells were dug at all? Today, springs are common in the vicinity of Mylouthkia and there is no reason to think they were not so in the Neolithic.

I went to visit the wells in September 2001. At least six were known by that date, several remaining unexcavated and appearing as dark rings of soil on the ground amidst the apartments of a hotel complex. Indeed, as I was shown around by a local archaeologist, tourists sat nearby under fake-straw umbrellas drinking cocktails. I didn't find that in the Jafr Basin!

Perhaps we need to think about social or ideological motivations for digging wells rather than simply the need to get sufficient water to drink. Indeed, excavation revealed two striking aspects about the wells. First, the wells show little sign of wear and tear, either from human action or through natural erosion. This suggests they were in-filled quite shortly after construction. They may have quickly become polluted, but the back-filling included materials suggestive of ritual activity: one of the wells contained the remains of five human skeletons and those of 22 goats – a small herd slaughtered just to place inside a well. So, all in all, the Neolithic wells at Mylouthkia appear to have had significance beyond simply providing access to a fresh water supply.

Another set of intriguing wells, slightly more recent in time, have been found at Atlit-Yam, a Neolithic village dating to between 7500 and 6000 BC on the coast of Israel.[33] In fact it is now 300 metres off the coast because the village was submerged by rising sea level – a consequence of the final melting of the ice age glaciers in the northern hemisphere. When occupied, Atlit-Yam lay adjacent to a lagoon formed by a sandstone ridge and a river; it was a farming and a fishing settlement, benefiting from a seasonal flooding of the lagoon that provided freshwater fishing opportunities.

Three wells have been discovered at Atlit-Yam. These were 1.5 metres in diameter and had been excavated to 5.7 metres, at which depth they would have hit the Neolithic water table. Unlike the Mylouthkia wells, their sides were lined with stone blocks to a depth of 4.7 metres, after which they were cut into bedrock. The archaeological excavations suggest that the wells had been abandoned because of rising sea level, contaminating the fresh water with salt. After this, the wells were used as rubbish pits until the settlement as a whole was abandoned.

Water management in the Pottery Neolithic

The Jafr Basin dam and the Cypriot and Atlit-Yam wells are difficult to date precisely but most likely come from the end of the PPNB and start of the Pottery Neolithic periods. The invention, manufacture and use of ceramic vessels may be the most obvious sign that a new approach to water had emerged. These replaced stone bowls and leather or wickerwork containers that had been waterproofed by covering them with bitumen collected from deposits at the Dead Sea. While there are many ideas as to why pottery vessels were adopted at this time, their role in simply easing the storage and transport of water should not be neglected.

One of the largest Pottery Neolithic sites provides us with the third earliest and the most impressive example of a well, dating to 6300 BC, that at Sha'ar Hagolan.[34] Located in the northern reaches of the Jordan Valley, close to the Yarmuk River and the Syrian border, Sha'ar Hagolan is one of the largest Neolithic villages ever found. It has a much higher degree of 'town planning' than elsewhere with a row of houses and courtyards backing on to a 'street'. An extensive area has been excavated with the well being found in a rather peripheral location. Rather than simply digging out the fill of this well, Professor Yossi Garfinkel, from the Hebrew University in Jerusalem, exposed a complete vertical cross-section so that its original construction could be clearly understood (Figure 2.3; Photograph 7).

The well had been made by initially digging a stepped pit through the natural sediment to reach the water table, located within a gravel deposit at a depth of 4.26 metres. The upper 2.5 metres of the well shaft was then lined with stone blocks. It appears that the water within the well eroded the sides of the shaft, perhaps suggesting that it had been used for a significant length of time.

Just as with the wells at Mylouthkia and Atlit-Yam, that at Sha'ar Hagolan is located close to a source of fresh water, in this case the Yarmuk River. While this does not necessarily exclude practical reasons for the well – its water may have been cleaner than that from the river – it may suggest that social or even ideological factors might have influenced the location of this Neolithic well. The well at Sha'ar Hagolan might have been for private use – the late Neolithic period being one when issues of ownership and social status were increasingly significant.

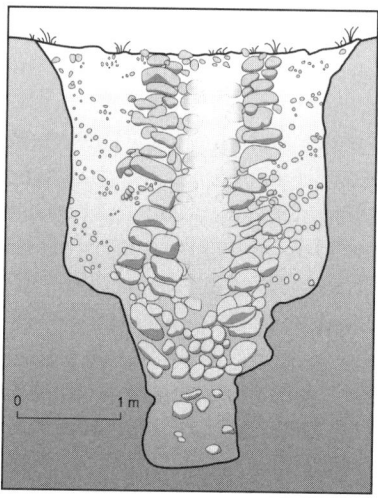

2.3 Section through the Neolithic well at Shaʻar Hagolan, Jordan Valley, Israel (after Garfinkel et al. 2006)

The Pottery Neolithic period is also when we find the earliest terrace walls, built to inhibit soil erosion and maximise water use within a field system. These are found at the Pottery Neolithic settlement of Dhra', located on a terrace above the Dead Sea, providing marvellous views across the rift but exhaustingly hot to visit – and no doubt worse to excavate.[35] Excavation was undertaken by Bill Finlayson and his colleagues, who in 2005 revealed a suite of nine terrace walls close to the buildings of the Pottery Neolithic settlement. The walls were oriented perpendicular to the slope and placed directly across bedrock outcrops in order to anchor them against the flow of water.

These walls, some of which stood almost one metre high and ran for more than 20 metres, indicate a significant investment of labour in the construction and maintenance of a field system. They would have functioned to collect and retain run-off water and the sediment particles it carried. The soil that accumulated behind the walls would have acted as a sponge for the water. Scatters of pottery and other domestic refuse were found in the vicinity of the walls, most likely from attempts at manuring to fertilise the soil.

Developments in the Bronze Age

Although they appear to have taken a peculiarly long time to appear, by 3600 BC we have dams, wells and terrace walls – three methods of water management that provide essential engineering infrastructure to later civilisations. Quite how pervasive these were at 3600 BC remains unclear: archaeologists are always struggling against poor preservation and the lack of discovery. But as the Bronze Age develops there are rapid advances in hydraulic engineering which appear critical to the emergence of the first urban communities.

The scale of development is most effectively appreciated by taking a visit to the Early Bronze Age settlement of Jawa, otherwise known as 'the lost city of the Black Desert'.[36] The ruins of this walled town date to the start of the Bronze Age around 3600 BC and are located in the arid basalt desert close to the Syrian–Jordanian border. It isn't easy to find. On my visit I knew that the site was at the head of Wadi Rajil but had immense difficulty in discovering that wadi, indeed any wadi at all, within the vast expanse of boulder-strewn desert in which I found myself wandering. Not surprisingly I was detained by a military patrol for several hours, thought to be acting suspiciously near the border. I did eventually find the site and was astounded by the size of its walls. They were built out of basalt blocks and hence disappeared into the landscape when seen from a distance, becoming another geological escarpment.

The desert was little different in 3600 BC from what it is today, an extraordinarily challenging landscape for human settlement, its floor scattered with basalt boulders that had once been volcanic lava spilling from the now extinct volcanoes of the nearby Jebel Druze. There is no surface water other than highly seasonal and short-lived flows within the wadis from the brief, although often intense, winter rains. Today the desert receives less than 10 centimetres of rainfall per annum; it is estimated to have been somewhere between 12 and 33 centimetres a year when Jawa was occupied in the Bronze Age. There are no springs and the aquifers are deep below the basalt bedrock, far too deep to be reached by wells. The only water supply for Jawa came from rainfall.

The site was excavated by Sven Helms in the 1970s. He estimated a population of between 3,000 and 5,000 inhabitants on the basis of the number of houses within the walls but thought the town may have flourished for as a little as 50 years. Establishing the length of occupation was made difficult because there was virtually no soil surviving at the site, it all having been eroded away by wind, and hence there was no stratigraphy to excavate. The absence of soil and vegetation cover, however, enabled Helms to undertake a meticulous survey of the town and the immediate area outside the walls, including subtle changes in the undulations of the ground. He found channels that were designed to extract water flowing in Wadi Rajil into at least ten ponds surrounding the outside of the town wall, with an estimated storage capacity of between 28,000 and 52,000 cubic metres (Figure 2.4). Water would only have flowed in the wadi during the winter months, and hence the storage system provided water for the long dry summer that would follow – drinking water for people and animals, and possibly for irrigation.

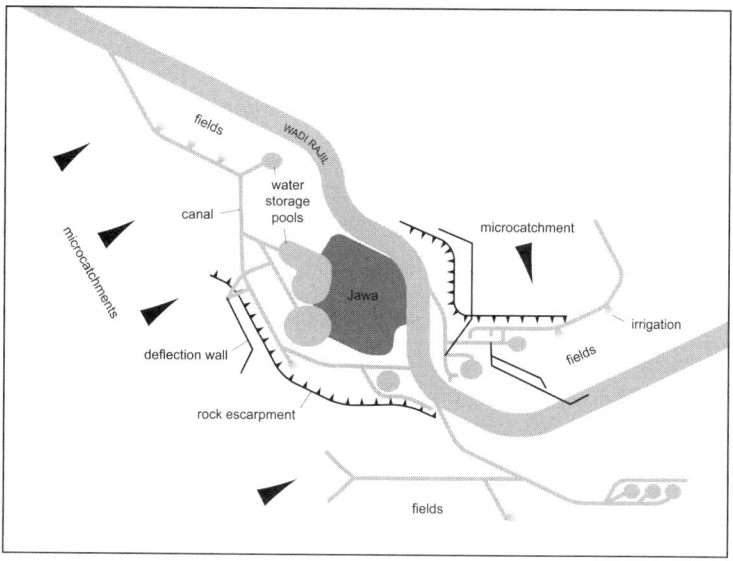

2.4 Schematic diagram of the water storage system at Bronze Age Jawa (after Whitehead et al. 2008)

A recent computer simulation study of the water storage system at Jawa as reconstructed by Helms was undertaken by Professor Paul Whitehead and his colleagues from the University of Reading.[37] They drew on a range of environmental data and models to reconstruct the Bronze Age rainfall regime and made estimates for the quantity of water required by people and animals and for irrigation. They found that Helms' population figures were feasible, provided that the stored water was not being used for irrigation and that rainfall was above 200 mm a year. Once it fell below this, as it is likely to have done at around 3000 BC, the sustainable population of Jawa fell rapidly, implying that the town would have soon been abandoned.

Whether or not there had been long-term occupation at Jawa, to have established a settlement at all within the Black Desert was a remarkable achievement. This had required not only a capacity for expert hydraulic engineering but also the confidence and ambition to undertake this in the most challenging of environments.

We gain the same impression from Levantine Early Bronze Age sites from around 3600 BC and later.[38] Two hundred kilometres south-west of Jawa there was the contemporary walled town of Bab edh-Dhra, located on the east side of the Dead Sea, close to the Pottery Neolithic site of Dhra'. While this is thought to have had a population of a mere 1,000 people, it has an extensive cemetery containing 20,000 tombs, many of them family tombs with multiple internments suggesting that the town had many generations of occupation. The ruins today are extraordinarily dry and dusty, as any visitor soon becomes aware; access to water must have been at a premium when living there in 3000 BC. The remnants of plaster-lined cisterns are found within the walls, while evidence for irrigation comes from the recovery of swollen flax seeds.[39]

One hundred kilometres north of Bab edh-Dhra, the occupants of the Early Bronze site of Tell Handaquq had built a dam to capture seasonal floods within the adjacent Wadi Sarar,[40] while a further 50 kilometres north at Khirbet Zeraquon a rock-cut underground tunnel had been excavated to provide a water supply to the town.[41] On the western side of the Jordan Valley, reservoirs were constructed within the towns of 'Ai and Arad.

As these few examples illustrate, methods of hydraulic engineering proliferate during the Early Bronze Age with each town employing those methods appropriate for its immediate environment. While such water management was essential for the growth of these towns, it wasn't sufficient to sustain them. For reasons that remain unclear, but which are probably linked to climate change, many of the Early Bronze Age towns were abandoned soon after 3000 BC, the settlement pattern largely returning to one of dispersed villages and farmsteads. The Bronze Age continues until 1200 BC with cycles of urban growth and then collapse, most likely reflecting inherent environmental constraints on the long-term sustainability of urban communities in this region. This is quite different to the Nile Valley to the south-west and Mesopotamia to the east, where the Early Bronze Age urban growth continued unabated to create the earliest cities and civilisations – as we will see in the next chapter.

Cycles of growth and collapse continue throughout the Iron Age of the southern Levant, with further developments in what can be legitimately described as hydraulic engineering. At Tell Deir Allah on the eastern side of the Jordan Valley, a dam and a system of canals were constructed to bring water from the nearby Wadi Zarqa for irrigation.[42] Visiting that tell today one can look across the well-irrigated fields of an agricultural research station and get some impression of how the Iron Age fields may have appeared.

In modern Israel, we find several large tells which had once been the site of dense urban communities, each of several thousand people, notably Meggidio, Hazor and Beer Sheba. They all had underground water systems and reservoirs.

Into history – and the Biblical spirit of water

Before we make our last visit of this chapter, we should pause and note that the Bronze Age and Iron Age archaeology is believed by many to have been the setting for the stories of the Old Testament.[43] Bab edh-Dhra is thought to be one of the 'five cities of the plain' referred to in Genesis 13 and 19; some identify it with Sodom and suggest that the nearby archaeological site of Numeira was that of Gomorrah. For the Iron Age some archaeologists openly use terminology taken from the Bible referring to the Philistines, Israelites and Canaanites for particular Iron Age cultures. It might be argued, therefore, that the stories in the Old Testament have a bearing on the lifestyles, and perhaps attitudes to water, in the Bronze and Iron Ages of the southern Levant.

The most striking point is simply that references to water are pervasive throughout the Old Testament. Within Genesis, water is described as having been created before plants, land and even light. God uses water to help his chosen people: he reveals a well to Hagar and her son when suffering in the desert (Genesis 21:10), he tells Moses which rock to strike with his staff to find water for the people he is leading to the Promised Land (Exodus 17:1). Just as the life-giving power of wells and springs is expressed, so too is the destructive nature of floods: God parts the waters of the Red Sea to allow passage for Moses and his followers and then releases the water in a flood to destroy the chasing Egyptian troops (Exodus 14:15). And, of course, there is the story of the Flood and Noah's Ark (Genesis 7:17).

The significance of floods, wells and springs as referenced in these stories is readily appreciated from the archaeological evidence itself. But the Old Testament also explains how water was used for cleansing and purifications – something we could never identify from the ruins of Bab edh-Dhra. There are many references to God's followers using water within ritual and religious ceremonies, such as washing animals before they are sacrificed and to purify a person after they have touched a corpse. In general, water is used to wash away all that is impure.

One might take a functionalist interpretation of such stories, arguing that they simply serve to emphasise good hygiene. Alternatively one might suggest that water had taken on a spiritual significance in the Bronze Age and Iron Age communities of the southern Levant – if indeed the Old Testament has any relationship to them at all.

Finally, in Jerusalem

A new threshold in hydraulic engineering was passed in Jerusalem in the year of 700 BC, or sometime very close to that, by the construction of 'Siloam's Tunnel'.[44] This is a 533-metre-long curving tunnel, constructed to carry water from the Gihon Spring, located outside the city wall, to the Pool of Siloam within the city (Figure 2.5). This is the first known tunnel excavated from both ends simultaneously with a meeting in the middle. It is claimed that the underground engineers and workmen were directed by sounds coming from hammering on the surface above. The tunnel is referred to in the Old Testament as having been constructed during the reign of Hezekiah as a means to thwart an impending siege of Jerusalem by the Assyrians.[45] It replaced a Middle Bronze Age channel dating to 1700 BC that also brought water from the Gihon Spring to the Pool of Siloam, but which was more exposed to a besieging army.

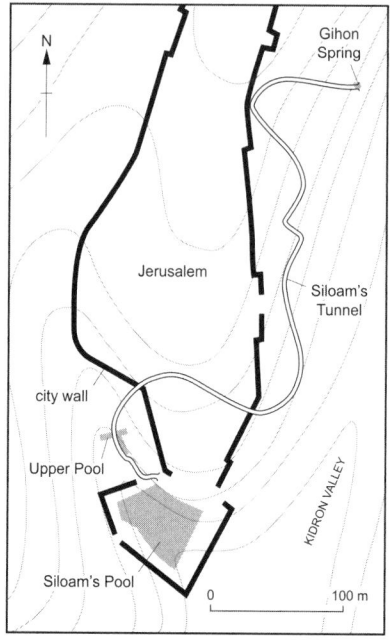

2.5 Siloam's Tunnel at Jerusalem

I have yet to visit Siloam's Tunnel but it is top of my list for sightseeing when I am next in Jerusalem, above both the Church of the Holy Sepulchre and the Dome of the Rock. Seeing this tunnel would complete my own journey through the evolution of water management in the Levant. That began at 'Ubeidiya, occupied one million years ago, and took me to the peripheries of the Levant – Göbekli Tepe in the north, Beihda in the south and Cyprus in the west – and to numerous sites in the centre of the Levant: Ohalo, Jericho, Netiv Hagdud, WF16, Jawa, Bab edh-Dhra and – still to come – Jerusalem.

At all of these sites I searched for traces of water management going beyond its mere transportation in cupped hands, folded leaves or even ceramic vessels. The ice age came to an end, plants and animals were domesticated, and villages that housed hundreds if not thousands of people were built, before any signs were found: the Neolithic and urban revolutions passed by without any need for water management. The first dams, wells and terraced walls only appear towards the end of the Neolithic period and became pervasive in the Early Bronze Age. Knowing what comes later – as the next chapters in this book will reveal – I am confident to describe this as a 'Water Revolution', one that changed the course of history.

3

'THE BLACK FIELDS BECAME WHITE / THE BROAD PLAIN WAS CHOKED WITH SALT'

Water management and the rise and fall of
the Sumerian civilisation, 5000–1600 BC

With water now domesticated, we leave the Levant and move 1,000 kilometres east to the alluvial plain of the Tigris and Euphrates. Here, Bronze Age towns of 4000 BC underwent inexorable growth to create the first civilisation of human history – the Sumerian. Those towns had grown from Neolithic villages which had arisen from either the arrival on the plain of Neolithic farmers, dispersing from the Levant, or by the adoption of Neolithic lifestyles by its indigenous hunter-gatherers. Our concern is with the role that hydraulic engineering played in the rise and then the fall of the Sumerian civilisation.[1]

With regard to its rise, canal building and irrigation may have been critical, providing water to increase yields for agricultural surpluses and transport links to enhance trade. With regard to its fall, excessive irrigation may have contaminated the fields with salts – chemical compounds containing sodium, potassium, calcium, magnesium and chorine. These would have reduced soil fertility leading to agricultural collapse, as implied by the extract from a Sumerian document written at around 1640 BC that I use as my chapter title.[2] As ever, there are disagreements between archaeologists to consider as we explore the role of water management in this first, and to many the most magnificent, of ancient civilisations.

Out of bounds

The arena for Sumerian civilisation was southern Mesopotamia, now known as Iraq (Figure 3.1). For reasons that do not require enumerating, Iraq has been out of bounds for archaeologists, especially those from Britain, for a number of years. As such, I have been unable to visit the landscapes and rivers that once formed the setting; neither have I seen the ruins of ancient Mesopotamian towns and cities where the action took place. The drama had been a remarkable fluorescence of human culture in the widest sense – architecture, trade, engineering, writing, religion, cosmology, politics, warfare and artworks of enduring beauty. Professor Guillermo Algaze, one of the world's leading authorities on the Sumerian civilisation, would go further by describing this as a 'veritable revolution in human spatial, social, political and economic organisation'.[3]

Some compensation for frustrated archaeologists and travellers unable to visit the area is found in the many remarkable artefacts from Southern Mesopotamia that are on display in museums throughout the world. The most impressive are the treasures from the Royal Tombs of Ur in the British Museum – Ur having been one of the great cities of the civilisation.

The first thing that caught my eye when entering Room 56 at the top of the stairs in the British Museum was the glint of gold. Necklaces, bangles and head-dresses made of gold and precious stones such as red carnelian and blue lapis lazuli were arranged in glass cases, looking as new as if they were in a Bond Street jeweller's shop. In fact they are arranged according to where they were found in the numerous burial chambers and death pits found in the excavations directed between 1922 and 1934 by Charles Leonard Woolley, who was later knighted for his services to archaeology.

The guards and servants of the deceased were dressed in the finest of regalia and buried with their master or mistress, some lying in rows and others arranged in groups, positioned as if playing their musical instruments. Their remains had been crushed by the collapse of the earth-covered ziggurat that covered them, and which had caused the helmets and head-dresses to remain in situ on their owners' heads. Other items provided for the afterlife had included games, gold ornaments, ceramics, carved stones and incised clay tablets.

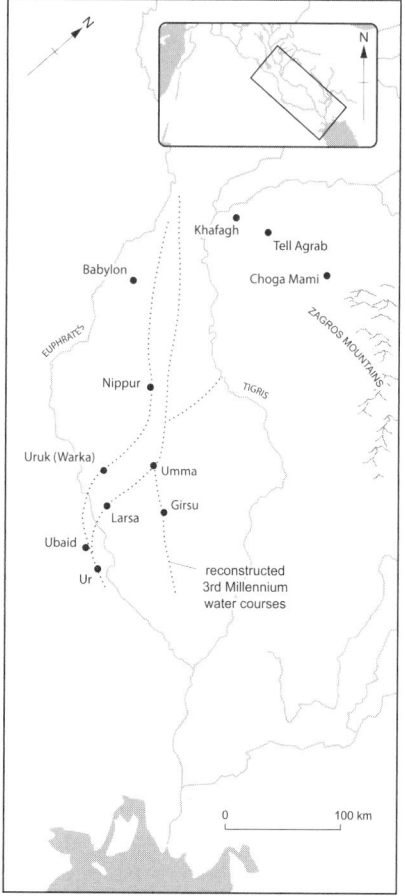

3.1 Archaeological sites and locations referred to in Chapter 3

Thousands of such clay tablets incised with the Sumerian cuneiform script are available for study in museums throughout the world,
as are collections of pottery, sculpture, jewellery and so forth. Such
collections exist because much of the late nineteenth- and early
twentieth-century Mesopotamian 'archaeology' had been funded by
those museums in a competitive race to acquire antiquities, a race that
in some circumstances led to little more than the pillaging of ancient
cites. One could not make that accusation against Charles Leonard
Woolley, whom many consider to be of the first 'modern' archaeologists using detailed recording methods. His work was funded by the

British Museum and the University of Pennsylvania's Museum of Archaeology and Anthropology, both of which now hold collections of the most significant artefacts.

Woolley may have found a lot of gold, but the 'golden age' for archaeology in Southern Mesopotamia was between the 1960s and 1970s. This was when large-scale, regional surveys were undertaken to map the distribution of settlements and questions focused on issues of social and economic change rather than the mere acquisition of antiquities.[4] We are still largely reliant on those surveys, although they are now supplemented by images from both satellites and the Space Shuttle.[5] In Northern Mesopotamia, covering Syria, western Iran and southern Turkey, we have more recent surveys and excavations. This northern region rose to prominence after 1800 BC, just when the Sumerian civilisation was in a fatal decline in the south. Why such a difference in the timing of cultural prominence? This will need addressing, but Northern Mesopotamia is of less interest to us, its developments being largely derivative of what happened in the south. To return to water management, irrigation is arguably the key to the remarkable cultural achievements in Southern Mesopotamia. And water – as the carrier of salts – is argued by some to have been the key to its later demise.

The land between the rivers

Mesopotamia is the name given by the Greeks to the land between the two great rivers of the Tigris and the Euphrates. They flow through a trough between the Arabian shield to the west and the Zagros Mountains in the east. The rivers rise in the mountains of south-east Turkey, the Tigris flowing for 1,850 kilometres through Turkey and Iraq and the Euphrates for 3,000 kilometres through Turkey, Syria and Iraq. They eventually merge to form the Shatt al-Arab waterway at the location of modern-day Basra and then discharge into the Persian Gulf. In Northern Mesopotamia, these rivers and their tributaries are deeply incised into the rolling steppe. Once beyond the rock wall of the Jebel Hamrin, an outlying outcrop of the Zagros that divides Northern from Southern Mesopotamia, they flow across a flat plain depositing their loads of sediment carried from the uplands.

The Tigris is fast flowing and unpredictable. The Euphrates travels a longer distance, losing 40 per cent of its water by evaporation when crossing the Syrian Desert. It braids into multiple channels and the silt it carries has a greater natural fertility. The Euphrates played the dominant role in providing water for the Sumerian civilisation.

These rivers have been flowing across the plain for many thousands of years. During the critical period for the emergence of Sumerian civilisation, the fifth and fourth millennia BC, they formed a single, dynamic network only fully dividing into their separate courses in the third millennium BC. A vast expanse of alluvium is the consequence of their flow, especially in the north of Southern Mesopotamia, where as much as ten metres has been deposited during the last 10,000 years. Banks of sediment, levées, have built up on each side of the river courses, gradually elevating the level of the water itself. The levées can reach up to three kilometres wide and would have originally carried lush riverine woodland of willow, poplar and fig, in which deer, wild boar and big cats were found. As the rivers approach the Gulf, the landscape turns into one of lagoons and marshes – or at least it used to before recent drainage and landscape degradation. Their combined courses in the Shatt al-Arab once reached the sea as much as 200 kilometres before they do so today, the sea having retreated because of falling sea level and silt deposition into the Gulf.

Northern Mesopotamia receives more than 250 millimetres of rainfall each year, crossing the threshold required for rain-fed agriculture; Southern Mesopotamia fails the rainfall test. It has long hot dry summers between May and October with less than 200 millimetres of rain falling during the cooler winter months. The climate is unlikely to have been significantly different during our period of interest. Some argue that there would have been higher winter rainfall across the Tigris and Euphrates watersheds, resulting in heavier winter and spring flows in the rivers. Others suggest that the summer monsoon, which today falls further south, may have had a more northerly track, bringing summer rainfall. This would have increased animal forage at the time of greatest need. But these climatic differences are quite minor and the fundamental pattern has remained the same: insufficient rainfall in Southern Mesopotamia to allow farming without irrigation.

Prior to the fluorescence of the Sumerian civilisation, the Neolithic and early Bronze Age communities of Northern Mesopotamia in the fifth millennium BC were quite similar to those of the Levant that we considered in Chapter 2. If anything, they were somewhat less advanced. They date back to at least 9000 BC and were preceded by a long record of Stone Age hunter-gatherer transient campsites. The first farming villages in Southern Mesopotamia are more recent than the Levant, 5000 BC at the earliest. This is not surprising because the irrigation-based farming required to develop the land is more challenging to develop. And yet it was here rather than in the Levant or Northern Mesopotamia that the first civilisation emerged.

Foundations: Ubaid and Uruk

In the sixth and fifth millennia BC, few if any cultural differences are discernible between the Levant, Northern Mesopotamia and Southern Mesopotamia. Small farming villages and pastoralist campsites were common, communities who also engaged in hunting and gathering, pottery production and other household crafts. In Mesopotamia those of the fifth millennium BC are referred to as the Ubaid Culture. Although the term refers to a common lifestyle and culture, in a narrow sense it refers to a distinctive pottery which is black or dark brown in colour with distinctive horizontal bands, and with animal and geometric motifs.

The Ubaid archaeological record is not well known; many sites are likely to be buried beneath later alluvium or the towns and cities of later periods. Others will have been destroyed by the changing courses of the rivers. Moreover, large areas of Southern Mesopotamia have simply never been surveyed. Few of the known Ubaid sites have received significant levels of excavation, especially with the suite of modern techniques and scientific applications now available. From what we do know, it is thought that Ubaid communities relied on household-based production of food, utensils and tools. They were most likely egalitarian in nature, in the sense that formal leadership positions were rare, and those which did exist would not have been hereditary. The focal point for some of the larger settlements may have been a 'temple', or central building, possibly acting as a place for the storage of grain and then its redistribution during times of shortfall.

There appears to be little difference between the Ubaid communities of Northern and Southern Mesopotamia; if anything, those in the north had more signs of durable wealth, such as jewellery made from lapis lazuli, which one might take as a sign of impending cultural change. In this region the Ubaid culture is either immediately preceded by or partly contemporary with the Samarran culture named after a distinctive style of pottery (from a site of Samarra) and dated between 6000 and 4500 BC. It is within this culture that the earliest irrigation system for Mesopotamia has been identified at the site of Choga Mami. This is located 150 kilometres north-east of Baghdad in the low foothills of the Zagros Mountains close to the Iranian border and was discovered by the (now) Cambridge-based archaeologist Professor Joan Oates during a survey in 1966.[6] Her 1967–68 excavations showed the site to be a small farming village growing barley and wheat, and with goat, sheep, pigs and cattle. A suite of irrigation ditches were found bringing water from a nearby river, providing the earliest known evidence for water management in Mesopotamia. In her 1969 excavation report, Oates described how 'the prosperous Samarra settlement at Choga Mami represents an intermediate stage between the early rain-fed agriculture of the northern plain ... and the full efflorescence of the 'Ubaid economy in the south, which must have been based on fairly large-scale irrigation'. That conclusion still stands more than 40 years later.

While Choga Mami and other settlements in the north experienced a gradual growth in their size and complexity, the Ubaid culture in Southern Mesopotamia experienced a dramatic 'take-off', denoted as the cultural period of the Uruk culture. Although our knowledge of the specific changes in settlement patterns and chronology of events remains limited, within a few hundred years around 3900 BC, the foundations for the Sumerian civilisation had been laid: urban communities, monumental architecture, public art, centralised rather than household production, extensive trade networks. Water management for both irrigation and canal transport provided the foundations for this earliest civilisation in the world.

The development of technology and crafts had also played a key role in this cultural transformation. The Uruk was a Bronze Age culture, the archaeological record enabling us to track the change from the use of copper alone (Chalcolithic) to inclusion of tin to make bronze. Metalworkers were critical to economic growth, producing artefacts that significantly increased farming efficiency, such as bronze sickles to replace those of stone, and which transformed warfare and transport. Another group of key workers were those involved in the woollen-textile industry. During the Uruk period this grew to be many thousands of people undertaking the sequence of tasks from the management of the flocks to plucking wool, sorting it by quality and colour, combing, spinning, dying and weaving.[7] The textiles were a key export commodity. Indeed, Professor Robert McCormick Adams, arguably the greatest Mesopotamian archaeologist of the 20th century, proposed that 'without the wool for textiles to be traded ... it is difficult to believe that Mesopotamian civilisation could have arisen as early and flourished as prodigiously as it did'.[8]

Metal- and textile workers were just two of the groups of craftsmen that now required support by agricultural surplus from the rural areas. There would have also been armies of potters, stonemasons, leather workers and so forth – as well as armies themselves. To these we must add administrators and priests and, of course, the rulers themselves – the kings. These claimed to rule in God's name (or rather Gods' names), some may have even claimed divine status themselves. We lack the names and biographies of the earliest rulers – writing had yet to appear – but see their images as bearded figures, wearing distinctive skirts and hats as depicted on their cylinder seals, carved vases and monumental architecture. They are presented in typically kingly poses: with bound captives, shooting lions and standing on the shoulders of Goddesses.

The Uruk period saw a proliferation of settlements with many villages in Southern Mesopotamia growing rapidly into towns. A hierarchical settlement pattern soon developed, with four tiers of settlement. The largest was the town of Uruk itself, sometimes known as Warka (its modern name), located in the far south of the alluvial plain. During the Early Uruk period this is estimated to have been up to 100 hectares in extent, more than twice as big as the next largest settlements at around 40 hectares. There are likely to have been numerous other settlements reaching up to 25 hectares, and then smaller scattered villages. Towards the end of the period, Warka asserted its dominant position, reaching 250 hectares in extent and with a population of at least 20,000 people – that probably being a highly conservative estimate. In the mid fourth millennium BC, Warka would have been the largest settlement in the entire world. The next largest towns, such as Nippur, would have been up to 50 hectares, followed by multiple small and large towns between 15 and 25 hectares in extent.[9]

This was a remarkable growth of urbanisation. It was dependent upon a continuous influx of new people into the towns: mortality rates are likely to have been high, and family size limited, with many people living as slaves, soldiers and in situations not conducive to producing many children. Indeed, the Southern Mesopotamian towns may have been dependent upon immigrants from Northern Mesopotamia because their expansion appears to mirror a contemporary contraction of towns in the north.

The towns had many attractions, one of which may have been security, in light of their enclosing city walls. Conflict appears to have been endemic, often depicted in scenes of warfare. This did not only happen between the towns themselves but also with periodic invasions into the alluvial plain, coming from the Zagros hills to the east and deserts to the west.

The Uruk culture was also expansionist: we find communities within Northern Mesopotamia – in Turkey, Syria and Western Iran – with a distinctively Uruk culture. Some are located within large settlements, suggesting enclaves of traders, while others constitute whole communities which are likely to have been Uruk colonies established at key geographic locations to control trade. That of Hacinebi Tepe in Turkey, for instance, was situated at one of the few natural fording places of the Upper Euphrates and Godin Tepe, also in Turkey, at a point that controls the overland route into the Iranian plateau.

By 3200 BC as much as 70 per cent of the population of Southern Mesopotamia may have been living in towns and cities – a population thought to be in the order of 200,000 (but this is notoriously difficult to estimate). The Uruk proliferation of new settlements had halted, with a consolidation into around 30 cities distributed across the plain. Each had a distinctive identity focused around its own city-god and temple, some of which were now positioned atop huge monuments – the ziggurats.

As well as being places of worship, the temples played a major economic role, owning swathes of land and being centres of production with their own contracted craftsmen and labour force. The presence of a powerful elite is now made explicit by marked differences in burial practices, the elite having elaborate graves with immense riches – as

Sir Charles Leonard Woolley discovered when he excavated the Royal Tombs of Ur. The poor were now coaxed and coerced into dependent relationships with the rich and powerful, who claimed to rule in God's name.

The city-states, surrounded by their networks of canals, appear to have been politically independent and yet inextricably interlinked by trade, sharing an economic base of mixed arable and livestock farming and a common culture; Professor Nicholas Postgate, the Cambridge University Mesopotamian specialist, describes the cities as having shared a consciousness of belonging to what he calls 'The Land' – the alluvial plain.[10] Writing had now become pervasive and from the cuneiform tablets we know the names of the cities and their rulers and have a narrative about a selection of events: Girsu, Umma, Larsa, Nippur and Ur are especially prominent in the texts. Many think that it was via the interaction between such city-states that the cultural achievements were attained – this coming from a rich mix of cooperation, competition and emulation.

Cycles of unity and division

This initial period of relatively independent but interactive city states lasted for almost 700 years, between 3000 and 2350 BC, and is referred to as the Early Dynastic Period. The next 700 years began a cycle of unification, followed by rebellion and incursion, a return to city-state independence, followed once again by unification by a single ruler but leading eventually to the demise of the Sumerian civilisation.

First came Saragon, one of the rulers who gained overall control of the city-states, being content to claim his authority by force of arms rather than divine right. He founded the city of Akkad to be his capital, a city yet to be located by archaeologists. The most marked sign of new order was the use of a new language in the written texts, one known as Akkadian. This used precisely the same symbols as the Sumerian language, just as, say, English and French use the same alphabet. Some texts were written in both languages which proved the key to their translation. Whether or not a change in spoken language accompanied that used in writing is entirely unknown. Saragon's Dynasty of Akkad lasted a mere 200 years. His descendants lacked his charisma and leadership abilities and his grandson, Naram-Sin, appears to have been notably deficient, being described in the texts as arrogant. As the city-states reasserted their independence, 'The Land' was also being invaded by foreign tribes, notably the Gutians coming from the Zagros Mountains. Soon Southern Mesopotamia reverted to a collection of largely autonomous city-states.

By 2150 Southern Mesopotamia was again unified, this time under the hegemony of Ur, a city located to the south of the alluvial plain. A leader known as Ur-Nammu rose to prominence in Ur and founded what is called the Third Dynasty; he led the conquest of neighbouring city-states, and then those further afield. Unlike the descendants of Saragon, those of Ur-Nammu were more effective, notably Šulgi, his son who ruled for 48 years. He eased commerce by imposing standardised units of weights and measures and introducing a new calendar. During the Third Dynasty of Ur, there was a remarkable proliferation of administrative documents reflecting an immense burgeoning of bureaucracy. Thousands of these were illicitly excavated between 1880 and 1920 and are now dispersed in museums throughout the world; thousands more must remain buried amongst unexcavated ruins.

The collapse of the Third Dynasty of Ur occurred in the centuries around 2000 BC. It led into a period of great instability as a succession of city-states tried to assert their power. Some built their own short-lived empires, although never as extensive as those which went before, notably the cities of Babylonia and Assur – both located in the north of 'The Land' rather than close to Ur in the south – so representing a clear geographical shift of power. Indeed, this final phase of the Sumerian civilisation (2000–1600 BC) is referred to as the 'Old Babylonian' period within which the First Dynasty of Babylonia with Hammurapi as its leader rose to some prominence.

Hammurapi is well known for having produced one of the first written codes of law. These inform us about the issues of greatest concern to the Babylonians and hence it is interesting to note the prominence of water, this being referred to in several of his laws. Law 53 states: 'if anyone be too lazy to keep his dam in proper condition, and does not keep it so; if then the dam breaks and all the fields be flooded, then shall he in whose dam the break occurred be sold for money and the money shall replace the corn which he has caused to be ruined.'[11]

At this time water was used as a weapon of war, as it may have been throughout Sumerian history. Abi-eshuh was the king of Babylonia between 1711 and 1684 BC and was in conflict with a rival known as Iluma-ilu, thought to control cities further south, possibly including Uruk itself. A written tablet, in this case an oracle rather than administrative text, records how in the 19th year of his reign Abi-eshuh blocked the flow of the Tigris. The water was redirected into an adjacent irrigation channel, which was unable to contain the deluge and, as a result, all of the surrounding fields were inundated. Abi-eshuh had succeeded in his plan to impede the movement of his enemy's forces.

We know from texts that Ur, and we must assume other cities, periodically lacked supplies of food, whether caused by falling yields or disruption to the distribution system is unclear. Texts also show that during this period it suffered chronic inflation, most likely a symptom of overall economic decline. In addition to the conflicts between city-states there were external threats from invading peoples, most notably the Amorites from the west, who soon gained control in many cities, adopting the traditional attributes of Sumerian urban dwellers in place of their previous nomadic lifestyles. There were also the Kassites coming from the Zagros Mountains, and finally the invasion of the Hittite army around 1600 BC. After that the documentary record from Southern Mesopotamia ceases – a sign of its final demise. But these final incursions were no more than a coup de grâce. Ever since 2000 BC the ancient cities had, according to Professor Nicholas Postgate, been 'dying on their feet' and becoming 'more or less moribund'. The centre of gravity – of political power, economic prosperity and cultural innovation – 'was moving inexorably northward'.[12]

Reading the tablets and the landscape

The rise and fall of Sumerian civilisation that I have summarised very briefly is partly derived from the archaeological surveys which attempted to read the landscape and reconstruct settlement patterns, but also partly from reading the cuneiform tablets. The latter became possible after 1857 when the script was fully deciphered to provide a window into the Sumerian world and mind. Indeed it was only at this relatively late date in the 19th century that the existence of the Sumerian civilisation became known.

Cuneiform (meaning 'wedge-shaped') writing appears to have been invented in the city of Uruk itself sometime around 3200 BC and gradually evolved throughout the course of the civilisation without losing its basic character. It began as a system of pictograms, which then evolved into mnemonic aids and finally into a system of signs that had a one-to-one correspondence with spoken sounds. The symbols were impressed and incised into tablets of damp clay, the majority of which were hand-held. This had disadvantages for the Sumerian scribes – long texts needed multiple tablets, while large tablets were difficult to hold for writing. But for archaeologists this medium was ideal, far better than, say, papyrus scrolls, because many thousands of tablets have survived. Some were baked hard in the sun after being stored or discarded while others were deliberately fired before being placed in the ancient libraries.

Just as the cuneiform symbols gradually evolved through time, so did the uses to which the writing was put. The dominant use was always for administration, primarily concerned with the movement of commodities, the sale and purchase of land, the measurement of yields and so forth. Another pervasive use was for making lists – a means of organising knowledge that is difficult to do in the head alone. The most famous list is that of the Sumerian kings, written in the early second millennium BC. There appears, as ever, a blurred line between recording historical fact and propaganda: some of the early kings are described as having reigned for thousands of years.

While some of the records of sales and lists of commodities might appear dull reading for us today, the impact of such documents on economic growth must have been profound. Writing constitutes a form of external memory, extending the capacity of the human brain and permitting levels of planning, problem solving and decision making that would not otherwise be possible.

During and especially after the Early Dynastic period, writing was put to rather more diverse usages, including legal documents, Royal inscriptions, letters, proverbs, hymns to the temples and epic stories about the Creation, mythic heroes and the Gods. It is from the latter that we gain further insights into the Mesopotamian mind with the Epic of Gilgamesh recounting a flood and indeed a whole cosmology within which water plays an absolutely central role.

The ability to write and read must have been restricted to a small echelon of people, the scribes who were most likely employed and educated by the temples. It would have provided them with not only high status but also considerable power, especially in light of the key administrative focus of tablets. The proliferation of documents in the Third Ur Dynasty might suggest literacy was becoming more widespread – King Šulgi himself claimed to be able to write.

With regard to reading the landscape, the key contribution came from Robert McCormick Adams, supported by his colleagues and students from the Oriental Institute of Chicago. Between the 1950s and 1970s he led teams that mapped the distribution and size of archaeological sites across the landscape, dating them by the type of pottery exposed on the surface. One of his findings was that archaeological sites were arranging themselves in lines over several kilometres. He surmised that these had once been located along the edges of canals, thus confirming the indications from written tablets and the implication of rainfall patterns that irrigation had been essential for farming. Fifty years after his surveys, satellite and Space Shuttle images can now trace the course of the canals themselves, many being impossible to see on the ground. These reveal a complex changing geography of river meanders, levées, canals and minor irrigation channels, many criss-crossing each other and near impossible to disentangle.

With regard to excavation, much of the key work was undertaken before Robert Adams's surveys. During the 1930s Thorkild Jacobsen, also from the Oriental Institute of Chicago, led research in the Diyala region of Iraq, excavating houses at the sites of Khafagh, Tell Asmar and Tell Agrab dating to between 3300 and 1800 BC. These remain as some of the most extensive excavations undertaken and, for our interests, revealed houses with in-built toilet and drainage facilities. Jacobsen's surveys in the Diyala area were the first to explore the relationship between settlement and irrigation by plotting the sites of each period on to maps and exploring whether the site alignments related to the courses of rivers and canals. In 1957 a combined project between Jacobsen, Adams and the Directorate General of Antiquities of Iraq had the specific aim of identifying ancient irrigation practices, drainage facilities and the potential salinisation of soils.

The result of that project leads us to a possible reason for the demise of Sumerian civilisation. But first we should consider whether water management had been the key to its emergence.

The significance of irrigation

As explained at the start of this book, one of the most provocative works of the middle of the twentieth century had been Karl Wittfogel's 1957 *Oriental Despotism*. This had claimed that irrigation had been inextricably linked to the presence of a centralised authority: Wittfogel proposed that only with the prior existence of a political and bureaucratic hierarchy could the planning, construction and maintenance of an irrigation system have been undertaken. In the 1950s the Sumerian civilisation appeared to be an exemplar: the economy had evidently been dependent upon irrigation and there had been kings with armies of bureaucrats to manage the system.

Robert Adams's surveys came to a different conclusion.[13] He found that many canals were associated with mere villages, dating to several centuries before the emergence of the Uruk towns and Early Dynastic city-states; there was no indication of a centralised authority for their planning or construction. If anything, it appeared that the existence of a canal and irrigation network may have been essential for the emergence of a centralised authority – the complete reversal of what Wittfogel had proposed.

This should not, in fact, have been surprising as there are numerous accounts of irrigation systems being constructed and maintained by tribal communities. One of the most important is the study by the anthropologist Professor Robert Fernea of the El Shabana tribe of the Daghara region of Iraq.[14] He believed that the traditional social and farming systems of this tribe were typical of those in Southern Iraq prior to changes in land management imposed by Ottoman rule in the later 19th century and then British rule after the First World War. The El Shabana had built and maintained irrigation systems without any need of a centralised authority. The canals had been developed in a piecemeal fashion and hence lacked the scale and ordered layout that one often assumes for irrigation systems. Sections of single canals were regularly cleaned of reeds and silt, repaired and extended by neighbouring communities. All of this was undertaken without any inordinate amount of effort and evidently without a centralised authority.

In this light, the Ubaid and early Uruk pre-state communities would not have had difficulty constructing their networks of irrigation channels and larger canals. Indeed, because of the natural braiding of the rivers, short off-shoot canals running for a few kilometres could be dug with minimal alteration of the water regime. The rivers were naturally elevated above the surrounding plain by the levées and hence by simply cutting an outlet through those banks, a sluice gate, water would flow into a channel by gravity. The water could then flow into a network of small channels surrounding the crops. In some cases more work had to be done, such as raising the level of the flowing water. This was sometimes achieved by constructing a barrage from reeds or baked brick and bitumen in the river channel. With irrigation, the rich alluvial soils became highly productive, notably for cereals with yields two or three times higher and with greater reliability than those from the rain-fed farming regions of Northern Mesopotamia.

While the construction of irrigation canals may not have been as challenging as Wittfogel had imagined, some degree of cooperation for their construction and use would have been required. Even a short stretch of canal two or three metres wide involved a substantial investment of labour, most likely requiring collaboration and coordination between several villages. Once it was built, a key issue was the amount of water taken from a channel by each set of farmers; if those close to the sluice gate at the river took too much water, there would be insufficient for those farmers further downstream. Nicholas Postgate suggests that the coordination of water extractions may have been undertaken by those referred to in the cuneiform texts as 'gugallum', which he translates as 'canal inspectors'.[15]

The levée banks provided the best soils – this being where the largest particles of sediment were deposited, resulting in good drainage. These banks supported an array of vegetables and fruit trees such as palms, which provided not only dates but wood and fibres for ropes and matting, together with apple trees, pomegranates, figs and vines. The nearby gardens and fields contained onions, garlic, legumes, herbs, spices and a variety of leafy plants.

The cereal fields were a little further from the river, being fed by the network of irrigation channels (Figure 3.2). Wheat and barley were the staples, being sown in the autumn and harvested in the spring. They were grown not only for making bread, but also for beer – the texts describe nine different types – and to provide animal fodder. Flax was also cultivated for both its linen fibres and its oil-bearing seeds.

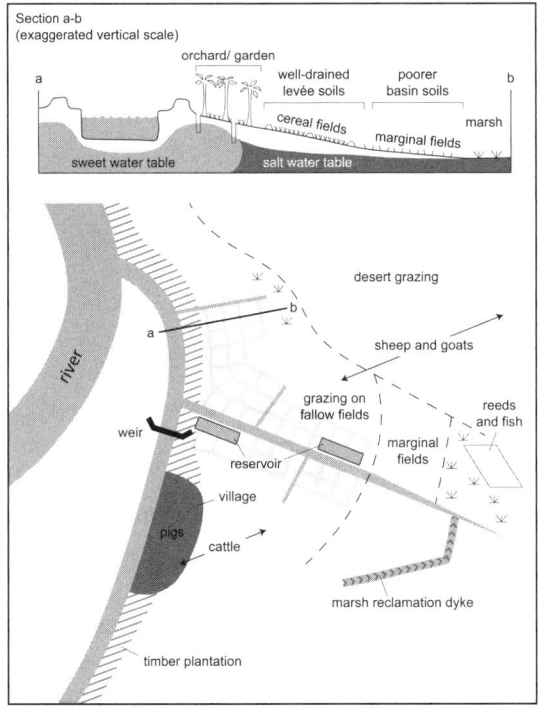

*3.2 Hypothetical layout of an agricultural complex
in Southern Mesopotamia (after Postgate 1994)*

Further away from the river the alluvium became dominated by
fine particles of silt with poorer drainage. Here the land was used as
pasture for sheep, goats and cattle, the latter being key for transport
as well as for their meat, milk and other products. Wool, and the con-
sequent textile manufacture, was of fundamental importance for the
Mesopotamian economy, while the sheep and goats also provided
meat, fat, milk and dairy products. Pigs were also kept for their meat.
With such a mix of arable and animal products, along with the hunting of wild game, catching of turtles, fishing and fowling, and collecting wild plant foods in the marshes, one can readily appreciate that
there was a prolific economic base to support urban communities.

To summarise, two key attributes were present: first, a sufficiently
productive farming base to generate surpluses to feed the non-
farmers: craftsmen, priests, soldiers, bureaucrats and the kings;

second, a mosaic landscape with different types of produce in different areas within easy reach of each other to stimulate the development of trade.

Coping with uncertainty, silt and salts

It may sound like a Garden of Eden, but the Sumerian farmers were in fact faced with many challenges. One of these was annual floods from the rivers, caused by snow melt in the mountains. This came too late for the spring sowing and too early for those of the winter season. Young plants were always in danger of being washed away, while the occasional massive floods threatened the sluice gates, regulators and even banks of the channels themselves. The flatness of the land meant that huge areas could be rapidly inundated with flood water and consequently a network of earthen dykes had to be built and maintained. The flood might, of course, be minimal or even not come at all; there might be particularly low rainfall, even drought. At such times the irrigation system would have to function as efficiently as possible, requiring that the channels and canals were constantly maintained. Life was full of uncertainty regarding water supply.

One means of adaptation was by water storage. Alongside some of the major rivers and canals, long narrow basins were constructed, referred to as nag-kud within the texts. Although some were as much as 90 metres in length and two metres deep, their capacity was insufficient to have had any impact on the irrigation of cereals, especially when one takes into account the high rate of evaporation. They may have been designed to supply water for the vegetable crops or have acted as settling tanks to remove silt from water for domestic use. Water would have also been drawn from wells dug through the levées or even lifted directly from the rivers themselves.

The consequence of a flood or a drought might be a substantial change in the course of the river channel itself. Within a relatively short period of time, a town or even a group of towns could find themselves in the midst of a desert after having grown up along the banks of a river. New generations of people would have been constantly shifting around the landscape as access to water changed over the decades. We find the evidence in the archaeological record: a pattern of the repeated abandonment and resettlement of towns and localities.

61

In addition to the uncertainty about rainfall and river flow, there was the constant challenge of silt. This would accumulate within the irrigation channels to eventually cause a blockage, requiring them to be regularly cleared to maintain the functioning of the system. It had to be shovelled on to the levée banks, where it would dry and then blow onto the fields. Reeds and other vegetation would also need to be cleared – the coordination of such work was possibly another role for the 'gugallum'. And then there was the accumulation of salts with the soil – to which we will shortly return.

Flood, famine and the gods

Mental adaptation to this uncertain and challenging world was by attributing the vicissitudes of nature to the will of the gods. In this regard the Sumerians were no different to many other ancient societies. The creation and control of the water supply by the gods is a central theme of their mythology. Parts of this were only written down towards the very end of our period of interest, and in some cases several centuries afterwards, but it is generally believed that these record the stories that had sustained Sumerian civilisation from its earliest times. The Enuma Elish, for instance, is 1,000 lines of text written on seven clay tablets in the seventh century BC and records the Creation. In the story the god, Marduk, plays the central role, creating the sources of fresh water in both heaven and earth and appointing himself as the controller of the weather. He used the body of the goddess Tiamut to create fresh water on earth, opening up the Tigris and Euphrates in her eyes.[16]

The most widely known epic story from Mesopotamia is that about Gilgamesh, King of Uruk, which famously includes an account of the Flood. The most complete version comes from 12 clay tablets written in the seventh century BC, but fragments of the same story survive from much earlier. Of most importance are the three tablets from the 18th century BC telling the epic story of Atra-Hasis, the Sumerian king of Shuruppak. This includes accounts of both the Creation and the Flood. The first tablet describes how mankind was created to alleviate the hard work the gods had to do, which included the digging and maintenance of canals. Having created mankind and watched it multiply in number, the gods were disturbed by all the noise and decided to find peace again by eradicating mankind. One of their attempts is described in the second tablet: 'Adad [god of weather] should withhold his rain / and below, the flood should not come up from the abyss / ... / let the fields diminish their yields ...' and '... the black fields became white, / the broad plain was choked with salt'. The famine, partly caused by the salinisation so poetically described, failed to eradicate humankind and hence a devastating flood was sent that would 'tear up the mooring poles' and 'make the dykes overflow'.

Productivity, trade and water transport

That the Sumerian farmers were able to cope with floods and droughts – sent by the gods or otherwise – is proven by the agricultural productivity of the landscape, attested within the written records. There are, for instance, cuneiform tablets from the cities of Umma and Girsu referring to the amount of grain loaded into boats for shipping to storehouses at Ur and elsewhere. These indicate that between 49 and 69 tons were being shipped within a single year. In one of these documents, a bureaucrat estimated that in one single trip to Ur 8,700 'gur' of barley would need shipping – the equivalent of 2,610,000 litres. Twelve boats were going to be needed, each carrying 76 tons.[17]

Agricultural productivity does not automatically lead to cultural fluorescence, let alone what might be termed a civilisation: it is a necessary but not a sufficient condition. Another such condition is trade, essential not only for deriving the raw materials for daily life – stone, timber, foodstuffs – but also the exotic items essential for specialists' crafts and ultimately for the elite to legitimise their power.

Southern Mesopotamia was no exception. The development of local and then long-distance trade was critical to its development. Indeed some academics, such as Professor Guillermo Algaze from the University of California, argue that it was the ease and extent of trade in Southern Mesopotamia that caused this region to have a cultural 'take-off' earlier than in the north or anywhere else in the Middle East. According to Algaze: 'where trade flows, its ramifications of increasing social complexity and urbanisation soon follow. Thus the precocious development of southern Mesopotamia throughout the fourth millennium BC comes as no surprise.'[18] The trade could flow largely because water was flowing: it was the canal system that enabled trade to flourish and civilisation to emerge.

Southern Mesopotamia either had a shortage of or lacked entirely many of the raw materials that were essential for the shift from farming villages to urban communities: roofing-grade timber, wood, copper, tin, silver, gold, limestone, granite, chert, semi-precious stones and so forth. These came from trade with all surrounding communities, but especially with those in the uplands to the north. Some of the items originated from great distances, such as lapis lazuli coming from what is now Afghanistan. There may have been a trade in slaves coming from the desert tribesmen to the west, while seaborne trade from around the Gulf brought in, among other products, copper from Oman.

In return for all of these goods, the Southern Mesopotamian traders provided manufactured goods, especially textiles, together with hides, dried fruit and salted fish. Fortunately, the flat alluvial plain and network of natural and manmade waterways made possible both low transportation costs and route ways between all major city-states. These allowed the movement of not only materials but also labour, information and ideas. A variety of vessels were used: simple log rafts mounted on inflated animal skins, coracles, canoes and bitumen-covered boats.

Guillermo Algaze argues that the benefits of trade became increasingly significant during the Ubaid period. The villages in the immediate vicinity of the major rivers and canals had orchard crops, vegetables and grain to exchange for the wool, hides, goat hair, meat and dairy products produced by those with animal herds living away from the fertile soils; both groups needed to exchange their produce with the people living around the lagoons and marshes near the Gulf, who had rich supplies of fish, fowl and reeds – the latter being a key building material. This trade in produce, Algaze argues, contributed to the emergence of social elites who then became aware of the economic and social implications of trade; they began to organise and control the trade, acquiring prestige goods as a means to validate their status. As their power and demands grew, trade had to become more extensive, reaching into foreign lands for both ever more exotic items and the basic building materials for their palaces and temples.

The domestication of the donkey in the middle of the fourth

millennium BC greatly facilitated the trading networks of the Southern Mesopotamian cities. Previously, people had to carry their own goods but pack animals greatly enhanced the amount that could be carried and distances travelled and were not restricted to waterways. As such the produce of 'The Land' could now be taken far beyond the bounds of Mesopotamia itself – and even more exotic items received in return.

The key trade routes within Mesopotamia itself were from the north to the south, using the flow of the rivers and then relying on donkeys to tow the boats back upstream. Of equal importance, however, were the transverse movements across the alluvial plain using the network of canals and minor rivers. By these means the city-states were all interconnected, enabling trade to have that immensely stimulating effect on economic growth, the spread of ideas and the emergence of social elites.

Here – according to Algaze – was a key difference to Northern Mesopotamia, where the villages and towns of the fifth millennium BC did not experience the same 'take-off' towards civilisations. In the north, the Tigris, Euphrates and their tributaries were deeply incised into the landscape, preventing the construction of transverse canals. Towns were relatively isolated from each other, having to conduct their trade overland, with the goods being carried by pack animals, wheeled carts or by people themselves. That was both slow and limited in capacity when compared to waterborne transport in the south; indeed Algaze estimates that waterborne transport had been 170 times more efficient than that overland.

We have therefore not two but *three* vital ingredients for the emergence of Sumerian civilisation in the fourth millennium BC. First, an alluvial landscape that when irrigated could produce high yields of grain to provide the staple foodstuff along with a diverse range of other plant and animal produce. Second, Southern Mesopotamia was also a mosaic landscape with a wide variety of produce and raw materials in different areas that encouraged communities to engage in trade. Third, there was a network of waterways, natural, modified and entirely manmade, that enabled this trade to be undertaken with relative speed and on a substantial scale.

Whether these three factors alone were sufficient for Sumerian civilisation to emerge is a moot point. One could also refer to the new means by which labour was organised: Algaze proposes that the towns 'domesticated people' to form mass labour forces in an equivalent manner to which Neolithic villages had domesticated animals. The significance of recording and then writing systems must also be stressed. But many would argue that the critical fourth ingredient must have been the presence of particularly ambitious individuals, those with either exceptional leadership and management skills or who were thirsty for power and prepared to use force to secure it – or perhaps all of these characteristics. We do not need to debate that issue here – in any case our knowledge of the early Ubaid period is inadequate to find the answer. All we need to acknowledge is that water management, for both irrigation and transportation, was vital to the emergence and flourishing of Sumerian civilisation throughout its 1,500-year existence. Now we can turn to its fall.

Salinisation: causes, consequences and avoidance

I am told that if one were to visit Iraq today, especially the land to the south of Baghdad, much of it would appear as a desolate, flat brown expanse, occasionally cut by canals and with irrigated fields.[19] Some areas would have a white crusty appearance from the accumulation of salts within the soil. These fields are no longer fertile: high salt concentration impedes seed germination and inhibits the absorption of nutrients by roots. Those plants which are relatively salt-tolerant, such as barley, provide low yields. Willows and poplars which had once grown on the levées have been replaced by dense thickets of the more salt-tolerant tamarisk. Could the process of salinisation in Southern Mesopotamia have begun several thousand years ago during the Sumerian civilisation? Could it have been sufficiently serious to halt its further development or even cause its demise?

The documentary record attributes the successive collapses of the Akkadian, the Third Ur Dynasty and the First Dynasty of Babylon to military and political conflict within 'The Land', and the incursion of invaders from the Zagros Mountains and the western deserts. But what lay behind such instability and the failure of any one city-state to gain and then sustain overall power? Could it have been an overall economic decline from falling yields?

The salinisation of soil is a persistent risk for any region which makes extensive use of irrigation. The salts are carried by water and remain within the soil after plant roots have extracted the moisture. Unless removed, salts accumulate within the root zone and eventually poison the plants. In regions where there is sufficient rainfall – such as Northern Mesopotamia – the salts are rapidly leached downwards, away from the roots.

Leaching can also be achieved by irrigating to excess, beyond the actual water needs of the plants. This, however, is only a temporary solution because the salt-laden water accumulates in the subsoil and begins to raise the water table. As this approaches the topsoil, capillary action comes into play, drawing water to the surface, where it evaporates, reinfusing the soil with salts. Ultimately, the only way to avoid salinisation is to introduce a system of ground water drainage either through a series of open ditches or by installing underground pipes. In either case, the salt-laden ground water must be disposed of in a manner to avoid the contamination of further land – ideally into the sea.

The nineteenth- and early twentieth-century tribal communities of Iraq, those prior to the imposition of British rule as studied by the anthropologist Robert Fernea, coped with the threat of salinisation by a system of alternate-year fallow.[20] Up to half of the land suitable for cultivation was simply left out of production each year. After harvest, the fields chosen for fallow were left for the growth of two desert plants, shok and agul, both having especially deep roots. These went dormant during the winter months and then thrived during the following spring. As they did so, they drew moisture from the water table and dried out both the surface and the subsoil to a depth of two metres, while also reducing wind erosion. Being legumes, they also replenished the soil with nitrogen. When cultivation started once again the following autumn, the subsoil was sufficiently dry for the irrigation water to leach salts away from the root zone.

Ultimately, however, salt accumulation in the subsoil could not be entirely avoided and even the shok and agul became unable to dry out the land. When that happened the farmers would have to remove the salinised topsoil and then leave the field for anything up to 50 or even 100 years to recover.

While the fallowing system is effective, it requires half of the potentially productive land to be left out of cultivation each year. As such, the traditional tribal system of land tenure was of critical importance.

The tribe as a whole maintained sufficient land for the alternate fallowing system, while farming plots were continually shifted around members of the group; kinship ties and obligations of hospitality ensured that the risks of low productivity and the costs of fallowing were shared throughout the tribe. Overall, by embedding the system of fallowing into tribal social organisation the threat of salinisation was minimised. As the geographer J. C. Russell – a colleague of Thorkild Jacobsen – once described, it constituted 'a beautiful system for living with salinity. The rural villages understand it in that they know it works, and they know how to do it and they insist on it.'[21]

When Ottoman and more especially British rule were imposed, the need for extensive fallowing and the significance of tribal social mechanisms were not appreciated. Systems of land ownership, often with absentee landlords, were imposed; labourers and farmers lacking in local knowledge were brought in to cultivate the land. The imposition of rents, taxes and debts meant that smallholders had to violate fallow to meet the short-term demands; this had the consequence of reducing their livestock because the fallow had supplied the animals' feed. The overall consequence was progressive salinisation throughout the 19th and 20th centuries, exacerbated by the construction of massive new canals.

The Sumerian salinity thesis

One of the outcomes of the Diyala Basin Archaeological Project led by Thorkild Jacobsen and Robert Adams was the proposal that salinisation had had a severe impact on Mesopotamian agriculture. Salinisation was, they argued, the key reason for the shift of economic and political power from the south to the north. In 1958 they published a seminal article in *Science*, the premier science journal, entitled 'Salt and Silt in Ancient Mesopotamian Agriculture.'[22] This argued that salinisation had indeed impacted on the Sumerian civilisation. They explained how salinisation had persisted into the present day, and needed to be fully understood in the light of new irrigation projects being planned by the Iraqi government.

The potential risk of salinisation in Ancient Mesopotamia was unavoidable: the Tigris and Euphrates carried salts dissolved from the sedimentary rocks of the northern mountains into the south, and these were taken into the fields when water was drawn for irrigation. As water evaporated, calcium and magnesium precipitated as carbonates, leaving sodium ions dominant in the soil solution. The key question was whether the extent of salinisation had become sufficiently serious in ancient times to have had a significant impact on Sumerian agricultural productivity. The Diyala Project examined the cuneiform texts and concluded that salinisation had been very serious indeed, most notably between 2400 and 1700 BC – precisely the period that saw the demise of Sumerian civilisation.

Jacobsen and Adams were able to point to a specific historical event that, they claimed, played a key role in raising the water table and degrading the soil with salts. As I noted above, the independent city-states were often in dispute with each other. One of the long-running disputes had been between the cities of Girsu and Umma, neighbours along a water course stemming from the Euphrates. They clashed over the fertile land that lay between them. Girsu gained control of this land but Umma, situated to its west and higher along the water course, was able to breach or obstruct the canals that brought the water for irrigation. With his protests ignored, Entemenak, the king of Girsu, built a new canal to irrigate his land by drawing water from the Tigris. This flowed to the east of Girsu and hence the new canal was entirely out of Umma's control. By 1700 BC this canal was supplying copious amounts of water to an extensive area to the west of Girsu, far more than the Euphrates had ever provided. Indeed the new canal itself was being referred to as the Tigris River. This, according to Jacobsen and Adams, resulted in over-irrigation. A decisive rise in the water table followed, leading to salinisation and a dramatic loss in fertility.

Jacobsen and Adams drew on three sources of evidence to support their case. First, soon after the reign of Entemenak, texts begin to describe patches of saline ground and refer to specific fields that had become infused with salt. Second, during the course of the Sumerian civilisation there had been a shift from the more productive wheat to the more salt-tolerant barley. In 3500 BC these had been grown in equal proportions, or at least that was the claim based on the impressions of grain in excavated pottery. Texts from the reign of Entemenak dating to around 2500 BC indicated that wheat now formed a mere sixth of the cereal harvest; further texts suggest that it continued to decrease in significance and by 1700 BC had been abandoned altogether. Third, there had been an overall decline in fertility that progressively reduced even the yields of barley and can be attributed to salinisation. Jacobsen and Adams cite texts dating to 2400 BC from Girsu that indicate yields of 2,537 litres per hectare – high yields even by modern-day standards. By 2100 BC, however, these had declined to 1,460 litres while texts from the nearby city of Larsa dating to 1700 BC cite 897 litres as the average yield.

With such a dramatic fall in yields, could sufficient surpluses have been produced to maintain the army of bureaucrats, priests, crafts-men, soldiers and merchants that urban life requires? How could the kings claim authority if the gods had lost confidence in them by causing the harvests to repeatedly fail? Jacobsen and Adams thought salinisation had been a decisive factor and the catalyst for the shift of economic and political power to the north of 'The Land', to Babylonia, and eventually to Northern Mesopotamia. The salinisation had led to a 'disastrous general decline' of fertility and so the Sumerian cities were either entirely abandoned and left in ruins or they dwindled to villages.

Against the salinity thesis

Theories in archaeology are never allowed to go unchallenged, espe-cially when they deal with such topics as the rise and fall of civili-sations. The key challenge to Jacobsen's and Adams's salinity thesis came after nearly 30 years had elapsed, in a long detailed argument published in the academically prestigious but rather inaccessible jour-nal *Zeitschrift der Assyrologie* in 1985.[23] This was by Marvin Powell, a specialist on cuneiform texts based at Northern Illinois University.

Powell took Jacobsen and Adams to task for an incomplete and often erroneous analysis of the textual record. A key issue was with the specific meaning of the cuneiform words that the salinity thesis relied upon. For instance, a word that is transcribed as 'gú-na-bi' could mean either 'yield' or 'revenue'; in some circumstances these might be identical but would normally be quite different and could give a misleading impression if read incorrectly. According to Powell, Jacobson and Adams erroneously combined information about yields and about revenues from different texts to derive their overall estimate of yields, such as that of 1,460 litres for 2100 BC. They also failed to take into account the different areas of land in cultivation at different times, and the fact that low yields might arise from factors unrelated to salinisation.

While acknowledging that numerous texts written after the construction of Entemenak's canal refer to soil salinity, Powell disputes that there is sufficient evidence to conclude a cause and effect relationship. He claims that we have insufficient information about the particular salt tolerance of the strains of wheat and barley that had been used, suggests that the evidence for a shift from wheat to barley is equivocal, and also suggests that even if it did occur this could have been for reasons unrelated to salinisation. Powell then argues that another error by Jacobsen and Adams had been their misinterpretation of the textual evidence regarding seeding rates. Without knowing how much seed was planted one cannot make an accurate estimation of the return. Not a single document exists that records both the amount of seed and the size of harvest from a specific plot of ground and consequently a great deal of conjecture has to be involved in stating a decline in agricultural productivity. Powell meticulously re-examines each of the texts drawn upon by Jacobsen and Adams pointing to numerous misinterpretations of cuneiform words, phrases and numbers.

The final aspect of Powell's critique is that Jacobsen and Adams failed to credit the Sumerians with sufficient ability to combat the threat of salinisation by the use of fallow and leaching – a further example of their misinterpretation of the documentary evidence. One incised tablet names a field as Gana Kimun, translated as 'Salt Ground Field'. This is described as 'dag giš-bar', which Jacobsen and Adams as well as Powell read as 'fallow'. Why, Powell asks, would a record have been made of a saline field as being fallow if this had not been in the process of reclamation? Similarly he argues that the term 'ki-duru', which Jacobsen and Adams had translated as 'moist ground', actually means land that has been 'soaked in water immediately before ploughing' and attests to the use of artificial leaching to combat salinity.

Overall, Powell does not deny that the Sumerians had faced the risk of salinisation. He simply questions the evidence that it had ever been sufficiently serious to impact on agricultural productivity, noting that the Sumerians appear to have used an alternate-fallow system and leaching in the same manner as recent tribal communities.

The salinity thesis defended: the failings of state government

The academic world had to wait a mere three years for a rebuttal of Powell's arguments. It came in the form of a relatively short note in an academic journal entitled *Geoarcheology*, appropriately co-authored by an archaeologist, Michael Artzay (University of Haifa) and a soil scientist, Daniel Hillel (University of Massachusetts).[24] Rather than challenging Powell's reinterpretation of the cuneiform tablets, Artzay and Hillel simply pointed to the inexorable nature of the salinisation process in environments such as Southern Mesopotamia. Powell had overlooked, they argued, the basic fact that without intensive forms of modern pump-driven underground drainage, irrigation-based agriculture in such environments is simply unsustainable in the long run.

But what of the application of alternate-fallowing – that which J. C. Russell described as a 'beautiful system for living with salinity'? Powell is surely correct that such fallowing would have been used in Southern Mesopotamia, even if the textual evidence he draws upon is as ambiguous as anything used by Jacobsen and Adams. Why could not this and perhaps the use of artificial leaching have sufficed to avert the impact of salinisation?

A persuasive answer was provided in 1974, in an article written by McGuire Gibson, also of the Oriental Institute of Chicago, entitled 'Violation of Fallow and Engineered Disaster in Mesopotamian Civilization.'[25] He suggested that the relatively recent impact of British rule in Iraq, which had destroyed the tribal arrangement of land management as reconstructed by Robert Fernea, had simply been the most recent example of a state government inadvertently causing economic decline by a process of agricultural intensification.

By the time that the city-states began to form at the start of the Uruk period in 3200 BC, communities had been practising irrigation-based farming in Southern Mesopotamia for close on 2,000 years – throughout the Ubaid period from 5000 BC. We have no evidence, but it would be surprising if some form of alternate-fallow system had not been in place to combat salinisation and to maintain the general fertility of the land.

The formation of the city-states involved a massive process of urbanisation. People who had once been subsistence farmers, or at least who would have taken up that role from their parents, now crowded into the new towns. They found new roles as craftsmen, administrators, merchants, soldiers and so forth. Agricultural surpluses were required to support them, as well as the manual labourers who built the new temples and palaces – most likely a combination of slaves and corvée labour. The rulers needed to fund their military adventures and trade for exotic goods, and did so in the tried and tested manner: by imposing taxes on the farmers. As such there was further pressure for increased agricultural production.

In such circumstances one can readily imagine short-term profits beginning to take precedence over the long-term management of the land. Just as happened in recent times, workmen lacking any knowledge of farming or vested interest in the land would have been set to work in the fields. They may have worked for a new generation of farmers, people arriving from the desert margins. Having been pastoralists, they were now attracted to the culturally vibrant alluvial plain but lacked an understanding of how to maintain soil fertility. With the disintegration of the traditional Ubaid tribal society, farming became intensive, providing for short-term need at the cost of long-term sustainability; the needs of the soil were ignored through ignorance as much as neglect.

The drive for greater productivity might be seen in the ambition of the irrigation schemes undertaken in the Akkadian period and after, during the Third Dynasty of Ur and the First Dynasty of Babylon. These were on a massive scale, such as the effective redirection of the Tigris by Entemenak of Girsu, and would have required a central authority to manage the projects. Canal construction in the Uruk period was rarely mentioned in the texts, presumably because it was undertaken at a local level; the new schemes were described within royal inscriptions.

McGuire Gibson suggests that this process of intensification is reflected in a surge of officials concerned with irrigation and its administration. We see this in the law codes with an array of regulations concerning water management, suggesting a loss of local knowledge. For instance the laws of Ur-Nammu (2115–2085 BC) refer to the penalty that a farmer will face if he floods the fields of another. According to McGuire Gibson, the proliferation of rules, regulations and administrators reflects attempts by the kings to halt the collapse in yields by devising ever more irrigation schemes or reforms in their administration. Ultimately, however, the intensive-farming system could not be sustained. As a result the power of the state government declined, urban life lost its attraction and the tribal-based farming system tentatively reasserted itself. Soon the cycle of city-state growth would begin again with urbanisation, taxation, a shift towards intensive farming, an initial increase of yields and then collapse as the land became infused with salt once more. This was not, however, a cycle but a downward spiral: ultimately there could be no return to the traditional system of alternate-fallow.

The salinity thesis originally proposed by Jacobsen and Adams in 1958 appears robust. They may have misinterpreted some cuneiform tablets and pushed the limited textual evidence beyond its reasonable limits. But the essence of their argument is too compelling to resist: by drawing salt into the soils, irrigation played a key role in the demise of Sumerian civilisation, just as it had made possible its original fluorescence.

4

'WATER IS THE BEST THING OF ALL'
– PINDAR OF THEBES 476 BC

Water management by the Minoans, Mycenaeans, and
Ancient Greeks 2100–146 BC

On 12 September 1903, the *British Medical Journal* (*BMJ*) carried the following description of a recently discovered flush toilet:

It consists of a small chamber about a metre in width and two metres in length, resembling the shape and size of a similar room in a modern house. Outside the lintel of the door is a flag sloped towards a hole, which opens into a short drain beneath the floor, and down which water could be thrown for flushing purposes, and also the household 'slops'. From the groove seen in the wall there appears to have been a wooden seat about 57cm from the ground.[1]

Captain T. H. M. Clarke, MB, DSO, RAMC, Medical Advisor to HRH High Commissioner of Crete and author of this report, went on to speculate about a 'curious projection' that he thought might have been used to support an earthenware basin or receptacle for the excreta. 'Judging by the shape of the cavity, this basin would have been vertical in front and sloping at the back, or, in other words, it would resemble in shape the 'wash out' of the closet of the present day, in which a certain amount of water is kept in the basin by a ridge, over which the excreta are carried by a flush of water.'[2]

The toilet being imagined in such vivid terms for the readers of the *BMJ* had just been discovered at Knossos, the supposed palace of King Minos dating to 1800 BC. It was then in the process of excavation by the Oxford archaeologist Arthur Evans. For the many thousands of tourists who visit Knossos today, the room is on display and described as no less than 'The Queen's Toilet'.

The Captain was delighted with this discovery. He had already explained how no special training is required to appreciate the 'graceful statuary, the huge amphorae, the beautiful vases, the marble throne of prehistoric kings, or the lifelike figures in the brilliantly coloured frescos' that Arthur Evans had found in such quantity during his three years of excavation. But, the Captain continued, writing for the *BMJ* readers, 'amidst the debris of the Minoan civilisation at Knossos a discovery has been made which will appeal with peculiar interest to the members of our profession – namely that a large and well-designed system of sanitation existed in the palace'.[3] There can be little doubt that for Captain T. H. M. Clarke it was the sanitation rather than the graceful statues and brilliant frescoes that marked the Minoan culture out as truly the first European civilisation.

The Minoan civilisation

The hole in the floor at Knossos is the earliest known 'flush toilet' in the world, dating to around 1800 BC. It was just one element of a sophisticated system of water management that developed on the island of Crete during the Middle Bronze Age and was essential to the great cultural achievements of the Minoan culture (Figure 4.1).[4]

Minoan Crete was by far the most elaborate of the Bronze Age cultures of Greece and the Aegean islands. Contemporary communities on the mainland and other islands never achieved the Minoan level of architectural and artistic accomplishment even though the preceding Neolithic in all regions looks virtually identical – small villages and farmsteads cultivating cereals and legumes, herding sheep and goats.

Neolithic villages are the earliest known settlements in Crete, there being no evidence for preceding hunter-gatherers on the island. We don't know where the Neolithic colonists came from, but ultimately their lifestyle derived from that developed in the Levant as I described in Chapter 2. This occurred either through the spread of ideas, technology and domesticated seed to indigenous communities in Turkey and then Greece, or by the dispersal of Neolithic farmers themselves.

The Bronze Age towns of the Greek mainland have been described as 'humdrum' in comparison with those on Crete.[5] Two factors appear critical to the Minoan achievement.

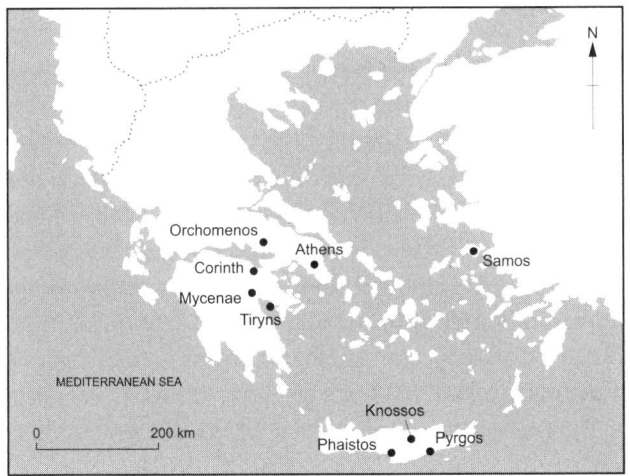

4.1 Map of Greece showing archaeological sites referred to in Chapter 4

First, the earlier Bronze cultures on the mainland were subject to waves of destruction, probably by the arrival of new populations; in contrast there was an uninterrupted period of cultural development from the Early (2700–2160 BC) to the Middle Bronze Age on Crete (2160–1600 BC), the latter being the period of great palatial buildings. Second, just at the time when maritime trade was flourishing, Crete was ideally positioned at the crossroads of Asia, Africa and Europe; this made it nothing less than the cultural centre of the ancient world.

The Minoan towns were substantial: between 80,000 and 100,000 people are estimated to have lived within the urban complex of Knossos. Large populations were found at the other Minoan centres while there was also an extensive rural population. At least four of the largest settlements were centred on so-called palaces, at Knossos, Phaistos, Malia and Zakros. The palaces shared a similar design: a large building with many rooms – more than 1,000 at Knossos – designed around the four sides of a rectangular central court, with another court built on its western side. The idea of a central court may have been influenced by contemporary palaces in Anatolia and Syria seen by Minoan travellers and traders, although they also appear to have been a natural development of the paved courtyards at the earlier Bronze Age settlements in Crete.

The Minoans traded with those on the mainland and Aegean islands; they also went much further afield and secured copper, tin, gold, ivory and fine stone from Syria and Egypt. There have been striking finds in Crete that testify to such trade, notably scarabs from the Middle Kingdom of Egypt and haematite Babylonian cylinders, most probably coming from Syria. With these material goods came knowledge, ideas and inspiration, including those about water management. These combined with locally developed techniques, building on those inherited from their Neolithic forebears, to create an infrastructure of hydraulic engineering that could sustain communities of kings, priests, farmers, traders, craftsmen and, most probably, slaves.

All of the palaces suffered from a severe earthquake at around 1700 BC. They were rebuilt to be much grander than before: multi-storeyed with staircases; processional pathways and monumental entrances; royal residences with throne rooms; the 'graceful statuary, the huge amphorae, the beautiful vases [and the] lifelike figures in the brilliantly coloured frescos' that had so impressed Captain Clarke in 1903; not to mention the faience beads, gold jewellery and figurines that continue to impress archaeologists, tourists and no doubt medical practitioners today.

Hydraulic engineering was required because rainfall was limited, as was the availability of springs. The climate of Bronze Age Crete seems unlikely to have been significantly different from that of today,[6] with its rainfall being heavily influenced by the three mountains of the island, Mounts White, Ida and Dikti. The mountainous west receives more than 1,800 millimetres of rain per annum while less than 300 millimetres falls in the east. The Minoan centres developed in the semi-arid regions which today receive less than 500 millimetres. Much of the rainfall is lost again almost immediately because of the hot sun and winds, causing an evaporation rate of more than 2,000 millimetres per annum.

A further critical factor, both today and in the past, is that the rainfall is uneven throughout the year: hardly any in the summer and heavy rainfall in the winter. As such, Minoan water management had to deal with three challenges: summer droughts, winter flash floods and simply ensuring that there was sufficient water to sustain the population throughout the year. There was perhaps a fourth: to instil a level of sanitation that was fitting for what was one of the most artistically elegant cultures that had ever appeared in human history – and which had to present itself as such to its overseas visitors.

Getting water in and out of Knossos

Knossos was the greatest of these palaces.[7] It was the centrepiece of an extensive Bronze Age town located on Kephala Hill, on the eastern side of the Kairatos River, located today five kilometres from the city of Heraklion. The site had been originally occupied soon after 7000 BC by Neolithic settlers; a mound of their debris accumulated as new mud-walled dwellings were built on the old. This was added to by the debris of an Early Bronze Age community. At around 1930 BC the mound was deliberately levelled and the first palace constructed. Initially a series of separate buildings around a central court, these were combined into a single piece of architecture and given storerooms, terraces, workshops, an impressive approach road and a monumental entrance. The Knossos palace had become not only an economic and administrative centre for the region but also one for religion and ritual – the boundaries between the sacred and profane being much more fluid than we are familiar with today.

Knossos reached its cultural apogee between 1700 and 1350 BC, after which it was abandoned and gradually became lost to human knowledge. Its rediscovery began in 1878 when Minos Kalokairinos, a Cretan merchant and antiquarian, dug into the side of Kephala Hill and exposed part of the storerooms. The Turkish owners of the site prevented him from continuing; they also thwarted an attempt by Heinrich Schliemann, the German businessman and archaeologist who had excavated at both Troy and Mycenae, to purchase the hill. But the site was eventually bought by Arthur Evans, a wealthy Englishman and Director of the Ashmolean Museum at Oxford. Evans began excavating in 1900; within three years he had exposed the remains of the palace – a rate of excavation that is as unimaginable as it is intolerable today. He continued working at Knossos until 1931, not only exposing the ruins but also undertaking his best-guess reconstructions (Photograph 8).

When visiting Knossos today it is virtually impossible to distinguish between what is original and what is reconstruction; few archaeologists would today rate Evans's 'best guesses' as having been good enough to have warranted such work. Enter the throne room, for instance, and you will see an alabaster throne-like chair that remains where it was originally located – there are excavation photographs showing it being exposed. But the surrounding paintings are as good as fiction. These depict two reclining griffins, one either side of the throne, looking terrifically magisterial. We know that the griffin was a mythological important creature for the Minoans as it is represented elsewhere. But those in the throne room were painted by a father and son team commissioned by Evans with hardly any evidence for such depictions. Equally, we must be extremely cautious about the names given to frescoes, objects and rooms that so often involve unwarranted interpretation – the 'Prince of Lilies' relief, the 'Snake Goddess', the 'Corridor of the Procession' and, of course, the 'Queen's Toilet'.

Fortunately pipes and drains are less open to fanciful interpretation (Photograph 9). The most striking evidence found by Evans was terracotta piping that had once formed parts of the water system at Knossos.[8] This piping came in sections 50–76 centimetres long, each tapered in design from 13 centimetres wide at one end to 9.5 centimetres at the other; they were designed to fit into one another, with the joint most likely having been cemented to be watertight. The tapering created water pressure within the pipes that would serve to flush any sediment along the pipe. These pipes formed a network below the palace floors. The pipes circulated water collected from rainfall, drawn from wells or coming from springs via aqueducts.[9]

The earliest aqueduct appears to have been from a spring at Mavrokolytbos, located 0.5 kilometres south of the palace at a slightly higher elevation. A combination of stone-cut conduits and terracotta pipes brought the water along the side of a ravine and into the palace. A much longer aqueduct may have brought water to Knossos from springs at Archanes, about 10 kilometres away, having branches to both the palace and the town.[10] Precisely who was served by these aqueducts remains unclear – the whole population or just the elite? The majority of water most likely arrived at Knossos within containers made of clay, leather or wood, carried by hand or on the backs of donkeys from the mountain springs.[11]

While getting sufficient water to Knossos was a challenge for most of the year, removing run-off water from the heavy winter rainfalls and disposing of that which had been used for domestic purposes was a necessity. For this there was a system of stone-built drains running beneath the buildings. These appear to have been partly fed by at least four vertical shafts from the upper storeys of the domestic buildings, perhaps catching the run-off from roofs. The drains were made of limestone slabs and lined with cement. One ran below the 'Queen's Toilet' and they all ultimately drained into the Kairatos River.

Collecting run-off at Phaistos

Run-off from winter rainfall could, of course, have been a valuable source of water to either supplement that from springs or as the principal source. One site where run-off may have been 'harvested' is Phaistos, the second-largest known Minoan centre. This is located on the south side of Crete, on top of a hill with dramatic views across the Messara plain with its sprinkling of villages to the south and the summit of Mount Ida to the north; to the west one can see the brilliant blue water of the Messara gulf just beyond the hill of Agia Triada, where a Minoan country house had been located.

Visiting Phaistos can be far more satisfying than Knossos: it lacks both the vast crowds of tourists and dubious reconstructions; indeed with the sultry smell of pine, constant chirruping of cicadas and hot Mediterranean sun it is an absolute archaeological joy. The palace layout is similar to that of Knossos with central and western court-yards flanked by buildings; storerooms with massive pithoe (ceramic vessels) that had been used for oil, grain, and olives; grand stair-cases and a theatre. The surviving buildings are a mix of first- and second-phase palaces, but the site appears to have an integrity that is somewhat lacking at Knossos. When walking around there is a constant reminder of a water system as complex at that of Knossos with stretches of terracotta pipes and stone conduits – although trying to piece together the water supply and drainage plan has not yet proved possible.

The most thought-provoking evidence for water management at Phaistos comes from the Western Court. This is a large paved area below the theatre which is crossed by a so-called 'raised processional walkway' (Photograph 10). When making a recent visit, the walkway was immediately evident, raised about 15 centimetres and stretching diagonally across the plaza. I had read that the plaza was thought to have been used for collecting run-off but I initially looked around bewildered because I couldn't see any means to do so.[12] Then it dawned on me – and I must have been suffering under the hot sun – that the courtyard as a whole was the harvesting device, sloping gently to the south and with shallow channels leading into round structures that were described in my guidebook as 'storage pits'. These were several metres in both diameter and depth and surely could only be cisterns (Photograph 11). Rainfall on to the theatre, the staircases and the roofs of the surrounding buildings would have drained on to the floor of the plaza and then filled the cisterns to capacity. The raised walkway may have been for elevating the Phaistos elite during ceremonial processions but was also a practical solution for crossing the plaza with dry feet.

Were the 'kouloures' cisterns? The evidence from Pyrgos

The presence or otherwise of cisterns in Minoan Crete is, in fact, an issue of controversy among archaeologists. There are five such circular structures at Phaistos. Four near-identical structures have been found at Knossos where they have been termed 'kouloures'. With the exception of one such structure at Knossos, these are all found set into or adjacent to the west courts of the palaces and appear to be associated with special walkways. Whatever they had contained must have been something rather special.

The idea that they were cisterns has been largely resisted because of the absence of a plaster lining. Alternative ideas have included depositories for sacred offerings, granaries, rubbish pits, settings for the planting of symbolically potent trees and soakaways for excessive surface water. Knossos also has another candidate for a cistern: a rock-cut space, 8.32 metres in diameter and at least 15 metres deep known as the Hypogeum located at the southern side of the palace. We only know this from the excavation records because, having been partially exposed, it was then reburied. This had a spiral staircase cut into its side and has been proposed as a guard house, a storeroom and a granary. Evans rejected the idea that it could have been a cistern because its sides were not plastered, but having been rock cut that would not appear essential.

A further controversial example comes from the Minoan palace of Zakros.[13] Here there is a spacious rectangular hall immediately next to the Central Court of the palace that has been termed the 'Hall of the Cistern'. Inside there is a circular stone-built underground structure that very much looks to be a cistern. It collected water from a spring, although it is full of brackish water today because of seawater intrusion. This structure is seven metres in diameter with cement-covered walls and was surrounded by five columns – nothing else like it is known in Minoan Crete. Several ideas have been proposed for this water-holding structure other than for merely the storage of water: a swimming pool, an aquarium, water on which a sacred boat was kept?

The possibility that some of these Minoan structures had been cisterns for water storage has been raised by the discovery of two further structures at the Minoan site of Pyrgos on the southern coast of eastern Crete.[14] This is a relatively small hilltop settlement when compared to Knossos and Phaistos, covering a mere 0.5 hectares, with steep ascents on the south-east, south and west sides; the hill lies to the immediate east of the mouth of the Myrtos River which would have been the only source of water if there had been no means to capture run-off, the local geology being unsuitable for the sinking of wells. Water would have had to be carried up the hill in buckets either by hand or using donkeys.

At around 1700 BC there was a substantial spurt of building activity at Pyrgos that included two large, circular structures (Figure 4.2). Gerald Cadogan, a leading specialist on Minoan archaeology and excavator of Pyrgos, is convinced that one of these at least is a cistern used to capture run-off water. Both have vertical walls and rounded bottoms lined with white lime plaster applied over a surface of small river pebbles on their base and roughly hewn stone blocks on the walls. Neither has evidence for a roof, but could easily have been covered by awnings.

The most likely candidate for a cistern is located five metres below the peak of the hill within a flat paved courtyard adjacent to a large building known as the Pyrgos 'Country House'. This comes from a later phase of building activity but is assumed to stand on the footprint of a preceding house, contemporary with the circular structure. Cadogan excavated just one quarter of this structure, finding it to be almost 3.5 metres in diameter and at least 2.5 metres deep. As there was nothing to prevent run-off flowing from the courtyard into the structure, it would have been of no value for storing grain. What else could it have been other than a cistern?

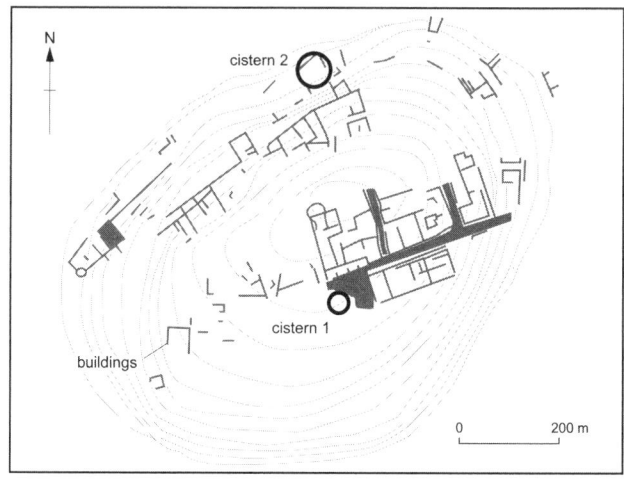

4.2 The Bronze Age settlement of Myrtos-Pyrgos, Crete, showing location of the two cisterns (after Cadogan 2007)

The second structure has three times the capacity, being 5.3 metres in diameter and at least three metres deep, but is of a more irregular shape. This was located just two metres from the summit, close to the edge of the settlement and partly built into the slope of the hill itself. It was well positioned to capture run-off from roofs and there appears to have been a number of open drains channelling water into the 'cistern'. But there appears to have been a design fault: the down-slope side of the structure had broken away in antiquity, its stones being splayed outwards. It looks as if the weight of water had been too great and it burst open, spilling water down the hill.

Cadogan asked why cisterns should have been built on the hill at Pyrgos when water was available in the river below. Perhaps there had been a shortage of labour to carry water up the hill? This seems unlikely when one considers the labour demands of the building programmes that occurred at the site. Perhaps the valley was infested with mosquitoes that carried malaria? If so, surely so would have been the standing water within the cisterns.

Cadogan is tempted by the idea of defence. Although archaeologists have traditionally believed that there was a 'Pax Minoica' throughout the island, providing continuous peace that enabled the cultural developments to flourish, he points to an increasing amount of evidence for destruction of settlements by fire. So perhaps the cisterns had been built because of the threat of siege. Immediately upslope from the cisterns are terrace walls assumed to be for defence, along with a tower-bastion. Ultimately, however, he favours a social explanation:

we see here a decision – unique in Minoan Crete at the time as well as later – to make a substantial water supply available in the hilltop settlement and provide (virtually) unlimited supplies of a vital commodity ... this would have had a wide propaganda impact on other settlements in Crete (perhaps leading even to competitive, warlike relations), while enabling the people of Pyrgos to enjoy the luxury of a self-sufficient abundance of water – a situation that is always something to covet in the Mediterranean.[15]

The cisterns – if that is what they had been – did not remain in use for long, the river becoming once again the only source of water. In the next phase of development at Pyrgos, the larger of the two circular structures became a rubbish dump, as did its immediate surroundings. Intriguingly, the cistern adjacent to what had become the Country House had also been deliberately filled in, but with carefully selected river pebbles: grey limestone stones with white veining. Cadogan thinks that this was symbolic: 'Thus, these pebbles and the former cistern that held them could have kept the rich memory of the history of the community and the vicissitudes of its water supply – and so acted, or could be used, as an important tool for social cohesion.'[16]

Mycenaean hydraulics, 1900–1150 BC

While the Minoan civilisation was flourishing in Crete, its eventual successor was emerging on the Greek mainland – the Mycenaean culture.[17] Named after the citadel at Mycenae, first excavated in 1876 by Heinrich Schliemann, this was a warrior-led society that matched the Minoan achievements in architecture, crafts and extravagant works of art, reaching its peak in the 13th century BC. It originated in South and

Central Greece, and then expanded across the Aegean Islands, taking over Knossos following its destruction by an earthquake in 1450 BC.

We know the Mycenaean world through extensive archaeological research, using state-of-the art scientific methods in the field and laboratory. But however much one might try, it is near impossible to avoid interpretation of the material evidence without drawing on Homer's *Illiad*, the story of the Trojan wars. Although not written until the seventh or eighth centuries BC, Homer's stories are about a heroic warrior-led society and cite many of the known palaces and citadels of the Mycenaean World. Schliemann was in no doubt that he had discovered the golden burial mask of Agamemnon, the hero of the Trojan War, when he excavated the most sumptuous of the so-called 'shaft' graves at Mycenae.

Archaeological reality and Homeric myth

'Sumptuous' is hardly adequate for what Schliemann found: ornaments, diadems, rings, armbands and a variety of vessels made in gold and silver; ostrich egg shells from Africa; bowls carved from rock crystal; bronze swords and daggers with the most exquisite decoration. Agamemnon or not, these were certainly the graves of warrior-kings. Such Mycenaean treasures will be familiar to many readers; and not just artefacts but architectural treasures such as the dramatic Lion Gate of Mycenae itself – the most memorable sight from my first visit to Greece as a schoolboy – and the monumental 'tholos' tombs found throughout Greece. Such cultural achievements were dependent upon the intensive cultivation of grains, olives and cereals on fertile alluvial soils. This produced the surpluses to support not only the craftsmen, the armies and the palace bureaucracies, but also the extensive trade networks that supplied exotic goods, which were essential to maintaining the status of the warrior-kings.

Just as critical as managing the food supply was managing that of water – to the crops as irrigation, to the towns as drinking water and across the landscape as a whole, because the Mycenaeans were the first people to undertake land reclamation within Europe. Indeed, to my mind, Mycenaean hydraulic engineering is a far more important measure of their cultural achievement than the mere fripperies of gold masks, rock crystal bowls and bronze daggers.

The Ancient Greek writers, those from Homer onwards, recognised the significance of Mycenaean water management, although they deferred from saying so explicitly. Washing is a central theme throughout Homer's *Illiad* and *Odyssey*. He describes the washing of hands before prayer and the purification of an entire army by water.[18] Travellers are cleansed physically and psychologically by bathing after a long and taxing journey. Bathing acted as a social ritual, as a rite of passage prior to being received into an unfamiliar household. Men were washed, anointed with oil and then dressed by unfamiliar women with no sign of embarrassment or fear of their predicament.[19] So when archaeologists excavated the Mycenaean site of Pylos, believed to be King Nestor's palace as described by Homer, and found a room containing a bath set in plaster, water pithoi and a pottery goblet that may have held a sponge it became near impossible to avoid embellishing the archaeology with Homer's story of how pretty Polykaste had washed Telemachos when he arrived at King Nestor's palace, and then anointed him with oil and dressed him in a tunic so that he looked like an immortal God.

Those who know their Greek myths will be aware that one of the twelve labours of Hercules had been to clean the stables of King Augeias within a single day. He did so by diverting the River Alpheios to flow through those mucky stables. Hercules is said to have conducted his labours while living at the citadel of Tiryns on the Argive Plain of Greece. This is one of the most impressive Mycenaean sites, one with massive walls, bathrooms and wells. Remarkably, archaeological studies of the town and its surrounding landscape have shown that its Mycenaean rulers did indeed accomplish that particular Herculean task: they diverted a river. This was not, as far as we can tell, to clean stables but to protect their town from flood inundation.

Mycenaean land reclamation

We will come to Tiryns shortly, but we should first note that throughout the Mycenaean world we find archaeological evidence for irrigation and drainage systems, for cisterns and bathrooms, for dams and dikes.[20] Rock-cut passages brought water to the citadels at Athens, Tiryns and Mycenae from underground springs. At Pylos, water was conducted around the palace and workshop area within terracotta pipes, this having arrived at the city from a distant spring via a kilometre-long wooden aqueduct. At Mycenae, a bridge across the River Chaos, appropriately named because the winter rains would have made it a raging torrent, was specially designed to prevent erosion of the river banks that carried a key roadway. Constructed from large, rough limestone blocks and laid in regular courses packed with small stones and clay, this had a 'false-arched inverted V triangular' culvert in the middle for the passage of the torrent – this serving to restrict the flow and hence its erosive force.[21]

These examples are not significantly different from what we had seen within the Minoan world. But the Mycenaeans also undertook extensive drainage schemes that amounted to the earliest land reclamation known in Europe. The challenge they faced was that the potentially fertile inland basins of Greece were subject to flooding. The extensive plains were surrounded by mountains and prone to inundation during the rainy winter seasons by run-off, gushing springs and rivers in torrent, and then again in the spring by melt waters from the snow-capped peaks. The only drainage was provided by sinkholes, fissures and subterranean passages in the limestone bedrock. Today, these basins are managed by sophisticated modern drainage schemes; archaeological research has shown that these are not significantly different to the schemes of the Mycenaeans between 1500 and 1200 BC.

Several basins show evidence for the construction of Mycenaean dykes and dams to contain flood water and in some cases to channel it away. The most impressive scheme was that devised by the Mycenaeans of the city of Orchomenos to drain Lake Kopais in Central Greece. Their achievement was noted by later writers: Strabo, the great traveller of the Classical period who wrote his *Geography* around AD 7, noted that 'They say that the place now occupied by Lake Kopias was formerly dry ground, and that it was tilled in all kinds of ways when it was subject to the Orchomenians, who lived near it. And this fact, accordingly, is adduced as evidence of their wealth.'

The research at Lake Kopais was undertaken by Professor Jost Knauss from the Institute of Hydraulics and Water Resources of the Technical University of Munich. He produced three volumes of technical reports on the hydraulic systems that he found through archaeological surveys.[22] Knauss discovered that the Mycenaeans had initially constructed numerous dykes – simple earth dams at various topographically advantageous points around the basin, most likely around 1500 BC. These protected areas of land were then used for agriculture or for settlement. Several of the dykes were found in the vicinity of the rocky island of Gla, on which an impressive Mycenaean citadel developed.

The system of dykes was not sustainable; the reclaimed agricultural land and new settlements became flooded again, possibly because of increased rainfall after 1500 BC. The Mycenaean response was to construct a far more sophisticated drainage system. A great canal, 25 kilometres long and 40 metres wide was built to lead water from the main inflowing rivers and springs directly to the natural sinkholes in the north-east area of the basin. This lowered the water level enabling the older dykes to function once again. It also provided a supply of freshwater to settlements during the dry summer season and played a key role as an inland navigation system. The old dykes were rebuilt as walls of limestone blocks embedded in clay and an additional canal served the citadel of Gla. Knauss estimates that the Mycenaeans moved more than two million cubic metres of earth and 400,000 cubic metres of stone.

Again, the system was not sustainable – this time because of earthquakes. One of the challenges the Mycenaeans had always faced was a degree of unpredictability about the drainage capacity of the natural sinkholes in the limestone bedrock. When one or more earthquakes struck at around 1200 BC, the sinkholes appear to have become entirely blocked and the whole landscape became flooded once again.

Although Ancient Greeks and Romans constructed their own drainage schemes, the problem was not resolved until the end of the 19th century, when a tunnel was driven through the bedrock to drain flood water into a neighbouring lake basin. Remarkably Johann Knauss found evidence that the Mycenaeans had attempted to do precisely the same, most likely immediately following the earthquakes. In the north-east corner of the basin they sank 16 vertical shafts into the bedrock, the deepest reaching 63 metres. They had then attempted to connect the bottom of each shaft to create a tunnel for the drainage of water. Had they done so, an inclined tunnel more than two kilometres long would have been completed and the lake basin permanently drained. But the scheme was never completed, possibly because of the victory of the Thebans in their local war with Orchomenos, after which they blocked all the shafts to ensure the settlements were drowned.

A Herculean task at Tiryns

The cultural centre of the Mycenaean world was the Argive Plain of the Peloponnese. This is surrounded by mountain ranges and faces the Aegean; it has a pleasant climate, protected position, rich alluvial soils, freshwater springs and a high water table enabling wells to be easily sunk, and so it is not surprising that the plain was peppered with Mycenaean settlements. The greatest of these was Tiryns, built on and surrounding a limestone knoll, located two kilometres from the shoreline today and one of three great Mycenaean centres in the region, joining Argos and Mycenae itself, located 20 kilometres to the north.

The walls surrounding the royal palace of Tiryns remain standing today, these having impressed Homer, who wrote about the 'the mighty walled Tiryns' (Photograph 12). He also frequently referred to the fine horses that came from the Argive Plain, reflecting its high fertility. The walls were built from such massive blocks of stone that a superhuman effort was supposed, one attributed within the Greek texts to the cyclops, the mythical primordial giants. And hence the typical Mycenaean masonry, consisting of huge uncut blocks of limestone fitted together without use of mortar and found throughout the Myceanean world, is referred to as 'Cyclopean'.

The walls enclosed the citadel within which the King's Palace had been located, while the town was located on the plain surrounding the limestone knoll. The site was first associated with Homer's city of Tiryns by William Gell in 1810 and the first excavations undertaken in 1831. When Schliemann arrived at Tiryns in 1876, fresh from his excavations at Mycenae, he initially supposed the site was medieval and came close to destroying it in the search for the Mycenaean city below. The first systematic excavations were by the Deutsches Archäologisches Institut between 1905 and 1929, and then between 1967 and 1986. While these excavations exposed the plan of the palace, found the throne room and identified multiple phases of rebuilding, they provided limited information about the town as a whole because that required excavation of the less glamorous remains surrounding the knoll. Such excavation has been limited but a campaign of work led by Professor Eberhard Zangger from the Geographic Institute of the University of Heidelberg between 1984 and 1988 revealed how the people of Tiryns had campaigned against a marauding force, one only defeated by a masterpiece of hydraulic engineering.[23]

The force was not an army from a competing city-state but the Manessi River that flowed across the plain from the eastern mountain ridge. When in flood this deposited silts and gravels across the farmland and within the town itself; mud-brick buildings were severely damaged if not entirely washed away. To fight back, the Mycenaeans initially tried to protect their settlement with levées. When these failed they took more drastic action – they entirely diverted the course of the Manessi.

Zangger led the Argive Plain Project. This explored the history of landscape change around Tiryns by making an extensive series of boreholes through the plain to establish the sequence of sediments, by inspecting aerial photographs and by piecing together many excavation reports that described the stratigraphic deposits surrounding the limestone knoll. Because parts of the town are now buried by six metres of sediment, this borehole project was itself a rather Herculean task.

Zangger found that during the initial phase of the Mycenaean town there had been extensive settlement around the coastal-facing south-west foot of the limestone knoll, the shoreline being only one kilometre away as the sea level was slightly higher than today. The river flowed through the middle of the town. Although its south bank was stabilised by an artificial levée to reduce the risk of flood damage, expansion of the town was made impossible by the constant build-up of sediment deposited by the river. When this area of the town was eventually abandoned, the Bronze Age houses, streets and marketplace were rapidly buried by several metres of silt, sand and gravel.

In the later phase of Mycenaean settlement the Manessi River had shifted its course to flow around the northern side of the knoll posing new risks to the settlement in this area. Quite soon, deposits of coarse alluvium had destroyed the buildings on the eastern and northern side of the citadel, these becoming buried below four metres of sediment. But then, apparently quite suddenly, the deposition ceased.

New houses were built on the former stream bed itself, the threat of flooding seeming to have disappeared entirely.

What had happened? Quite simply, the Mycenaeans – most likely under the orders of the King of Tiryns himself – had diverted the river away from their town by constructing a massive dam across its channel just as it entered the plain. The dam had to be placed just after the confluence of three mountain streams that fed into the single river and hence had to withstand a massive force of water. It was 10 metres high and 100m long, built from packed earth and stone and secured with walls built with 'Cyclopean' masonry. Without an outlet channel the build up of water behind the dam would soon have been overwhelming. And so a 1.5-kilometre-long artificial channel was dug to divert all of the water into another riverbed, along which it could flow to the sea, entering safely a few kilometres south of Tiryns (Figure 4.3).

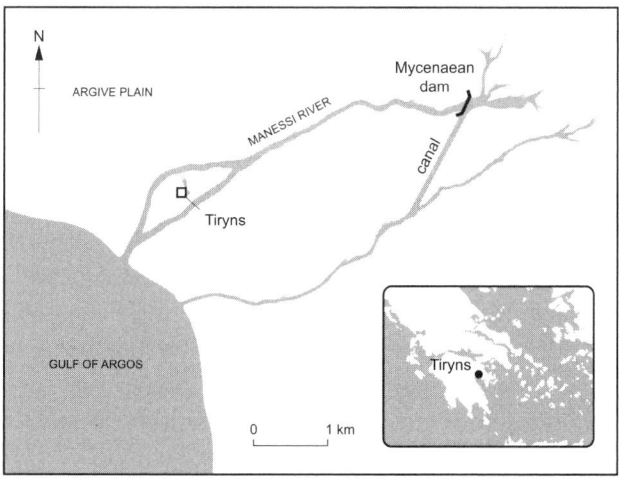

4.3 The original course of the Manessi River, its new course during the Late Mycenaean period around the northern side of Tiryns, and the Mycenaean dam and canal built to divert the river (after Zangger 1994)

Tiryns may have been protected from flooding but remained prone to other natural forces. Around 1200 BC it was struck by at least two major earthquakes which caused the collapse of many buildings within the citadel and damaged its massive walls – archaeologists would eventually excavate the crushed bodies of a woman and child beneath the rubble of a collapsed building. Zangger suspects that, irrespective of the dam, the earthquake might have caused a flash flood leading to a catastrophic debris flow – perhaps that which buried the buildings of the town to the south-west of the citadel.

Whether or not these were the same earthquakes that also had a devastating impact on Mycenae, a mere 20 kilometres to the north and whether these events caused the collapse of Mycenaean civilisation are questions beyond the scope of this book. Our concern is with the manner in which the people of Tiryns protected their town by diverting the course of a river – just as Hercules, who is believed to have lived at Tiryns for many years, is said to have done in one of his labours.

Classical Greece: the water supply to Athens

Several hundred years after the collapse of the Mycenaean civilisation and after a period now referred to as the Dark Ages, Athens and other city-states such as Sparta and Corinth emerged as the key political, social and economic centres of Greece. During the Archaic Period (750–480 BC) the infrastructure was created to enable these city-states to flourish, along with the expansionist policies of Classical Greece as a whole, reaching its height by the fourth century BC. With its semi-arid climate and few rivers, hydraulic engineering was a key element of that infrastructure enabling public baths, fountains and drainage systems to become defining elements of the classical urban culture.[24]

By the sixth century BC Athens had emerged as the most powerful and culturally sophisticated of the city-states, with its acropolis and Parthenon remaining today as the icon of Greek civilisation. At its peak, Athens had a population of 250,000, all with significant water requirements. Most of the population lived in the city below the acropolis. By the fourth century BC fountains, frequently decorated and elaborate in design, gushed at all the crucial points within the city, providing clean public water, meeting places, washing facilities, and of course the sight and sound of water. Many of the more wealthy houses had their own private bathrooms with bathing tubs and 'flushing' toilets.

The city was served by a complex system of wells, cisterns, aqueducts and drains, providing different types of water for different activities: spring water to the fountains for drinking, less potable water to the cisterns for washing, and the careful recycling of used water to clean floors, 'flush' toilets and water the plants.[25] Clay pipes distributed water around the city with an underground network leading to all districts. Similarly a network of drains led the water away. Wells were dug deep into the sponge-like karst ground. Run-off was collected from the roofs of individual houses and stored in underground cisterns providing a necessary back-up supply in times of drought.

The Athenians were fortunate because springs were relatively common; indeed there were eight situated on various parts of the acropolis itself. Its geology consists of layers of soft stone interspersed with hard layers of impermeable marl; water naturally exudes from between the layers, forming springs and enabling wells to be created.

Such springs may explain the long history of settlement on the acropolis, extending as far back as the third millennium BC and beyond. The Mycenaeans fortified the acropolis in the last part of the second millennium and built several flights of steps down to a spring that is now called the 'Mycenaean fountain' but there is very little evidence left of the Mycenaean great hall, or megaron, that once stood on the top. In the Classical period of the late seventh century BC a wall was built around the largest spring, the Clepsydra, and gradually the temples and other buildings were constructed to create the acropolis that we see the remains of today. In about 510 BC the water supply was supplemented by a long-distance aqueduct, the Peisistratid, proudly built by the family of the same name, bringing water 7.5 kilometres from Mount Pentilicus and Mount Hymettus. Another aqueduct brought water to the city from a reservoir on Mount Lycabettus. Also during the sixth century BC, cisterns collecting run-off, wells and drainage channels were built on the acropolis.

Just as we have seen on Minoan Crete, and will later see elsewhere in the ancient world, getting rid of excess water was often as important as providing a supply. As Athens expanded during the sixth century BC the need for artificial drainage became paramount. In particular, the low-lying Agora, originally an area of private housing that was developed into the centre of Athenian government, was prone to flooding following heavy rainstorms. The Athenian solution was only discovered by accident during excavations in 1932. Torrential rain during that excavation eroded the ground and exposed a massive

Athenian drain that had been constructed to run south to north under a street on the west side of the Agora. It had been carefully built with a stone floor, walls and roof, providing a one-metre-square channel for water flow and was appropriately designated as 'The Great Drain'. Constructed early in the fifth century BC, the drain was later extended with southern and eastern branches to provide further drainage to the city as it continued to expand in the fourth century BC.[26]

The Corinthians and Eupalinos' tunnel on the island of Samos

Each of the Greek city-states had to respond to its unique environments as regards how to ensure a constant and sufficient water supply for its population. The engineers of Corinth, for instance, made the most of the karst limestone landscape that extends across much of mainland Greece and is riddled with natural cavities and channels. Corinth was built around a limestone hill, the Acrocorinth, and the engineers tapped into this sponge-like outcrop and its network of natural tunnels and caves for their underground aqueducts and reservoirs.[27]

What may have been the most complex feat of hydraulic engineering in Ancient Greece was accomplished in 530 BC on the island of Samos, located just off the coast of southern Turkey: Eupalinos' tunnel.[28] The main city on the island, also known as Samos, was separated from the major spring, known as the Agiades, by the 300-metre-high Mount Castro. Polycrates, the tyrant, had fortified the city and wanted to secure a water supply to ensure that it would continue to grow as an intellectual hub, it being the home for philosophers, musicians, artists and mathematicians. These included Pythagoras, after whom the town was later renamed as Pythagoreion.

In 520 BC Polycrates commissioned the engineer Eupalinos to excavate a tunnel through Mount Castro to create a direct aqueduct route to the spring. The result is remarkable for having been successfully excavated from both ends, demonstrating a level of mathematical and engineering expertise that remains astounding today. The 1.5 × 1.5-metre-wide passage was 1,036 metres long and ran at a height of 55 metres above sea level, cutting through the middle of the mountain (Figure 4.4). From the north, the tunnel starts off straight but then zigzags, possibly to avoid a natural fracture or rift in the rock. From the south, it follows a consistently straight path until halfway when it suddenly turns to the right to meet the other tunnel – an engineering technique that made it more likely that the tunnels would cross. Miraculously the difference in height when the two tunnels crossed was only 60 centimetres.

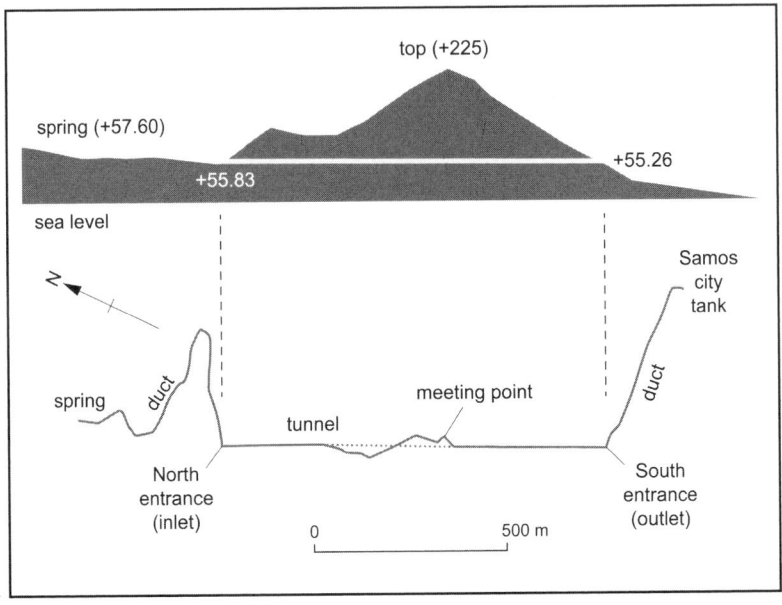

4.4 Eupalinos' tunnel on the Island of Samos (after Apostol 2004)

Modern mathematicians and engineers have puzzled over how Eupalinos could possibly have achieved this feat of engineering with only basic knowledge of geometry and fairly primitive tools. There were three key problems to be solved: firstly, how to ensure that the direction of each tunnel led it to meet the other on a horizontal plane; secondly to ensure that the perpendicular alignment enabled the two tunnels to meet at the same height; finally how to make the water flow in the right direction. The answer to the third question is fairly clear from what remains today: there are two tunnels, one above the other. The aqueduct tunnel lies directly below the main tunnel and was clearly built with more care after the initial passage had been made. Water ran through this lower tunnel in open-topped clay gutters and it was large enough to allow access for maintenance and an increased flow of water at certain times of the year. In some places it can be seen from the upper tunnel but at other places it is covered by slabs of stone.

The two remaining problems that Eupalinos faced were less easy to solve, even for modern-day engineers. Many attempts have been made to guess at and reproduce his methods. In 1965 two historians of science from the UK, June Goodfield and Stephen Toulmin, noticed that the tunnel passed directly under the only part of the mountain that could have been climbed from the south, even though this made it further away from the town of Samos. Having climbed to the top they were disappointed to discover that it was impossible to get a good view of both ends of the water course. It would, however, have been possible if Eupalinos had built a seven-metre-high tower out of wood and stone on top of the mountain. This would have provided a view of the Agiades spring and the seashore. By using a series of fairly primitive sighting tools it would then have been possible to place markers at each end of the tunnel and at fixed extended points in order to act as direction guides for the tunnels themselves.

Another feasible theory, though hard to imagine, first documented by Hero, a Roman engineer in the first century AD, was applied to Samos by Hermann Kienast of the German Archaeological Institute in Athens in 1995.[29] This argued that Eupalinos started by working his way round the mountain from the proposed entrance of the tunnel to the other side, following a constant elevation above sea level, that is, a contour of the hill. This could have been done with a primitive spirit level made from water lying in a strip of clay pipe. Posts or poles could then have been inserted into the ground, marking a constant height until an appropriate point at the opposite side of the hill was reached. This would have given Eupalinos a fairly accurate horizontal point of entry and exit for the tunnel. Then by carefully measuring and plotting the same route in steps of right-angled turns it is possible (though difficult) to calculate the overall distance south and east to get from entry point to exit point and then map a triangle of the route. By using Pythagoras' triangle principle the distance of the hypotenuse could be calculated and triangles of the same proportions could be marked out at each end of the tunnel enabling the workers to follow the correct direction by tunnelling straight and referring back to a guide point at the end point of the triangle outside the tunnel. By using the drainpipe spirit level they could keep the tunnel horizontal and potentially meet in the centre.

A further suggestion of how Eupalinos could have kept the tunnel level was put forward by Tom M. Apostol in 2004.[30] He and his colleague Mamikon Mnatsakanian suggested that a better way of devising a spirit level would have been to use a long straight pole hung from its centre by a rope. By lifting it slowly it would have been easy to see which end would clear the ground first and therefore the height of the surface at each end could be adjusted or measured. In order to plot a course round the mountain at a constant height above sea level temporary pillars could have been built, say eight metres apart, with level tops, measured using the pole spirit level. Having created two horizontal pillars it was then a fairly easy task to use them as 'sights' to indicate a distant point on the same horizon up to 100 metres away, much like aiming along the barrel of a gun. By building another two pillars at the end point and repeating the process, a more accurate horizontal path could have been mapped round the mountain in six or seven stages. This would have given a fairly accurate horizontal marker of the height of each end of the tunnel. But it would not have helped to plan the direction of the tunnel.

No archaeological evidence has been found for any of these engineering methods and there would obviously have been some margin of error in all of them. Needless to say the tunnel remains a marvel of technical and engineering achievement even today and, whichever method he actually used, Eupalinos achieved the result with remarkably little deviation from the planned route.

Philosophers and historians writing about water

Just as we found with the Old Testament (Chapter 2), the Sumerian cuneiform texts (Chapter 3), and the writing of Homer regarding the Mycenaeans, the nature and management of water are key themes within the literature of Classical Greece. Pindar stated the universal truth that 'Water is the best thing of all' in his ode 'Olympian I', written in 476 BC.[31] The writings of philosophers and historians provide us with insights unobtainable from the archaeological evidence alone.

The philosopher and scientist Thales (624–c.546 BC) claimed that water was the Primordial Substance, the so-called 'Arche', writing that it was the most beautiful substance. In his *Metaphysics* (983), Aristotle (384–322 BC) wrote:

Thales … says the principal [element] is water [for which reason declared that the earth rests on water], getting perhaps from seeing that the nutriment of all things is moist, and that heat itself is generated from the moist and kept alive by it [and that from which they come to be is a principle of all things].

In his *Politics* (vii, 1330b) Aristotle stated that 'cities need cisterns for safety in war'. Many Greek cities are indeed positioned on hills with access to water, perhaps for this reason. In his book on *Meteorology*, Aristotle wrote:

In the earth the water at first trickles together little by little, and that the sources of the rivers drip, as it were, out of the earth and then unite … When men construct an aqueduct they collect the water in pipes and trenches, as if the earth in higher ground were sweating the water out … Mountains and high ground, suspended over the country like a saturated sponge, make the water ooze out and trickle together in minute quantities in many places.

Plutarch (AD 47–127) describes the laws set by Solon, the lawgiver of sixth-century-BC Athens, illustrating the political and social implications of water use in a large city:

Since the water supply of rivers, lakes and springs was inadequate, and most people dug wells, he [Solon] passed a law that wherever there was a public well within a hippocon [i.e. 710 metres] this was to be used; where it was further away, one could dig one's own; but if, having dug ten fathoms [18 metres], one did not find water one was permitted to fill a hydria of six choae [20 litres] twice a day from one's neighbour's well; for Solon thought it right to help a man in need, but not to encourage laziness. (Plutarch, *Solon* 23).

Stories abound from Greek literature regarding the use and misuse of water in battles and sieges. One of these stories centres on the scientist Thales, probably of Phoenician descent, and later celebrated in the Temple of Delphi as one of the 'Seven Sages of Greece'. As a scientist he is credited with the discovery of the electrical properties of amber but by this time he had also gained fame as an innovative mathematician (referred to by Euclid in his *Elements*), an astronomer, inventor and engineer. It was his skill as an engineer that led him to accompany King Croesus of Lydia in his battles against Cyrus of Persia in 547 BC. Unable to cross the River Halys with his large army of foot soldiers, Croesus commanded Thales to divert the river. Heroditus writes in his *Histories* (Book 1, 75):

Beginning some distance above camp, he dug a deep channel, which he brought round in a semi-circle, so that it might pass rearward of the camp; and that thus the river, diverted from its natural course into a new channel at the point where this left the stream, might flow by the station of the army, and afterwards fall again into the ancient bed. In this way the river was split into two streams, which were both easily fordable.

A century later, under the reign of Pericles, the Athenians were trying to extend their power and influence in mainland Greece while also attempting to fight off the Persians in Egypt. According to Thucydides (1.109) the Athenians, who were trying to hold their land in Egypt, were forced to retreat to the small island of Prosopitis in the Nile delta, mooring their boats in the harbour. They were besieged there for eighteen months until finally, in 454 BC, the Persians dug canals in the delta to divert the river away from the island, effectively joining the island to the mainland. The Athenian ships were, therefore, marooned in the mud of the harbour and the few soldiers that escaped via Libya persuaded Pericles to abandon Egypt to the Persians.

Around the same time, on mainland Greece, an Athenian alliance of forces was in conflict with Sparta in the later stages of the Peloponnese wars. According to Thucydides (5.47.9), in 418 BC there was a standoff in Mantinea in which the alliance took refuge on high land and would not be drawn out. The Spartans, desperate to fight, resorted to water engineering to solve their problem. They are said to have diverted water from the Sarandapotamos River into the smaller Zanovistas River, causing the flooding of the Mantinea land and forcing the Athenian alliance out to battle. Sparta won this victory and the Athenians retreated.

Greek science: pumps and clocks

One of the most enduring technological inventions of the Ancient Greeks was 'Archimedes' screw', a rotating spiral in a tight fitting cylinder that is still used today to raise water or grain from one level to another.[32] Archimedes (287–212 BC) was from Syracuse in Sicily and was one of the greatest ancient mathematicians, along with being a physicist, engineer, inventor and astronomer. His famous pump was developed from his study 'On spirals', but he was also innovative in his investigations of the volume and equilibrium of 3D shapes, the properties of circles, the density of objects, and the properties of floating and sinking, and also his attempts to calculate the size of the universe in terms of grains of sand. According to Plutarch, Archimedes died during the Roman conquest of Sicily while he was working on a mathematical diagram. Although his safe passage had been ordered by the Roman generals, Archimedes was killed by a soldier because he would not leave his work quickly enough. The last words attributed to Archimedes are 'Do not disturb my circles!'

Another of the earliest inventors who used water was Ctesibius from Alexandria in Egypt (285–222 BC).[33] He apparently spent his early years working with his father as a barber, during which time he invented a counterweight system to raise and lower the mirror in his father's shop. This led to a series of inventions involving the compression of air, hydraulics and pneumatics. Like Archimedes, he invented a water pump. This consisted of two vertical cylinders, each with a plunger that worked reciprocally by a rocker arm with a valve that opened to draw water into one cylinder when the piston was raised, and then when the piston was depressed to close the valve and force water up into a higher tank, or to create a jet of water. This same principle led Ctesibius to create the first pipe organ.

One of Ctesibius' greatest inventions was the development and refinement of the water clock or 'clepsydra' (captured water or 'water thief'). Simple water clocks had been used for some time. A water-filled container with a small hole at the bottom for the water to drip out was used in Alexandrian courtrooms where a defendant was entitled to speak for a certain regulated time. The device could then be adjusted and the amount of water was regulated according to the crime. Ctesibius was dissatisfied with the process of measuring time by watching water drip through a hole in a container. This was useful for measuring a finite period of time but was not accurate because of the changes in water pressure in the bucket and could not be sustained.

Ctesibius set about refining the principle of the water clock, building on his previous knowledge of hydraulics and using valves, water pressure and siphons. By incorporating two water containers, water drips steadily from one constantly full container into another, keeping the pressure uniform throughout. But instead of dripping out of the second container, the water is collected there and rises steadily. The passing of time is marked by a rising float on the surface of this container which is attached to a marker on a vertical bar at the top indicating the passing of the hours. When the water has risen to the maximum height of the container it flows down an overflow pipe into a siphon which activates the whole process to begin again in much the same way as a modern flush toilet works. The marker at the top would therefore drop and a new day would begin. Ctesibius' water clock remained the most accurate way to measure time until the 17th century and was used in the 16th century by Galileo to time his experimental falling objects.

The Hellenistic world

Classical Greek culture, including its achievements in hydraulic engineering, spread throughout the Near and Middle East and into North Africa, blending with that of indigenous communities to create what we know as the Hellenistic world.[34] One vehicle for that spread was the army of Alexander the Great, the young king of Macedon, a northern Greek state, who had been tutored by Aristotle up until the age of 16. His military campaigns between 334 and 323 BC created arguably the greatest empire of the ancient world, one that stretched from the Adriatic coast in the west to the Indus Valley in the east.

In the two centuries between Alexander's death in Babylon at the age of 30 in 323 BC and the imposition of Roman rule throughout the Near East and mainland Greece in 146 BC, Hellenistic kingdoms thrived. Some were created by colonists from Greece, others by the local adoption of Greek culture fused with that of the east. In the next chapter we will find how the desire for water within Hellenistic culture, especially the significance of its display in fountains and pools, penetrated to one of the most arid regions of the ancient world and one of its most remarkable kingdoms: the Nabataeans.

5

A WATERY PARADISE IN PETRA

The Nabataeans, masters of the desert, 300 BC–AD 106

A visitor's first enthralling glimpse of the splendours of Petra is the yellow stone-cut façade of the 'Treasury' framed by the dark walls of the siq and a clear blue sky above. After a long hot coach or car journey to Wadi Musa in southern Jordan, and then the tiring, winding walk down the narrow two-kilometre siq itself travellers suddenly find themselves awestruck at such an architectural marvel (Photograph 13). This is only the start of a physically exhausting but mentally exhilarating walk around the rock-cut tombs, the theatre, temples and other ruins of Petra, the capital of the Nabataean Kingdom, 300 BC–AD 106. Fortunately there are always donkey rides available for the climb back to the Treasury, and horse and traps for the siq itself – although it is far better to walk because this makes the ice cream back in Wadi Musa taste all the better.

Visiting Petra is thirsty work. I always run out of water and am forced to buy small bottles at extortionate prices from stalls within the ruins. It is nevertheless a small price to pay for visiting one of the most spectacular of ancient cities, one situated within a basin formed by the course of the Wadi Musa and surrounded by a starkly beautiful mountainous landscape. It is a dry place. Just as in Nabataean times, today's rainfall rarely exceeds 75 millimetres per annum; it is often much lower and always highly localised. The sun is so hot that standing water quickly evaporates. The hundreds of thousands of tourists who flock to Petra each year quench their thirst on bottled water, and most have the knowing comfort of a limitless supply in their hotels at the end of the day. What about the 30,000 people who had lived within the ancient city 2,000 years ago?

No bottled water for them. They had it so much better, and probably for free. They had fountains of water. They had flowing channels of water and overflowing basins of water; not only these, but cascading waterfalls leading to still, silent pools of water. Such a quantity of water within Petra is almost unimaginable in today's ruined city, so parched and baked by the sun. However, this unlikely sight is precisely what the Nabataean kings had wanted: to shock the travellers, traders and emissaries visiting their city with the sight of such an abundance of water that it seemed they could even afford for much of it to spill wasted across the floor. For those ancient visitors, the first thrilling sight of the Treasury from the siq – a natural fissure in the rock – would have been nothing compared to that of the watery paradise they were about to enter just around the corner. That really would have been awesome.

Any modern-day visitor with even half an eye for archaeology can see the signs of Nabataean water control all around: channels and cisterns cut into rock faces, remnants of terracotta pipes, dams across side wadis (Photograph 14). However, their significance for the city is rarely fully appreciated as such mundane remains are outshone by the elaborate rock-cut tombs in strikingly coloured stone – yellow streaked with pinks, reds, blues, whites and browns. Much more evidence of the Nabataean use of water remains buried, some having once been temporarily exposed by excavations and then resealed again. The most recent discoveries have surprised even those who are the acknowledged experts on Petra, indicating that extravagant water displays had once occurred with the city.

Of all the ancient cities we are visiting in this book Petra provides the most startling contrast between the past and present. Its watery past is one of its most delightful secrets, while the city provides a profound lesson for us today: an ingenious example of how fairly simple hydraulic-engineering techniques can be applied to create an abundance of water in even the most arid of environments. The hydraulic engineers were the Nabataean people, whose skill can be observed not just in Petra, but throughout their desert kingdom (Figure 5.1).

The Nabataeans: from pastoralists and bandits to cosmopolitan city dwellers

The Nabataean culture emerged in southern Jordan during the late fourth century BC and was an amalgamation of indigenous Iron Age and nomadic tribal groups that had spread northwards from Arabia.[1] They left few written records themselves; the majority are short inscriptions and invocations that provide little more than lists of names. We must rely, therefore, on their archaeological remains and the documents written by either visitors to southern Jordan or, more often, those writing at a distance who had heard about the Nabataeans, second, third or even fourth hand.

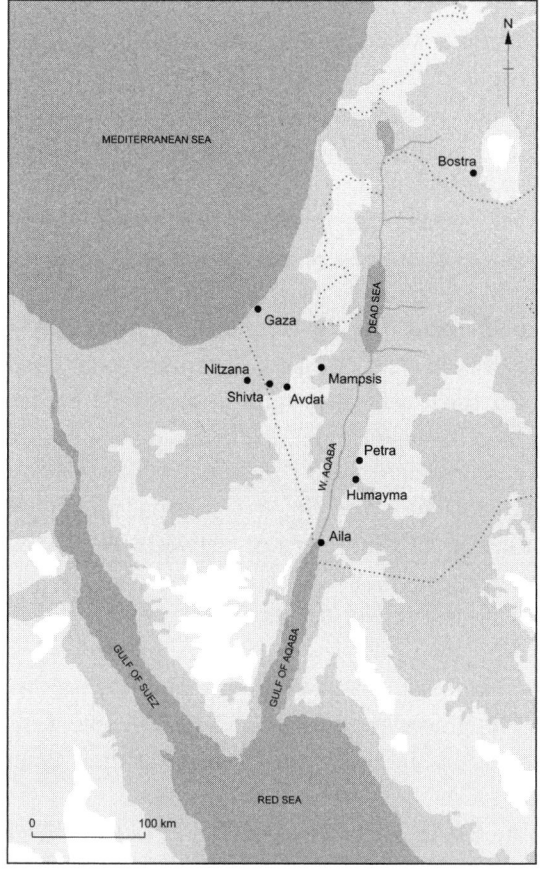

5.1 Nabataean settlements referred to in Chapter 5

While we must be cautious about any such sources, that of the Sicilian writer Diodorus of the later fourth century BC provides an appealing picture of people who had mastered the desert:

... they take refuge in the desert, using this as a fortress; for it lacks water and cannot be crossed by others, but to them alone, since they have prepared subterranean cisterns lined with stucco, it furnishes safety. As the earth in some places is clay and in others is soft stone, they make great excavations into it, the mouths of which they make very small, but by constantly increasing the width as they dig deeper, they finally make them of such size that each side has a length of one plethrum [27 metres]. After filling these reservoirs, they close the openings, making them even with the rest of the ground, and they leave signs that are known to themselves but are unrecognizable to others.[2]

Many such stone-cut, bottle-shaped cisterns can be visited in southern Jordan and the Negev today. They became a pervasive part of Nabataean culture during the next four centuries and a key element of their water management.[3] When Diodorus was writing, the Nabataeans are likely to have been primarily dependent upon herding sheep, goats and camels, along with some raiding of the trade caravans that passed through southern Jordan. Perhaps it was the latter that required them to disappear into the desert, as Diodorus described, like thieves into the night. But the Nabataeans soon transformed themselves into the middlemen of the old-world global trading network. By doing so, they secured immense wealth – and hence the architectural glories of Petra that we so much enjoy today.

Southern Jordan was at the nexus of not just one but several major trade routes: from the Gulf of Aqaba northwards to Syria; from southern Yemen to the Arabian Gulf; and yet another that stretched across the Negev to the eastern Mediterranean. More specifically, the Wadi Musa – where the Nabataean capital of Petra would develop – provided the most accessible route-way through the mountain barriers between the high plateaus to the east and the Wadi Araba to the west.

A multitude of goods passed along these trading networks: myrrh, balsam and frankincense came from the coast of southern Arabia and were in demand throughout the Near East and Eastern Europe, especially the frankincense for ritual use, cosmetics, perfumes and medicines. Pepper, ginger, sugar and cotton came from India; copper, silver, gold and bitumen came from within the southern Levant, the latter from the Dead Sea and in great demand within Egypt for embalming. All of these were transported by caravans that had to pass through the Nabataean homeland; the Nabataeans soon switched from raiding such caravans to charging tariffs and supplying shelter, food and – of course – water.

It is not surprising, therefore, that within a couple of hundred years of Diodorus' observations we gain a quite different view of the Nabataeans. In the first century BC, Strabo, the Greek geographer, described how:

The Nabataeans are prudent, and fond of accumulating property. The community fines a person who has diminished his substance, and confers honours on him who has increased it ... The houses are sumptuous, and of stone. The cities are without walls, on account of the peace that prevails among them. A great part of the country is fertile, and produces everything except oil of olives ... The sheep have white fleeces, their oxen are large; but the country produces no horses. Camels are the substitutes for horses.[4]

The transformation of the Nabataeans from desert nomads to cosmopolitan city dwellers and wealthy farmers began apace in the second century BC, involving the transition from a tribal-based society to a monarchy. The first named king is Harithath I, more usually known by his Greek name, Aretas I, dating to 168 BC. Ten kings followed until finally Rabb' el II (AD 70–106) brought the kingdom to an end by conceding to Roman annexation.

Petra was an obvious choice for their capital. This is a basin centred on the confluence of five wadis, most notably Wadi Musa coming from the east and Wadi Sayyigh that runs westwards towards Wadi Araba. The basin covers 100 square kilometres and includes narrow valleys and broad plains within a rugged limestone and sandstone landscape, with prominent outcrops of dark porphyry. The geological formations are so striking that it is easy to imagine the Nabataean architects and stonemasons being inspired.

A second geological offering for Petra was security in the form of a long, narrow siq as the entrance to the basin. But it was a third geological gift that was perhaps of most importance: springs. Petra has several of these, making it the ideal location for the capital of an ambitious kingdom dependent on international trade for its wealth.

The extent of the kingdom gradually expanded from southern Jordan into surrounding lands, this being partially dependent upon the fortunes of the neighbouring kingdoms, notably the Jewish state of Judea. The Nabataean Kingdom reached its greatest extent in the first century BC under the rule of Aretas III (85–62 BC), extending eastwards into north-west Arabia, westwards into the Sinai and the Negev, and northwards to Damascus. The kingdom had ports on the east coast of the Red Sea that received maritime trade from India.

The Nabataean success was not solely dependent upon their for-tuitous geographical location at the nexus of multiple trade networks. Two other factors were critical. One was their sophisticated water management – as this chapter will describe; the other was the shrewd political judgements of the Nabataeans to engage in peaceful relation-ships rather than conflict with their neighbours. Political and mili-tary battling between the kingdoms of Syria and Egypt left a power vacuum in Transjordan, the region to the east of the Jordan Valley that the Nabataeans were eager to fill. When the Roman general Pompey sent troops against Petra in 62 BC, the Nabataeans simply chose to buy peace, paying 300 talents of silver, rather than engaging in costly conflict.

Relations with Rome

With its new-found wealth, Petra was transformed into a city fit for a king. The style of its monumental architecture reflected the position of Petra at the centre of the many cultural influences of this region mixing Hellenistic and Egyptian designs with those of the local Arab traditions. Nabataean religion did the same, blending ideas, ritual and deities from Egypt, Syria, Canaan, Assyria and Babylon with elements from the Greek and Roman pantheon to create something distinc-tively Nabataean – something for everyone. By the first century BC far-reaching contacts had led to the introduction of coinage and new elements of hydraulic engineering, notably vaulted roofs for reser-voirs and aqueducts.

Such developments were essential because the Nabataean

population had grown substantially; Petra alone had now reached a population of 30,000 persons, many of whom were permanent residents. The Petra water system had to meet three demands: the domestic needs of urban dwellers; irrigation for the farmers within the Petra basin; and the requirements of the trade caravans.

The Nabataeans' most powerful neighbour was Rome, with a seemingly ever-expanding empire. In 64 BC Rome acquired Syria but remained tolerant of the Nabataeans, whose culture and wealth continued to flourish. By the first century AD, however, the transformation of trade routes had begun that would soon lead to the Nabataean Kingdom's demise. The Romans had learned to bypass the Nabataean-controlled land routes by using seaports on the west coast of the Red Sea and directing trade through Egypt. They also encouraged trade through Wadi Sirhan, running directly from Saudi Arabia into Syria, where the city of Palmyra would ultimately replace Petra as the entrepôt of Asian goods into Europe. The Nabataean king Rabb' el II (AD 70–106) responded by shifting his Nabataean capital from Petra to Bostra in southern Syria in AD 93 but the growth of Roman power and influence had become overwhelming. In AD 106 the Nabataean kingdom was incorporated into the Roman Empire and designated as Arabia Petraea.

Petra and the Nabataeans continued to prosper – for a while. The Emperor Trajan annexed the Nabataean kingdom to gain a comprehensive hold on the land and sea trade of the eastern empire. He invested in the region by constructing the Via Nova Traianus, a paved road that linked Bostra with Alia (present-day Aqaba) on the Red Sea Gulf. But Nabataean prosperity could not be sustained as centres of economic and political power continued to shift inexorably to the north and east. The decline was exacerbated by a major earthquake in AD 363, causing damage to many Nabataean settlements and widespread economic disruption.

The later history of Petra and its surrounding region reflects that of the eastern Mediterranean as a whole. During the Byzantine period between the fourth and sixth centuries AD, Petra became home to a substantial Christian community. Several rock-cut tombs were converted into churches and monasteries, although Petra continued to be an economic centre. It was damaged by further earthquakes, notably in AD 551. In the seventh century AD the region came under the rule of the Umayyads and gained some recognition again in the 12th century during the time of the crusades, but Petra and the Nabataean culture as a whole gradually slipped into historical obscurity. In 1822, Johann Ludwig Burckhardt, a Swiss explorer, became the first Westerner to visit Petra in over six centuries, only managing to do so by travelling in disguise and making his notes in secret.

Today, hundreds of thousands of tourists visit the ruins each year; archaeologists and historians continue to expose and seek to understand the cultural achievement of the Nabataeans. The majority of such research has been concerned with architectural styles and trade relations – vast amounts of necessary time spent cataloguing shards of pottery, coins and inscriptions. For me, however, the most impressive work has been that of Professor John Oleson of the University of Victoria in Canada, who studied the hydraulic engineering of the Nabataeans for more than 25 years.[5] His work and that of others has shown that the Nabataean capacity to manage the sparse water supply provided the foundations for their more widely noted cultural achievements in the fields of architecture and trade.

Nabataean water management

The Nabataeans made use of, and perfected, all of those methods of water management which had been developed in the Levant since the time of the final Neolithic communities of 8,500 years ago, those for its collection, transport and storage. To these they added some more methods which they had learned about from their extensive trading contacts.

Here we must remind ourselves that Petra was just one settlement among many and that the majority of Nabataean people were farmers and pastoralists. They flourished within a landscape that would often get no more than 10 millimetres of rain within a year, and on average between 25 and 75 millimetres. Except in rare circumstances, the geology and water table were not suitable for wells and so they primarily depended upon capturing the run-off from rainfall, transporting that to their fields and storing a substantial quantity for the long summer dry season.

Throughout the Nabataean territory, extensive use was made of stone walls to create terraces across hillsides; the walls trapped the run-off water and soil particles. Similar walls were built across wadis to slow down the water and enhance infiltration, just as the Neolithic inhabitants of the Jafr Basin had once done. The Nabataeans also utilised containment dams, as had been seen in the Bronze Age at Tell Handaquq and Jawa, to trap and hold pools of water. The largest was at Mampsis in the Negev, built of mortared stone blocks with a vertical face upstream and sloping face downstream for buttressing, and thus creating a pool of 10,000 cubic metres of water.

Bottled-shaped, rock-cut cisterns were the mainstay of water storage throughout the Nabataean period, these being formed just as Diodorus described and often lined with plaster. Open-air reservoirs were problematic because of the hot sun and winds leading to high rates of evaporation and the risk of pollution from animal droppings, insects and so forth. The shortage of timber and unsuitability of the available sandstone slabs, which were too small and friable, prevented the roofing of reservoirs until transverse arches were introduced in the first century BC. That originated as a Hellenistic technique, with John Oleson suggesting it had been seen by Nabataean traders when visiting Delos and other arid Aegean islands. Another development of Nabataean hydraulic engineering was the introduction of aqueducts to transport water from springs. These were at ground level, made from blocks of stone placed together end to end and held by framing walls.

The scale of such works suggests that royal patronage may have been required to meet the costs. John Oleson suggests that the titles bestowed on the kings may reflect their support for agricultural and water management projects. King Aretas IV, for instance, was known as 'He who loves his people', and King Rabb' el II as 'He who brought life and deliverance to his people'. There must have been an administrative system in place to manage the hydraulic systems. Here the sparse written records are of limited help, rarely going beyond such invocations. There is, however, an inscription at the Nabataean cult centre of Khirbet at-Tannur referring to a person with the title 'Master of the spring of La Ban', who was presumably a governmental official.

In the Negev desert

Some of the best-preserved Nabataean archaeology is located in the Negev Desert, across which caravans travelled to Gaza and Egypt. The

Nabataeans established at least six substantial towns in the Negev, these developing into Roman and then Byzantine settlements during the course of history, the most notable of which are Avdat, Shivta and Nitzana. Their ruins were first documented towards the end of the 19th century, but it was only during the 1950s that extensive networks of Nabataean field walls, terraces and farmsteads were recorded, leading to one of the most outstanding experimental archaeology projects of all time.

A team of Israeli archaeologists led by Michael Evenari began to explore the Negev desert during the 1950s. They were initially surprised and confused by the many terraces and stone walls they found, describing these as appearing to run 'unintelligently across stony barren hillsides'.[6] During the course of the next two decades they gradually mapped and came to understand these walls and thousands of associated stone mounds as the remains of the Nabataean system for collecting run-off to enable them to farm the desert.[7]

That the Nabataeans were able to grow not only wheat, barley and legumes, but also almonds, olives, dates, figs and grapes in the intensely arid conditions of the Negev Desert had been known since the mid-1930s. Excavations at Nitzana, located in the far south-west of the Negev, recovered a remarkable collection of papyri written during the sixth and seventh centuries in Greek and Arabic. Although dating to at least 400 years after the annexation of the Nabataean Kingdom by Rome, the records gave many names of Nabataean origin and described agricultural practices that were likely to have remained unchanged for many centuries. The documents were primarily accounts of economic, legal and financial transactions, but they provided information about land ownership, the size of farms and fields, and the diverse array of plant and tree crops being cultivated.

Michael Evenari and his colleagues mapped and began to understand how the field walls and stone mounds had acted to collect run-off and to divert flash floods on to fields. The sophistication of the systems they found can be appreciated from their description of what they named Yehuda's farm, located near the Nabataean town of Avdat:

This system comprises a terraced area of about 2.2 hectares which received its run-off from a catchment of about 70 hectares. This catchment area was artificially divided into a number of smaller watersheds by several run-off conduits, each of which was connected to a specific field within the farm unit. Some of the units began high on the plateau above the farm and collected additional run-off from there. It seems as if the ancient settlers first divided up the slope to channel the run-off into the fields and then extended these conduits to the plateau in order to catch more water. Since every field had its own catchment area, the run-off during a rainstorm was automatically divided among the fields.[8]

This particular farm had a catchment more than 30 times the size of that actually being farmed; the average they found was about 20 times. This meant that every square metre of hillside had been allotted and hence there was a complicated system of catchment boundaries. Not surprisingly, therefore, the Nitzana documents indicated how each of the farmers had their water rights guaranteed by law.

Not content with merely mapping and speculating about how the run-off systems would have functioned, Evenari and his colleagues embarked on a most remarkable experiment: reconstructing a fully functioning Nabataean farm within the Negev. They spent much of 1960 rebuilding a farm close to Avdat. Fourteen ancient terraces were reconstructed, along with hillside conduits to precisely replicate how run-off had been captured during the ancient times. They didn't rely entirely on ancient technology but imposed a modern flood water distribution system using concrete channels and steel pipes. Nevertheless, in essence, they rebuilt a Nabataean farm.

The account of how Evenari and his colleagues then successfully cultivated a diverse range of crops, including all those described within the Nitzana documents, securing substantial yields even in the years categorised as drought, is described in their 1982 book *The Negev: The Challenge of a Desert*. This is a marvellous account of how they learned through trial and error, experiencing the same uncertainty over when the rains would arrive, and hence when to sow, as had their Nabataean forebears. It is an enormously important account, the clearest demonstration that we still have a great deal to learn from the ancient world.

When reflecting on his three decades of experience in the Negev, Evenari described how he had learned that:

The ancient farmer fitted his artificially created agricultural ecosystems into nature and used landscape and topography to his best advantage without damaging his environment. He neither caused erosion nor brought about salination of his agricultural soils. By using the run-off he tamed the flood torrents and prevented the damage that uncontrolled floods usually produce.[9]

This wasn't a case of looking at the past through rose-tinted spectacles: Evenari had lived this farming system. Moreover, he acknowledged that mistakes were sometimes made:

The ancient desert farmer made mistakes and had to pay for them ... Their agricultural system using large catchments was a miscalculation leading to erosion, silting up, and destruction because this approach to cultivation was overambitious. The farmers responsible for it wanted to achieve too much and so disturbed the equilibrium of the desert.[10]

Humayma and the longest Nabataean aqueduct

None of the run-off technology is especially complex; what is impressive is the scale at which it was implemented and how the various methods were integrated into a single system. Before we look at how this reached its apotheosis at Petra, we should visit the rural settlement at Humayma, located between Petra and Alia – today's Aqaba at the Red Sea gulf – which was the focus of John Oleson's studies for more than three decades.[11]

Humayma was founded around 80 BC by King Aretas III with the dual purpose of supporting the caravan routes and being a farming settlement. The location appears to have been carefully selected not only for relatively fertile soils but for its particular topographic and geological features (Photograph 15). For a Nabataean engineer these provided the opportunity to capture substantial run-off water into cisterns and reservoirs, supplemented by fresh water channelled from springs. When the Romans later built a legionary fort at the same locality soon after the Provincia Arabia had been established in AD 106, they made no alterations to the Nabataean hydraulic-engineering scheme other than to construct an additional reservoir within the fort itself. Indeed, some of the Nabataean cisterns and reservoirs are still in use today, supporting the modern Bedouin population of Humayma.

Our understanding of the water management system at Humayma is due to the survey and excavations undertaken by Professor John Oleson and his colleagues since 1986.

The village centre at Humayma was surrounded by a hilly periphery forming a 250 square kilometre catchment from which run-off was captured by terraced walls stretching across fields and wadis, and feeding 48 cisterns and three containment dams. A rural population would have lived throughout this area, grazing their flocks in the hills and growing cereal and vegetable crops within the fields; a significant proportion of their time would have been spent maintaining the terrace walls and cisterns.

Oleson estimates that with an average rainfall of 80 millimetres a year, this catchment would have annually received 20 million cubic metres of water. The cisterns had capacity to store around 5,000 cubic metres of such water, while no more than 3,500 cubic metres would have been trapped behind the dams. He suggests that 80,000 cubic metres would have been held in the soils maintained by the terrace walls to sustain the Nabataean crops. Consequently, even with this investment in walls, cisterns and dams, the vast majority of rainfall run-off was lost to human usage, disappearing into the geology or simply evaporating into the air.

The village centre was supplied by an aqueduct that brought water from three springs at the base of the al-Shera escarpment, 'Ain Qana, 'Ain Sara and 'Ain al-Jammam, and covered a distance of no less than 26.5 kilometres (Photograph 16). The aqueduct fed two tanks that supported the rural population and then supplied a reservoir in the village centre, with a capacity of 633 cubic metres, available for use by those Nabataean families who had settled at Humayma and also by those who still lived a nomadic existence. There would have been a continuous flow of water. Some of the overflow appears to have fed a bath-house. One can easily imagine that such a constantly refreshing pool would have been especially attractive to traders and travellers passing through the region.

The village centre was also supplied by run-off water that filled two reservoirs, each with a capacity of 488 cubic metres. These were fed by a field of about one square kilometre that gently sloped towards the village centre and had a single natural outlet, from which channels directed the run-off to the reservoirs. These had slab roofs supported by transverse arches, one of which remains largely intact today. The reservoirs appear to have been for public use, perhaps supplying the caravans as well as the residential Nabataean population. They are surrounded by 13 cylindrical cisterns also fed by the run-off; these are likely to have been owned by families or clans for their private use. Not surprisingly, people were drawn to live close to the reservoirs and cisterns: during the 800 years of Humayma's existence around 30 houses were built surrounding these cisterns.

John Oleson has calculated the total amount of water storage provided by the tanks, reservoirs and cisterns at Humayma as 4,355 cubic metres in the village centre and 4,315 cubic metres in the rural areas. By then using estimates of water consumption by people and their animals, he proposed that the Humayma countryside had supported a population of 163 persons, 163 camels and 1,469 sheep/goats, while the village centre could have supported 654 persons, 20 camels and 180 sheep/goats.[12] Anyone visiting Humayma today will find these surprisingly large figures.

Within Petra

Water management at Humayma was successful because it used water from both run-off and springs, each system being able to provide a back-up to the other should one fail. Charles Ortloff, a California-based hydraulic engineer with an interest in the ancient world, argues that such 'redundancy' was also the key to the water supply system in Petra.[13] He has undertaken a detailed study of the archaeological remnants of aqueducts, water channels, cisterns and reservoirs that had once supplied the people of Petra, concluding that water conservation was practised within Petra on a much larger scale than in any other contemporary city.

Piecing that archaeological evidence together is not easy. After annexation by Rome, much of the city was renovated and new buildings erected; the same happened in the Byzantine period when some tombs were converted to churches. In addition, throughout its history, and especially during its period of decline and then relative abandonment, the city was subject to erosion by flash floods and the general wear and tear imposed by sun and wind. As a consequence, much of the Nabataean hydraulic system only survives as fragmentary remains and there is a great deal still to learn. But we do know sufficient to be confident that Petra represents one of the most elaborate hydraulic-engineering achievements of the ancient world.

Water and ritual in the siq

Springs provided a continuous flow of water to the city of Petra. The most important was 'Ain Musa, which is still being used to supply the modern settlement in Wadi Musa, and the ever-escalating water demands of tourists.[14] This is where the Old Testament (Exodus 17:7) claims that Moses struck the ground with his staff to generate a spring. An aqueduct carried water the eight kilometres from the spring to the city centre, following the course of Wadi Musa, and beyond to exit in Wadi Siyyagh. When first built, this aqueduct was an open channel, up to 2.5 metres wide and one metre deep, which followed the two-kilometre route through the siq. This aqueduct now lies underneath the Nabataean and Roman paved floor; it was replaced during Nabataean times by a neat channel carved into the northern rock face of the siq lined with terracotta piping. Four settling basins were incorporated along its course to remove silt from the water and stabilise the flow.

The remnants of the stone-cut channel and terracotta pipes are quite visible at waist height today and remarked upon by many of the tourists walking down the siq. Professor Leigh-Ann Bedal, an archaeologist from Pennsylvania State University, who has excavated in Petra, reminds us that any visitors to the city in ancient times would not have seen the pipes at all, these being sealed below mortar into the stone channel. But there would have been another sensation which we sorely miss today and that may have had a profound impact on those who had recently crossed the arid desert to reach Petra: Bedal writes that 'the sound of water that flowed, unseen, inside the covered channel and pipeline, would surely have been audible to passers-by, as a continuous murmur reverberating off the towering walls of the narrow siq'.[15]

The music of moving water may have had a meaning for the Nabataeans and their guests beyond the promise of quenching their thirst. Although we lack explicit written evidence, it seems likely that water was infused with religious significance within the Nabataean world, and especially in Petra. At the entrance to the siq are three huge 'god-blocks', non-figurative representations of the Nabataean gods who appear to stand guard over the water supply to Petra. There are 25 of these within the city, many appearing at key localities associated with the water supply. There are innumerable religious niches and engravings throughout the course of the siq at which the sound of running water may have been integral to whatever rituals had taken place.

The most striking relief sculpture is that of two pairs of camels, facing each other and led by a male guide. Only the lower halves of the camels survive; the legs are well preserved, carved in the round with the water channel flowing behind. Archaeologists disagree as to whether or not the camels were depicted carrying goods. The usual assumption is that this is a depiction of the vital caravan trade but Leigh-Ann Bedal believes the camels are part of a ritual procession, possibly being led to sacrifice, and that an altar had once sat between them[16] – directly in front of the murmuring water.

To make use of the siq for not only the 'Ain Musa water channel but for regular and safe access to the city, the Nabataens had to divert the course of the Wadi Musa river, whose intermittent floods after winter rains would have been a violent torrent. To do so they built a dam and a 40-metre-long tunnel to divert flood waters into Wadi Muthlim – also overseen by a 'god-block'. An inscription indicates that the dam had been built by either King Malichus II (AD 40–70) or King Rabb' el II (AD 70–106)[17]. Today's tourists walk over the dam to enter the siq, but one can choose to follow the tunnel and Wadi Muthlim into the city centre of Petra, this taking one round the southern side of Jebel al-Khubtha, the mountain into whose northern face the Royal Tombs were constructed.

The significance of the dam and tunnel for access along the siq to the city of Petra was illustrated in 1963: the tunnel had been long blocked and a group of French tourists lost their lives to a Wadi Musa flash food that had nowhere to go but down the siq.[18] The tunnel is now working again, but water still penetrates the siq through various gullies and can reach ankle-deep. Many small side wadis and channels have remained blocked by original Nabataean dams, some of which can still be seen with a bit of scrambling off the tourist track.

From the Treasury to the city centre

At the end of the siq the terracotta pipes fed into a water basin at the foot of the so-called Treasury building, but which was most likely a Royal Tomb, possibly of King Aretas III. This is an extravagant rock-cut chamber, including elements of Egyptian and Hellenistic design – the latter being unsurprising because Aretas III was otherwise known as Phillhelene, the 'Greek lover'. The surveying and masonry skills so evident from such tombs are relevant for the hydraulic archaeology because building aqueducts and cisterns would have required the same technical skills, although far less ostentatiously displayed.

From the so-called Treasury, pipes took water along the route that tourists walk today (Photograph 17): along the outer siq, past the tombs and theatre, and round the base of the north face of Jebel al-Khubtha and to the centre of the city. After the Roman annexation, the water fed into a grand Nymphaeum, marked today by the presence of a huge pistachio tree. This looks so old that it could be Nabataean itself and provides some welcome shade. Quite what had been here before the Roman build remains unknown; it seems likely that this place had once been marked by a Nabataean fountain or at least an open tank of water. From this point the water kept flowing alongside the main street of Petra that the Romans turned into a colonnaded street, and which would have been bordered by shops, marketplaces and houses – today there are just mounds of unexcavated rubble. Finally the water channel emptied into Wadi Siyyagh at the far western end of the Petra basin, which led to the Wadi Araba.

The aqueduct from the spring at Wadi Musa was, therefore, the central lifeline of Petra, flowing through the city along a path that can be easily traced today. But as the city grew in size, this by itself became insufficient and was supplemented by a complex system of reservoirs, cisterns, channels and pipelines, either exploiting other springs or run-off water from the surrounding hills. Population growth resulted in people living in higher areas, which could not be supplied by the low-lying channels. Reservoirs were therefore created within the mountains and paths contouring the rugged hillsides were surveyed and constructed.

Charles Ortloff describes one such reservoir known as the Zurraba, fed by run-off water and which supplied a large basin within the mountain plateau of Jebel al-Khubtha. The water flowed to the basin through terracotta pipes set within a rock-cut channel which wound its way along mountain sides at an elevated height. Channels from the basin led to cisterns at the base of the mountain, which in turn supplied the domestic needs for those living in the vicinity of the Royal Tombs, along with the water for any ritual requirements. While the channel from 'Ain Musa provided a continuous flow of water, that from the Zurraba reservoir was 'on demand' – it could be switched on and off at will. Ortloff describes this as having been the 'back-up' system, able to supply large quantities of water at short notice, such as would be required by the arrival of a large caravan in the city.

There were a multitude of cisterns and dams in the rugged hillsides around the city, capturing further run-off water, which was fed via channels to areas of urban housing and fields, supplementing supplies from the Zurraba and other reservoirs. Springs located on Umm el Biyara – the great mountainside that one directly faces when walking down the colonnaded street and which translates as 'Mother of Springs' – are also likely to have been tapped. One of these springs had a stele carved on to the adjoining rock face accompanied by a dedication to the Nabataean goddess Al-Uzza. She was the consort to the patron god of Petra, Dushares – 'He of the shara' (the mountains) – and appears to have been associated with Aphrodite, the Greek goddess of love, and Atargatis, the Syrian fertility goddess. Leigh-Ann Bedal suggests that Al-Uzza may have been the deity of water at Petra;[19] it seems likely that access to some, perhaps all, of the water sources was sanctioned by religion. Water, or at least some type of fluid, was certainly involved in religious ritual in light of basins in the Treasury and at the so-called 'High Place of Sacrifice' – one of many platforms on ridges and hilltops overlooking Petra where rituals had taken place.

Returning to more practical matters, the dams within the hillsides functioned not only to contain water that could be used for irrigation, but also to protect the city from flood water. The annual rainfall may have been limited, but much of it would have come in a small number of events during the winter rainy season; with the ground baked hard by the sun, such rain would have quickly resulted in torrents flooding into the city with the potential to do substantial damage. So the hydraulic engineering at Petra and the invocations to Al-Uzza were for the dual purpose of ensuring sufficient water was supplied when required and for protecting the city from winter floods.

Water demands south of Wadi Musa in the vicinity of the so-called 'Marketplace', theatre, temple and housing area were substantial. These were partially met by an overflow of water from the Royal Tomb area and partially by water from reservoirs in the Wadi Farasa area of Petra, and by that from springs – 'Ain Braq and 'Ain Ammon. There are many channels leading to the buildings within this area but it is difficult to decipher the precise patterns of flow when so many are now fragmentary and disconnected because of erosion, partial excavation and remodelling of the water supply system in antiquity.

Some of the most complex channels are associated with the so-called 'Great Temple' area (Photograph 18). This is a 'so-called' Great Temple, because the name is simply that given to a series of buildings in the city centre by the first archaeologists to work there, various German teams in the 1920s. They ascribed names – 'Great Temple', 'Marketplace' and so forth, but did not engage in any excavation to ascertain the appropriateness or otherwise of these names. It now seems likely that the 'Great Temple' was the Royal Palace of the Nabataean King Aretas IV; this would explain the substantial system of pipes that had carried freshwater into the complex, a large underground cistern, and the subterranean channels that had carried run-off water away.[20]

The Garden-Pool complex

It is, however, in the area to the immediate east of the Great Temple, the so-called 'Marketplace', where the most ostentatious display of water appears to have occurred. This overlooked the colonnaded street and is amid temples and civic buildings. Its name was acquired because it was a large (65 × 85 metres), open and seemingly un-built-upon area within the city centre of Petra; a marketplace seemed the obvious explanation to a western archaeologist of the 1920s. Excavations led by Leigh-Ann Bedal in 1998 revealed that this area had in fact contained a monumental pool surrounded by an ornamental garden with a central brightly decorated pavilion.[21] She found that a rocky slope had been terraced to leave a 16-metre-high vertical cliff face, in front of which the pool had been constructed, with the water retained by a massive east–west wall.

The pool had been 43 × 23 metres and 2.5 metres deep, able to hold 2,056 cubic metres of water, and lined with a pinkish-white concrete-like mixture. In the middle there had been a pavilion, perched on a sandstone pedestal. Several courses of wall had survived, along with patches of white plaster and painted stucco in dark red, orange and bright blue. There had been a pier against the north side of the pavilion, and another against the south side of the east–west wall. Water was transported around the pool in channels and through pipes, while the east–west wall had also supported an aqueduct. The wall had a central tank from which water had flowed into the terrace with its garden.

By standing on either the pavilion or the east–west wall one would have gained a view across a shimmering expanse of still blue water in the midst of the harsh, arid, mountainous landscape; beyond that there would have been a verdant garden with olives, figs, grapes, walnuts, violets and a diverse range of other plants,[22] but even that would have been overshadowed by the view and the sound of a waterfall. Leigh-Ann Bedal describes the remains of a tank on top of the cliff from which water would have cascaded into basins and then been channelled into the pool. This was an outrageously extravagant water display, but seems to have just been one of several deliberately created waterfalls within the city centre. Another fell from a carved niche in the northern edge of the Jebel al-Khubtha, some of the water being collected into a series of terraced basins and a cistern, but a great deal of it went to waste.

Bedal believes that the garden-pool complex was designed and constructed in conjunction with the Great Temple towards the end of the first century BC in the reign of Aretas IV. The garden-pool design with its island pavilion is strikingly similar to that at the summer fortress of King Herod in Judea built between 20 and 23 BC, although the Petra construction is much smaller in size. Such pools had become status symbols within the Hellenistic world and were particularly favoured by Herod. He constructed them at all of his palaces in Judea, probably having been influenced by the extravagant displays of water in Rome when he visited in 40 BC. Bedal suspects that Aretas IV built his garden-pool complex to establish his status within the region; he would have been familiar with those of Herod from either his own travels or by hearsay from traders. There may even have been dynastic marriage between the Herodian and Nabataean royal families.

The pool at Petra had a long history of modification. Bedal identifies nine chronological phases; the pool was only definitely out of use by the time of the AD 363 earthquake.

A valediction from Strabo

With Leigh-Ann Bedal's discoveries, the meticulous analyses of the fragmentary archaeological remains by Charles Ortloff, and the long-term research by John Oleson, one can perhaps have confidence in leaving the last word to Strabo:

The capital of the Nabataeans is called Petra ('the Rock'). It is situated on a spot which is surrounded and fortified by a smooth and level rock, which externally is abrupt and precipitous, but within these are abundant springs of water both for domestic purposes and for watering gardens. Beyond the enclosure the country is for the most part a desert, particularly towards Judea.[23]

6

BUILDING RIVERS AND
TAKING BATHS

Rome and Constantinople, 400 BC–AD 800

I doubt if it will be news for you that something was going on with water in the Roman world. Go almost anywhere in the former expanse of the Roman Empire – in Europe, North Africa, Asia Minor or the Middle East – and you will find evidence for the Roman manipulation of water: aqueduct bridges, wells, reservoirs, dams, cisterns, fountains, drains, toilets and, of course, Roman baths.[1] The bridges are often of monumental scale, exemplified by the Pont du Gard in France – almost 50 metres high and 275 metres long – that carried the aqueduct of Roman Nîmes across the River Gard.[2] Likewise the baths: those of Caracalla in Rome are so vast that 9,000 workers had been employed daily for five years for their construction.[3] Just as impressive is the sheer pervasiveness of the hydraulic archaeology: wherever one goes, from northern Britain to southern Egypt, there is compelling evidence for a Roman need for water. Moreover, we are only seeing a fraction of the evidence: thousands of kilometres of subterranean channels remain hidden from us, weaving their way around hillsides and valleys. All built to quench a Roman thirst, not necessarily for the water itself but for that which water brings: power, pride and prestige.

Throwing it away

Both the rural and urban populations of the Roman world certainly needed a reliable water supply and hence it was necessary for water to be transported, stored and distributed. It was needed in the construction industry to drive water mills for grinding grain into flour, for mining by using powerful jets of water and for agricultural irrigation. Everyone from the aristocratic elite to the legions at the frontier would have required some level of personal hygiene, and hence their toilets and bath houses. But was it really necessary to control water in such a grandiose manner? Did the Pont du Gard and the Baths of Caracalla have to be so imposing and magnificent in their design? Was continuously flowing water from fountains in every single Roman town really necessary to quench the thirst of the citizens? [4]

Of course not. While water had to be controlled in order to serve basic human needs, it was carried out in a manner that supported those with power: it legitimised and extended their authority. Professor Trevor Hodge, author of the most detailed study of the Roman water supply system and an expert on every engineering detail of aqueducts, baths and water mills that one could possibly imagine, was in no doubt. In his seminal 1992 volume, Hodge explained how, with few exceptions, Roman cities developed by depending on simple wells and cisterns for the water supply. The great aqueducts came long after the cities were established, for what Hodge described as a mere luxury: to supply the baths. Luxury perhaps, but baths were at the heart of Roman social life and their extravagant use of water was a political statement about power, wealth and identity. As was the water that flowed continuously from city fountains and the massive aqueduct bridges that remain as the most imposing monuments of the Roman Empire. Hodge believed the Roman aqueducts sent one unmistakable message to all around: 'Water? Why, we've got heaps of it, we just throw it away.'[5]

Other academics have gone further: in a review of the water system of Rome, Professors Bono and Boni stated that 'in ancient Rome, water was considered a deity to be worshipped and most of all utilised in health and art. The availability of huge water supplies was considered a symbol of opulence and therefore an expression of power.'[6] Much of the water was required for bathing, which Professor Inge Nielsen, the pre-eminent scholar on Roman baths, has described as being 'as vital as eating, drinking, making love and laughing' in the Roman world.[7] Even though baths were architecturally complex and expensive, they were among the first buildings to appear in the colonies and newly conquered towns;[8] all towns built the largest and most luxurious baths possible as a matter of civic pride. Likewise an aqueduct, if at all possible, and especially an imposing arched bridge – a structure that so perfectly encapsulated the Roman virtues of being solid, grand, practical and civilised.[9]

At the baths

With regard to identity, one could simply be neither Roman nor Romanised without bathing. This involved far more than merely getting clean by a sequence of sauna, hot bath, cold bath and massage: the baths were the centre of Roman cultural life. The largest baths catered for sport, music, art and literature, while all were the focus for social gossip. Cicero wrote that 'the gong that announced the opening of the public baths each day was a sweeter sound, than the voices of the philosophers in their school'.[10] But a letter sent by Seneca to Lucilius that captures the atmosphere within a provincial Roman bath of the first century AD also makes bathing sound a little less desirable than Cicero might let us imagine:

I have lodgings right over a bathing establishment. So picture to yourself the assortment of sounds, which are strong enough to make me hate my very powers of hearing! When your strenuous gentleman, for example, is exercising himself by flourishing leaden weights; when he is working hard, or else pretends to be working hard, I can hear him grunt; and whenever he releases his imprisoned breath, I can hear him panting in wheezy and high-pitched tones. Or perhaps I notice some lazy fellow, content with a cheap rubdown, and hear the crack of the pummelling hand on his shoulder, varying in sound according as the hand is laid on flat or hollow. Then, perhaps, a professional comes along, shouting out the score; that is the finishing touch.

Add to this the arresting of an occasional roisterer or pickpocket, the racket of the man who always likes to hear his own voice in the bathroom, or the enthusiast who plunges into the swimming-tank with unconscionable noise and splashing. Besides all those whose voices, if nothing else, are good, imagine the hair-plucker with his penetrating, shrill voice, – for purposes of advertisement – continually giving it vent and never holding his tongue except when he is plucking the armpits and making his victim yell instead. Then the cake seller with his varied cries, the sausageman, the confectioner, and all the vendors of food hawking their wares, each with his own distinctive intonation.[11]

The phenomenon of Roman baths and bathing was inherited from the Greeks, but took on a quite different character.[12] The Greek baths, their 'balaneion', were developed during the fifth century BC but lacked the prestige of the Roman version. Indeed they were generally thought of as immoral places mostly for the old and infirm. Far more status was attached to the 'gymnasia', where not only sports took place but also education, socialisation and cultural events. The Roman upper classes who visited Athens had become familiar with the gymnasia. These appear to have inspired the great imperial baths, or thermae, of Rome, where a similar range of activities took place – as well as bathing itself.

Until aqueducts were constructed, baths were constrained in size by having to draw water from wells or cisterns. The archaeology of the Stabian baths at Pompeii shows the transition. When first built in the fifth or fourth century BC they were supplied from a well by a bucket; later, a treadmill lifted water into a tank, and then a larger treadmill was used before an aqueduct was built at the end of the first century BC. This enabled the baths to expand and to become a massive complex of rooms of cold, tepid and hot baths, pools for swimming and an exercise area, all ornately decorated.[13]

Water wasn't the only resource needed for a Roman bath; fuel was required, initially for charcoal braziers and then for the hypercaust systems introduced in the first century AD, as well as for boilers that heated the water. The state-owned baths were fuelled by wood from the state forests, while fuel costs were a considerable expense for the private baths.

A visit to the baths was an indispensable part of daily life for all Romans, whatever their social class.[14] The significance of bathing for maintaining not only the Roman body but also the psyche is indicated by the sheer number of bath-houses and the amount of literary and epigraphic references to baths and bathing. The baths were supposedly strictly segregated by sex, but writers such as Cicero frequently lamented that this was not always adhered to. Not surprisingly, baths were haunts for thieves, with the upper classes often taking a slave with them to mind their belongings. Small entrance fees were the norm – one record notes that this was a quadrans, far less than was required to buy a loaf of bread. By the second century AD, baths throughout the empire had taken on a standard architectural form, distinguished by apses and curved shapes; Inge Nielsen describes the architecture of the bath-house, and by implication the practice of bathing with all of its social and cultural associations, as forming an entrenched cultural construct.[15] The same could be said for aqueducts: they appeared wherever the Romans went, more than 300 were built in Gaul alone.[16]

Although the cultural associations with bathing were clearly important, one cannot escape the fact that seeing, touching and becoming immersed in water was the ultimate rationale for the Roman bath, just as it was for the fountains and nymphaeum within the towns. Why should water have had such an impact on the Roman mind and civilisation?

We must, of course, be careful. Water is essential for life; even the most extravagant works of Roman hydraulic engineering delivered practical benefits, especially for urban populations. The water drained from the baths, for instance, was sometimes used as irrigation for agriculture and gardens.[17] Gross generalisation about Roman attitudes to water are only helpful when accompanied by specific historical, cultural, economic and environmental studies that expose the complexity of the Roman world. So to explore the complex mix of utilitarian value and political statement, this chapter will consider the water supply of the two greatest Roman cities, the capital of the Western Empire, Rome, and that of the Eastern Empire, Constantinople.

Water supply of Ancient Rome

At its height, the one million inhabitants of ancient Rome were served by eleven aqueducts, built over the course of six centuries from the end of the fourth century BC.[18] Each one had a name: the Aquas Appia, Anio Vetus, Marcia, Tepula, Julia, Virgo, Alsietina, Claudia, Anio Novus, Traiana and Alexandrina (Figure 6.1). There was no master plan, simply a gradual development of the water supply system as demanded by an expanding urban population, and especially a more affluent population.

It wasn't only an enhanced supply system that was required: just as we have seen in Classical Athens and will later see elsewhere in the ancient world, the drainage of water from cities is equally essential. The most striking evidence from Rome is the Great Sewer, *Cloaca Maxima*, which is one of the earliest sewage systems in the world and still partly operates today. It was probably constructed around 600 BC and served to drain the unhealthy swampland around the foot of the Palatine Hill – the very centre of the city – caused by periodic flooding of the Tiber River; the drain also emptied human effluent from Rome's burgeoning population into that same river.[19]

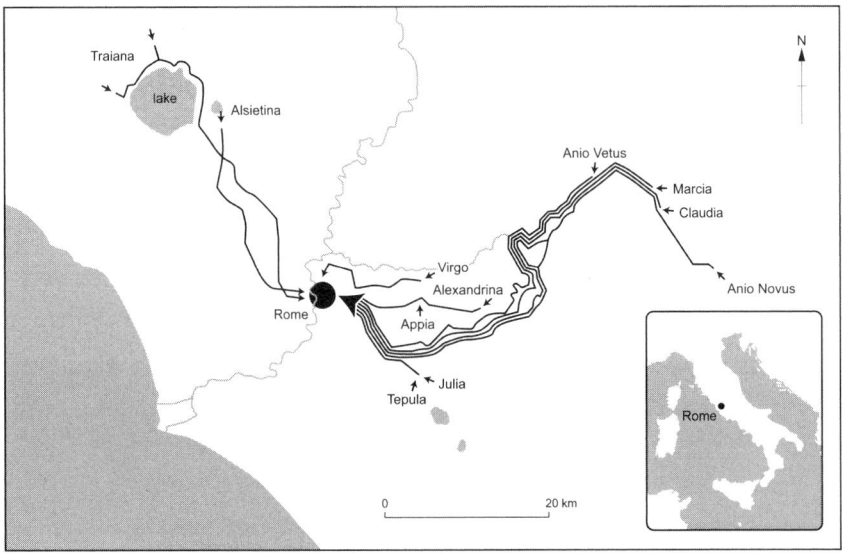

6.1 The aqueducts of Rome

Only the most dedicated hydraulic archaeologist will be looking for ancient sewers on a visit to Rome; but everyone can enjoy the ancient aqueducts, remnants of which can be easily seen within the city today. There are, for instance, the monumental ruins of Porte Maggiore, which is the only remaining section of the Aqua Claudia within the city and now straddles a busy crossroads in the centre of Rome. The Arch of Drusus is tucked away in a quiet side street and carries the only surviving fragment of the Aqua Antoniniana, an extension of the Aqua Marcia that crossed the ancient Appian Way to supply the Baths of Caracalla. But only one of the underground aqueducts leading into

Rome, the Aqua Virgo, continues to function, having been repaired in Papal and modern times.

The best place to get an impression of the extent of the aqueducts in Rome is eight kilometres to the south of the modern city in the 'Parco degli Acquedotti' where the remains of seven aqueducts converge on their route to the ancient city (Photograph 19). The most impressive remains are of the Aqua Claudia carried on majestic arches which still stride proudly across the Italian countryside. The park provides us with a glimpse of the landscape as it must have been in Roman times. Fields of wheat and wild flowers run underneath the arches and walkers follow the aqueduct as it marks the ancient route of the Appian Way. They rest from time to time at the base of the monumental walls and shelter from the hot sun. In the distance are the lower slopes of the Appenines that provide the springs for the aqueducts and the steep gradient to send the water on its route into Rome.

However grand these ancient ruins still appear, Rome certainly did not have the largest or most sophisticated aqueducts; these were constructed in southern France, Asia Minor and North Africa. Nor do we have a particularly good understanding of how the water supply system worked in Rome once the water left the aqueducts at their 'castellae', the distribution tanks they fed: medieval and modern-day Rome have partly destroyed and partly buried the evidence, leaving us guessing as to how water flowed around ancient Rome after it had arrived into the city.

Rome does, however, have two special features that make it central to our understanding of the Roman obsession with water, and especially water for bathing. First, being the original, largest, wealthiest and most diverse Roman city, it is where the Roman relationship with water was defined and found its most elaborate expression. Second, it had Sextus Julius Frontinus, AD 40–103.[20]

Frontinus was appointed as the *cura aquarium* – the city's water official – during the reign of the emperor Claudius, in around AD 95. He was the president of a three-person board, all of whom were of senatorial rank. The fact that Frontinus had previously been governor of Britain (AD 74–7) and a proconsul of Asia Minor, that he had written on military theory and was a distinguished lawyer, indicates how highly the position of *cura aquarium* was held within the administrative and political hierarchy of Rome. This was not only because water was required for public and private needs, but also because Rome was plagued with fires, the largest having happened a mere 40 years before Frontinus' appointment during the reign of Nero.[21] Others who held the *cura aquarium* post had similarly distinguished careers, such as M. Valerius Messalla Corvinus who been made *cura aquarium* in 11 BC after a prominent legal and military career. Frontinus was appointed partly to reorganise the service of the water system because it had become corrupt. Fortunately for us, he took his job most seriously, deciding to get to grips with the details of its history and management himself rather than relying on reports from his staff. As such he wrote a treatise entitled *De aqua ductu*.

Written in AD 97 this work recounted the history of the aqueducts and investigated the capacity of the water supply system. Frontinus measured the amount of water provided by each aqueduct and how this was divided up between the various uses of water: for public fountains and basins, for the imperial buildings, for private usage. His absolute figures remain subject to debate, with modern-day estimates for the overall water supply to Rome varying between 500,000 cubic metres and more than a million. But his relative measures for the capacity of the aqueducts, his evaluation of their water quality and descriptions of its usage, along with his accounts of how and when they were built, are simply invaluable for anyone concerned with water in the ancient world. Although his writing is largely typically dry civil servant speak, there are a few expressions of evident pride: 'Just compare', he instructs us, 'the vast monuments of this vital aqueduct network [with] those useless pyramids, or the good-for-nothing tourist attractions of the Greeks.'[22]

Aqueducts of Ancient Rome

The city of Rome grew and thrived for several centuries without any need for aqueducts: it was naturally well watered with abundant springs and a high water table. Drawing water from wells appears to have been perfectly adequate for all of the city's essential needs throughout the period of the Roman Empire, but in 312 BC the first aqueduct was constructed and named the Aqua Appia. This brought water into the city from springs 16 kilometres to the east and appears to have supplied the commercial district of the Forum Boarium. It ran underground for almost its entire length and was, according to Frontinus, considered a remarkable achievement at its time of construction.

Just like the aqueducts that were later built in Rome, and indeed the majority throughout the Roman world, the Aqua Appia worked by gravity – a gentle slope from the source to the castellum, this being a small reservoir from where water was distributed throughout the city. The arrival of an aqueduct was often marked with a fountain – a tradition that has continued for Rome's modern-day aqueducts constructed in the 20th century.[23] Water in the Aqua Appia had flowed through stone-lined conduits, usually about a metre wide and 2.5 metres deep and lined with cement for waterproofing and to create a smoothly flowing surface. The conduits were either capped with flat stones or had a vault, and were buried around half a metre below ground. They had to be large enough to allow access for cleaning, because they were prone to calcium carbonate incrustations and the accumulation of sediment, although this was often captured by settling tanks distributed throughout their courses. In some parts of the Roman world, pipes made from lead, terracotta or wood were used, as were siphons rather than bridges to cross valley floors. While the crusty carbonate deposits within lead pipes may have impeded their efficiency, they had a beneficial effect by preventing the flowing water becoming contaminated with lead and hence poisoning the population.[24] But this was rarely an issue because the large majority were conduits made of stone. They often had to be supported across steep valleys by bridges and led through ridges by tunnels. They are best thought of as artificial rivers.[25]

Forty years after the Aqua Appia, a second aqueduct for Rome was built, the Aqua Vetus. Although this too was almost entirely underground it was far more ambitious, bringing water 69 kilometres from the Upper Anio Valley and supplying a much more extensive area of Rome via 35 castellae into which its waters flowed. Frontinus notes that its waters were liable to become muddy after storms and hence may have been reserved for industry, irrigation and animals. The Aqua Vetus took two years to build and was funded by booty from the Pyrrhic War.

There was an attempt to build a third aqueduct in 179 BC but Marcus Licinius Crassus simply refused permission for it to cross his land, the course designated by the aqueduct engineers, and so the project was scrapped. Thirty years later the demand for water arising from an expanding population had become irresistible. The Senate empowered Quintus Marcius Rex to take the required action, providing booty from wars with Carthage and Corinth to cover necessary costs. He built the Aqua Marcia in 144 BC and repaired the two existing aqueducts. As Frontinus was later to explain, repairs were constantly required because of the crusty carbonate deposits laid down within the channels and disruptions caused by roots and earth tremors.

The new aqueduct, the Aqua Marcia, was the most ambitious yet and the first to involve arched bridges to carry the channel across steep-sided valleys and into the city. This also drew on a source in the Upper Anio Valley, but one more distant than that used by the Aqua Vetus, carrying water for 91 kilometres, only seven kilometres of which was above ground. The city was evidently proud of its new supply, erecting an honorary statue of its engineer near to where it terminated at the Capitol. Frontinus records opposition to its extension on the grounds that it would encourage further urban growth in an area that was a *loca publica* – nimbyism in the ancient world.

The next aqueduct, built a mere 19 years later in 125 BC, was named the Aqua Tepula after its tepid waters, which were drawn from springs just a few kilometres from the city at the foot of the Alban Hills. No archaeological traces of the aqueduct are known and it may have been designed as a supplement to the Marcia for an industrial water supply. There then followed a hiatus in aqueduct building while Rome was involved in civil wars.

A transformation of the Roman water supply system was begun in 33 BC under the leadership of Marcus Vipsanius Agrippa.[26] By this time, the existing aqueducts needed substantial repairs while the administration of the system also required attention. Demand for water was continuing to grow – there were no fewer than ten public baths in Rome. Agrippa rose to the challenge. Using his own wealth, acquired as booty from war, he repaired the Appia, Vetus and Marcia aqueducts, increased the capacity of the Tepula and built two new lines, the Aqua Julia and Aqua Virgo.

The Aqua Julia drew on a source just a few kilometres away from that used by the Aqua Tepula; its channel made use of the arches that had been built to carry the Aqua Marcia. The Aqua Virgo is supposedly named after a young girl who showed soldiers to a spring about 22 kilometres outside Rome. This is the only Roman aqueduct still functioning today, having been repaired and renovated many times, and it supplies the waters to the famous Trevi Fountain in the centre of Rome.

While Agrippa's repair work and new builds were undertaken to meet the needs of an expanding population, he appears to have been just as concerned with displaying Rome's – his – achievement: within one single year 300 ornate sculptures were added to the fountains. The 'needs' were not of an entirely practical nature: the Aqua Virgo was built to supply Agrippa's new bathing complex in the Campus Martius, including his stagnum (pool connected to his baths) and euripus (swimming channel). This provided free access for all Roman citizens – the majority of other baths required an entrance fee.

Agrippa created a permanent staff for the administration and maintenance of the water system, which became the basis of the imperial staff headed by the *cura aquarium* that was established after Agrippa's death in 11 BC; he bequeathed no fewer than 240 skilled workmen to Augustus. Indeed Agrippa's reorganisation sustained the city until the fall of Rome to the barbarians in AD 537. Nevertheless, the water system was massively extended during the imperial period to meet ever greater demands for free-flowing water. One such demand was by the emperor Augustus, who commissioned the seventh aqueduct in 2 BC, the Aqua Alsietina, primarily to supply his naumachia, drawing water from the Alsietina Lake 25 kilometres from Rome. The naumachia was Augustus' artificial lake on the right bank of the Tiber intended to allow the re-enactment of naval battles. It was sufficiently large to contain 30 vessels with rams, containing 3,000 men (excluding rowers), along with a number of smaller boats.

This is merely one indication of the opulence of imperial Rome. As the city expanded so did the demand for water. While Augustus had already doubled the capacity of the Aqua Marcia by tapping a second source, his successors doubled the overall supply to Rome by means of two new aqueducts, the Aqua Claudia and Anio Novus (Photograph 20). These were built as a single project, initiated by Caligula in AD 38 and finished by Claudius in AD 52, and funded by the fiscus – this being the personal treasury of the emperor, there no longer being any booty from war. The Aqua Claudia drew on springs within the Upper Anio Valley close to those used by the Aqua Marcia, but because of a more extensive use of bridges the aqueduct was a mere 64 kilometres long. One of its bridges was 40 metres in height and no fewer than 1,000 arches had once taken the aqueduct across the Campagna – the low-lying countryside around Rome – of which 350 survive today.

The Anio Novus drew water from the River Anio 86 kilometres from Rome and delivered water of much poorer quality than that delivered by the Aqua Claudia. The majority of it was, however, mixed within a shared castellum. They also shared an aqueduct bridge, by far the largest built for Rome, on which the Anio Novus channel was stacked above that of the Aqua Claudia.

Before the fresh clean water of the Aqua Claudia was mixed with the dirty water of the Anio Novus, a branch aqueduct, the Aqua Caelimonatani, diverted a channel to the area of Rome known as the Domus Aurea. This appears to have been primarily to supply Emperor Nero's new baths of AD 62. These included a stagnum, a vast nymphaeum and an ornamental fountain, and were the first to impose the architectural plan that became standard for the large imperial baths in which the thermal areas (frigidarium, tepidarium and caldarium) were situated on a long central axis, and other areas, such as the palaestrae, attached along the sides.[27]

We know that the Claudia and Novus doubled the water supply to Rome from the writings of Frontinus. His *De aqua ductu* describes how this combined line served no fewer than 92 castellae in the city, out of a total of 247; the Marcia was the second-largest source, supplying 52 with its especially high-quality water. Frontinus tells us that one third of the water from the aqueducts was distributed outside the city itself. Sixty per cent of this went to private consumers for their villas, gardens and irrigation needs; they had to pay for the water unless they had been granted water rights by the emperor. The remaining 40 per cent went to imperial properties. Inside the city 20 per cent was taken for imperial usage, 40 per cent went to private consumers and 40 per cent was for public needs: fountains, baths, amphitheatre, markets and so forth.

Frontinus tells us not only about this distribution, the history and the need for constant repair, but also about the illegal tapping of the system. It seems that the whole of the waterworks staff colluded by supplying water to those who paid them, such as by adding an additional channel to their properties or enlarging the pipes leading from reservoirs. The system was also exploited by the landowners through whose land the aqueducts ran: they illegally tapped the channels and planted trees in what was agreed as a 'reserved' zone around the channel.

Frontinus is thought to have died in AD 104 and hence did not see the last two of the great aqueducts of Rome. The Aqua Traiana was built in AD 109 to provide additional water to the district of Trastevere on the left side of the Tiber, bringing water from springs near Lake Bracciano 56 kilometres from Rome. The headwaters and the highly decorated subterranean nymphaeum of Aqua Traiana were only discovered in 2009, buried beneath thick vegetation and pig pasture to the north-west of Rome. Its waters arrived at a considerable height, providing sufficient pressure to drive water mills. The Aqua Traiana, probably crossing the River Tiber near the modern bridge of Ponte Sublicio, also supplied the Baths of Trajan – a massive complex that was built on what had been the site of Nero's palace and became the canonical model for large imperial baths with associated arcades, gardens, libraries and exercise regions.

The Aqua Traiana was celebrated by the minting of a coin showing a river god atop flowing waters. It was an impressive aqueduct and the penultimate one built in ancient Rome: with one exception, all later emperors confined themselves to repairs and improvements of the existing system. The exception was Alexander Severus (AD 222–35), who built the Aqua Alexandrina (around AD 230) to supply his baths, drawing on a source 22 kilometres from Rome. One can readily understand why another aqueduct may have been required: by the start of the fourth century AD, Rome is estimated to have had 11 large baths, 856 public baths, 15 monumental fountains, 1382 fountains and basins, and two naumachiae.[28]

Other emperors chose to supply their new baths by diverting water from existing aqueducts. Most notable was Marcus Aurelius Severus Antoninus Pius Felix Augustus (AD 188–217), otherwise known as Caracalla. In AD 212 he diverted a channel from the Aqua Marcia, calling this the Aqua Antoniniana, to feed his baths – the magnificent Baths of Caracalla. These are worth a brief visit to further explore the phenomenon of bathing in the Roman world.

The Baths of Caracalla

The Baths of Caracalla were also known as the 'Villas of the Plebeians' because they were built in the working-class area of Rome;[29] ironically, they were the most ornate and lavishly decorated baths with multi-coloured marble floors, mosaics, stucco paintings and hundreds of statues surrounding an enormous bathing complex.

Caracalla, so named because of his predilection for wearing a hooded garment of that name from Gaul, was a rather questionable character. Ancient authors described him as insane, bloodthirsty and fratricidal – the latter on account of the death of his brother with whom he had initially shared the title of emperor. Caracalla suffered from delusions that he was the reincarnation of Alexander the Great and spent the majority of his reign with the legions in Britain, Gaul and Germany, leaving his mother in Rome to attend to running the empire.

His baths, inaugurated in AD 216, covered a vast area, having required several million bricks, 252 columns and 9,000 workers, employed daily for five years.[30] The supplying aqueduct emptied into 18 cisterns to guarantee a continual water supply to the baths and fed an extraordinarily complex plumbing system. At the centre of the baths was a building that contained the frigidarium with four separate cold-water pools, a tepidarium with two pools and a caldarium with seven pools for hot water, along with dressing rooms, saunas, massage rooms and those for heliotherapy and depilation – plucking of hair (Photograph 21).

The frigidarium alone was quite magnificent with massive cross vaults resting on colossal columns of grey Egyptian marble, rows of arcades, walls covered with polychrome marble plinths, marble floors, fountains and statues. Its monumentality was little different to that of a great basilica; it provided the inspiration to the architects of the Chicago railway station and the Pennsylvania station in New York.

Below the baths themselves was a system of underground passages used for maintenance, some large enough for carriages to pass through. These led to rooms where the wood for the heating system was stored, for the ovens, for cauldrons and even a water mill powered by the surplus water.

A 'mythraeum' was also found underground – a room for the

worship of the Egyptian god Mythra. Caracalla and his family tended to religious syncretism, blending elements from different traditions, and this appears to have been the largest mythraeum in Rome. Its key feature was a hole in the centre of the floor. Some believe this was to catch the blood from sacrificed bulls which was also supposed to splatter across the white toga of whoever was being initiated into the cult. At least they were in the right place to then get clean because for all of the exercise, eating, talking and ritual functions of the baths, at the centre of the complex were simple pools of water.

Today, the ruins are easy to identify as they still reach 30 metres high and are set upon a huge platform of earth. The arches that would have led to the underground service passage, where bustling shops and food stalls would have been located, now lie quiet behind metal grilles.

Climbing the ramp up to the ground level of the baths themselves, one is immediately struck by the scale of the building. An outer wall encloses an area of quiet gardens with meeting rooms to the west and east and two libraries to the south. This outside area would have been very much as it is now – a peaceful, quiet garden for contemplation and relaxation before or after the visit to the baths themselves.

I strolled through the rooms with sunlight falling in shafts on to the brightly coloured and almost complete mosaic floors. Huge chunks of highly decorated mosaics rested against the walls where they had fallen from the upper floors and walls. Seagulls were nesting on the sunny platforms of the arches and swooped between the curves of the upper floors. Back outside, I found that the small stadium was still in use – an inter-schools sports event with a young audience using the slope to get a better view in the same manner as an audience in ancient Rome would have done.

A new Rome – but with something missing

The Baths of Caracalla are said to have received between 6,000 and 8,000 visitors a day. As they bathed, Rome was in terminal decline; Constantine, Roman emperor between AD 306 and 337, turned his back on the city and chose Byzantium – later to become Istanbul – to be his capital. Located at the gateway between Europe and Asia and close to the frontiers of his empire, this was now a more appropriate capital than Rome itself, being closer to the trading entrepôts and the imperial armies. Originally founded around 660 BC as a Greek colony, Constantine redesigned the city into an image of Rome. He built public works, created an administrative system and imported works of art from throughout the empire. Soon the city could boast a forum, walls, baths, a hippodrome, an Imperial Palace and all other critical infrastructure, monuments and displays required by the capital of the greatest empire on earth.

The new city of Constantinople was consecrated on 11 May, AD 330. It became a magnet for people, traders and wealth, resulting in an exponential growth of the urban population. While Constantinople expanded, Rome declined. In AD 395 there was a formal split into the Eastern and Western Roman Empires; the latter fell to the barbarians, with its last emperor, Romulus Augustus, being deposed by a Germanic chieftain on 4 September, AD 476. The city briefly re-entered Roman hands during the Gothic Wars when in AD 537 Belisarius retook the city. It was immediately surrounded by the Ostrogothic army under the leadership of Vitiges and put under a siege that lasted one year and six days. This put an end to the Baths of Caracalla as the population abandoned the outlying areas to gather in the centre of Rome. The Ostrogoths blocked the aqueducts to deprive Rome not only of its drinking water but also its bread as this water had powered the mills. They made a thwarted attempt to use the Aqua Virgo as a passageway into the city before Belisarius finally defeated the Ostrogoth army. Nevertheless, Rome was 'history' and Constantinople had already become the greatest city of Europe, perhaps the world; it was to remain so for more than 1,000 years.

To achieve its status, the emperors, architects and engineers of Constantinople had to address one challenge that was quite different to that faced in Rome: providing a supply of water for even the most basic needs.[31] Rome had springs relatively close at hand and their discharge was quite stable throughout the year. Constantinople, however, was built on geology that lacked a ground water supply; any springs were distant and irregular. Its first aqueduct had been built while the city remained as Byzantium during the reign of Hadrian, AD 117–38, most likely from his direct orders during his visit to Thrace.[32] This brought water 20 kilometres from hills to the north-west of the city, from the region known today as the Forest of Belgrade. By the fourth century AD this supply had evidently become inadequate to supply the needs of the ever-growing population and the demands of an imperial city with its requisite elaborate fountains and baths.

Now the engineers had to look further afield for water, to the springs and aquifers in the hills of the Stranja Daglan, running parallel to the Black Sea coast north of Constantinople in what we now call Thrace. The discharge from those springs was highly seasonal – abundant following heavy rain but limited during the dry season and most likely zero during periods of drought. As such, the hydraulic challenge was not just transporting the water, when available, into Constantinople, but storing sufficient water for those periods when the springs ran dry. There was also a second reason for storage: for much of Constantinople's history it was besieged with a constant risk of having its water supply interrupted, just as had happened in Rome during the siege of AD 537.

The longest Roman water supply line

The remarkable achievement of bringing water to Constantinople has only been fully appreciated within the last five years. *The Longest Water Supply Line* is the title of a 1996 book by Professor Kâzim Çeçen, a hydrological engineer.[33] He was the first to undertake a systematic mapping of the archaeological remains of an aqueduct system that stretched, he proposed, for no less than 242 kilometres, beginning at springs near the modern-day town of Vize and taking water all the way to Constantinople. His work was extended by Professor Jim Crow of the University of Newcastle and his colleagues between 1998 and 2008. They added a great deal more detail, combining archaeological survey using high-precision satellite and terrestrial techniques, a comprehensive study of all historical sources, chemical analysis of calcareous deposits surviving with the aqueduct channels to identify the source of the water, and creating a Geographical Information System for the region to integrate and analyse the archaeological, topographic and hydrogeological evidence.

Crow and his team found that Professor Çeçen, had been wrong: the long-distance channel had not been 242 kilometres but a remarkable 551 kilometres – this being the length of its sinuous way around the hills and steep-sided valleys of the Thracian countryside to cover what is a mere 120 kilometres by a straight line from Vize to Constantinople (Figure 6.2).

As with the majority of Roman aqueducts, the channel had been constructed from stone blocks placed into a cut in the ground; these were waterproofed by lining with lime-mortar and the channel was capped with more stone. To traverse the countryside – thick woodland and steep, twisting valleys – no fewer than 60 aqueduct bridges were constructed along with numerous tunnels, some thought to be over 1.5 kilometres in length.

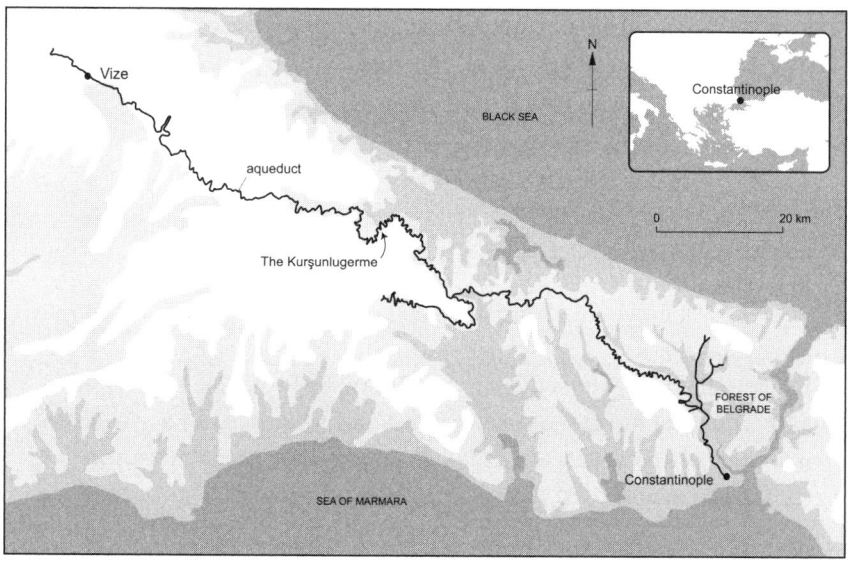

6.2 *The long-distance aqueduct from Vize to Constantinople, showing location of the Kurşunlugerme aqueduct bridge (after Crow et al. 2008)*

This was an extraordinary engineering feat, described by Jim Crow as 'one of the most extraordinary legacies from the ancient world'.[34] It appears to have been built in two phases. First, during the reign of the emperor Valens (AD 364–78), an aqueduct was constructed from the springs at Danamandıra and Pınarca, about 100 kilometres into the Thracian countryside. The water arrived into the city across a monumental aqueduct bridge with 89 arches; this still stands within Istanbul today, known as the Aqueduct of Valens or the Bozdoğan Kemeri, and spans one of the busiest city-centre highways. The arrival of water into the city was celebrated in an oration by Themistius to the Emperor Valens that extolled the value of water above that of statues and 'precious objects' (Photograph 22).[35]

In a second oration, Themistius characterised the flowing water as Thracian nymphs:

And they, faster than wings and thought, either fly high up through the air or run underneath steep jutting hills, in the earth and in the air, resembling a bunch of grapes as regards their backs, more than a thousand stadia uphill and downhill, neither running upwards nor downwards, and neither being held down nor being held in. And roofed over they come together and they have arrived here before the gates and they camp in the open waiting for the originator in order that with him as host they might settle in their temple, in which they dance together Hephaestus and Asclepius and Panacea.[36]

– their temple being the nymphaeum now buried below the concrete of present-day Istanbul.

During the fifth century the channel – effectively a man-made river – was given a more distant source, one near the town of Vize. This resulted in the vast 551 kilometres of aqueduct channel, passing across 30 stone bridges and through many kilometres of tunnel to reach Constantinople. It was built during the reign of the Theodosian emperors, AD 406–59 to meet a huge increase in water demand: records inform us that the city now had eight thermae, four nymphaea and no fewer than 153 private baths. We are also told about four cisterns for water storage and know that two of these, known as the Aetius and Aspar, were quite massive open reservoirs. They were located between one line of the city walls that were built by Constantine I, started in AD 324, and a second set of walls built 1,500 metres further out of the city by Theodosius II (AD 408–30) in an area that appears to have been devoted to agriculture. As such, these reservoirs seem to have been for irrigation and providing water for animals.

The fifth-century concern with water is evident from the number of laws that were passed at this time to control its use.[37] These covered such issues as how the diameter of a pipe to supply an estate was determined by the size of the estate; the penalties for taking water from an aqueduct rather than a reservoir; and how all tax payers had to contribute labour and materials for the repair of the water systems.

The Kurşunlugerme

When in Istanbul I went in search of one of those bridges that had carried water the 551 kilometres from Vize to the city. Driving north from the city for a couple of hours into the densely wooded Thracian countryside I came to the tiny village of Gümüşpınar. That was the closest place on the map to the Kurşunlugerme, the name of an aqueduct bridge that Jim Crow and his colleagues had described as the best-preserved and most monumental. I knew it was located within a steep-sided, thickly wooded valley, several kilometres from any road, but I lacked specific directions. Outside the mosque in the centre of the village I asked an old man if he knew of the Kurşunlugerme, immediately resorting to a picture and sign language. He recognised it immediately and began talking at great length – in Turkish and completely unintelligible to me. When my incomprehension dawned on him, he simply climbed into my car and directed me back the way I had come to a water-bottling station – Gümüşpınar Spa. After a few minutes' discussion with a couple of workmen, the old man passed me on to them and they drove me in their four-wheel-drive vehicle down a steep muddy track for about 30 minutes to their water-pumping station in the bottom of the valley. The driver gestured along the valley and there through the tops of the bare branches I saw the arches of the Kurşunlugerme. A few minutes' walk along a tiny winding path through brambles and across stepping stones took me to its base, where my neck craned upwards to see the vast stone construction, built to maintain a flow of water 35 metres above the valley bottom. A stream ran noisily through one of the arches; moss, ivy, brambles and even trees had taken root within the masonry (Photograph 23).

The Kurşunlugerme was the largest aqueduct bridge built outside of Constantinople, spanning the 149-metre-wide valley and carrying two water channels coming from springs at different heights.[38] It had three tiers: the bottom one had three great arches, six metres high, with a stream running through just one of these; the middle tier had six arches and a ledge along which the lower water channel ran; and the upper tier had eleven arches and supported the upper channel. It was built from grey-white limestone blocks held together with pink mortar with brick inclusions and with iron clamps. Having Jim Crow's detailed description to hand, I looked at the inside of the upper tier arches to see long cracks caused by earthquakes; this had destroyed the west face of the bridge, but the structure had remained standing.

I also looked for the decorations. The Kurşunlugerme was the most decorated bridge in the Roman world, with more than 40 symbols carved into its stone. These were primarily crosses and chrismons of varying degrees of elaboration, a chrismon being the 'monogram of Christ' formed from the Greek letter Chi ('X'), imposed on the letter rho ('P'). Other symbols include a flying eagle with a serpent and wreath in its talons, interpreted as a symbol of the emperor's triumph over his enemies. Of the numerous inscriptions, one that is associated with a cross and an eight-armed wreathed chrismon reads: 'The cross has conquered. It always conquers'.

Why should there have been so many decorations on this bridge? By a meticulous study of the specific location of the decorations on the bridge and by a comparison with decorations on fortifications, basilica and other buildings, Jim Crow and his colleagues suggest two reasons. Some of the decoration was, they argue, apotropaic in character – it was made to ward off evil and bad luck. Many crosses and other symbols are at positions of potential structural weakness or where reinforcements might be necessary, such as at the chamfers at the base of the buttresses. These are not particularly visible and seem likely to have been made during or soon after construction simply to protect the bridge from bad luck, such as earth tremors or especially violent river surges after storms.

The most elaborate symbols, such as the eagle and serpent, and the majority of the inscriptions are positioned quite differently – in places where they can be easily seen and read. Moreover, unlike other bridges within the aqueduct system, there was a well-constructed set of stone stairs leading from the valley floor to permit even closer inspection. But who, Jim Crow and his colleagues ask, would have been able to read the Latin and understand the symbolic meaning of the imagery? The Stanja Forest region would have been sparsely populated in Roman times, and few of its people would have been literate.

Crow suggests that there is likely to have been a grand ceremony upon the completion of the bridges of the aqueduct, involving a procession by the emperor himself to visit the Kurşunlugerme to celebrate the new flow of water into his city. The inscriptions would have been read out loud, the orations acting to both celebrate the engineering achievement and, of course, extol the emperor.

There may have been annual commemorative visits to the Kurşunlugerme, at least while the environs of Constantinople were sufficiently safe for travel. That certainly wasn't the case after the sixth century AD, when ceremonial events appear to have become restricted to within the city walls. With barbarian armies not far outside and now threatening the supply of water into the city by cutting the aqueducts, attention had to be paid to water storage.

Barbarians and droughts

Having its aqueducts cut by barbarians was just one of the threats to Constantinople's water supply; another was drought. These concerns led to a proliferation of cisterns, built for both public use and private supply to mansions, palaces and churches. The first recorded attempts to cut the water supply occurred in AD 487 as part of the extraordinarily complex politics and power struggles of fifth-century Constantinople. One episode involved Theodoric Strabo, head of the Thracian Goths, leading a revolt against Zeno, the Byzantine emperor, and involved him capturing Sycae, the region of Constantinople located to the east of the Golden Horn estuary and known today as Galata. Theodoric Strabo cut the water supply line to the rest of the city.

That was effectively an internal dispute within the city. In AD 626, however, the whole city was besieged by barbarian armies coming from Thrace. The 'long-distance aqueduct' was cut and not restored until AD 766.[39] The fact that the city survived for the intervening 140 years partly indicates that the much older Aqueduct of Hadrian coming from the Forest of Belgrade must have had a substantial and reliable flow. Of equal, if not more, significance was the proliferation of cisterns and reservoirs that had been built within the city during the sixth century as a means of water storage – as many as 70. Much of this work can be attributed to the Emperor Justinian I. While his greatest architectural achievement was undoubtedly the Hagia Sophia, the finest basilica in the world, for those admirers of hydraulic engineering his other 'basilica' is almost as impressive: a vast cistern built between AD 527 and 541.

Formally known today as the Yerebatan Sarayı this was built underground in a building complex that was once known as the Basilica and had been used to teach literature and law.[40] A small cistern had been built here during Constantine's reign (AD 306–37) but this was massively enlarged by Justinian to provide a reliable supply of water to the Hagia Sophia and Imperial Palace, ensuring that there was sufficient storage capacity to maintain the supply throughout the dry season. With a capacity of 80,000 cubic metres he certainly achieved this and has left us an architectural wonder with 336 eight-metre-high pillars supporting an underground monastery-beam-style ceiling. The pillars had been recycled from many locations and hence varied in their stone and design, while the walls were covered with a waterproof plaster.

However many and large the cisterns of Constantinople may have been, droughts appear to have remained a continuing problem. One of these is referred to within a historical text as occurring in November AD 526 and led to fights around the cisterns. The worst drought appears to have occurred in the year AD 765, described by the writer Theophanes as a year in which 'even dew did not fall from heaven and water entirely disappeared from the city cisterns and baths were put out of commission'.[41] He went on to explain how this caused the emperor, Constantine V, to restore the long-distance aqueduct from Thrace:

he collected artisans from different places and brought from Asia and Pontos 1,000 masons and 100 plasterers, from Hellas and the islands 200 clay-workers, and from Thrace itself 5,000 labourers and 20 brickmakers. He set taskmasters over them including one of the patricians. When the work had thus been completed, water flowed into the city.

In the cisterns and taking a bath

My ticket to visit the Basilica Cistern was bought from within a small brick building next door to the Hagia Sophia. At its back, a descending staircase took me into a vast underground cavity and to an immediate understanding of why the name Basilica has been used: the 336 columns were arranged in twelve straight rows creating symmetrical lines of perspective whichever way I looked; it had all the appearance of a religious monument. It was certainly presented as a cathedral for water with soft choral music playing through loudspeakers and artful electric lighting. The columns stood in a shallow lake of seeping ground water which reflected the lights and marble columns. Huge carp swam lazily, interspersed with goldfish.

A boardwalk led me round the cistern, joining a procession of other tourists all astonished that such a building could have been made for water alone. Towards the deepest part of the cistern, two columns were supported on huge blocks carved with the face of Medusa, obviously also recycled from somewhere else. When heading back to the entrance, past the inevitable 'cistern café', there was a partial power cut, perhaps a deliberate special effect for the tourists. In the dim light, the reflections of the marble columns became all the more striking and it became easy to imagine that up above ground Byzantine Constantinople rather than modern Istanbul was going about its business (Photograph 24).

Later I walked across the city to find the Aetius and Aspar cisterns, which had been open reservoirs. I took a circuitous route to first pass by another underground cistern located behind the Hagia Sophia that has been converted into an upmarket restaurant and then on to see the Aqueduct of Valens. Having stood its ground throughout Constantinople's turbulent history and as modern Istanbul was built all around, this impressive bridge now straddles a fast-flowing multi-lane highway, the Ataturk Bulvari. The high-rise buildings located across the Golden Horn seaway were framed between two of its arches through which buses and taxis streamed as if they were children's toys.

I stood awestruck by this monument to engineering genius, knowing that this was both the end of the 551-kilometre line that had brought water from Vize and also the start of the distribution system within the city. Somewhere below the modern-day concrete that paves the Istanbul University area of the city are the remains of a great nymphaeum where the aqueduct terminated. To me the Aqueduct of Valens appeared the equal of the Hagia Sophia – a monument to meet man's most basic need for water rather than his need for faith.

The Aetius reservoir was also impressive, although it could easily be missed by any uninitiated tourists or indeed residents of Istanbul because it is now used as a football stadium. Located close to a gate of the Theodosian city wall and next to a busy shopping street, one might easily imagine that it is a deliberately built stadium because the football pitch and seating fits so perfectly inside the walls. It was a massive reservoir, 244 × 85 metres and 10–15 metres deep, with walls more than five metres thick. Walking for another 20 minutes took me to the Aspar cistern, now used as a park with tennis courts, a children's playground and picnic benches. This reservoir had been just as deep but square, 152 × 152 metres, and apparently even more challenging to fill with water than had been the Aetius. This must be why it had been called the Xerokipion, which means 'the dry garden'.

There were further cisterns for me to see in Istanbul. I wanted to get to the western side of the city for the remains of the seventh-century cistern, built to supply a set of military barracks and a palace, which had not only its walls surviving but also internal staircases. This cistern had at one point been converted into elephant stables. Indeed, the diverse uses to which Byzantine cisterns had been put are striking: not only a tourist attraction, restaurant, football pitch, park and stables, but also a nightclub, art gallery, hotel foyer and marketplace.

But I had had enough of cisterns for one day. I needed to finish my day in Istanbul and my explorations into the water supply of Rome and Constantinople by getting a little closer to the Roman experience of water – by taking a Turkish bath.

The Roman bath experience was inherited by the Islamic world. The Arabs embraced the institution of the bath and through them it returned to Europe; it was passed to the Ottomans and then became the modern-day Turkish bath. It returned to Europe in the Middle Ages via the Islamic baths in Spain and evolved into the spa experience that thrives throughout the world today.

I went to the Cemberlitas baths in Istanbul, located close to the Grand Bazaar and built in 1584. Lying on a vast, heated-marble slab below a domed roof, I closed my eyes and listened to the sounds of trickling water and the gentle conversation of men relaxing at the end of a long day. They spoke in Turkish, probably on topics not dissimilar to those once discussed in Latin within the Baths of Caracalla in Rome. I could pick out a few words, more than I would be able to do during my next sojourn into the ancient world of water management. For that, I would be listening to Chinese.

7

A MILLION MEN WITH TEASPOONS

Hydraulic engineering in Ancient China, 900 BC–AD 907

From the top of the five-storey Qinyan Tower I finally grasped the scale of Li Bing's achievement. I could not fail to do so because it took centre stage in a stunning panorama that reached from the Tibetan mountains in the west to the distant modern tower blocks of Chengdu city in the east. Between them was the Dujiangyan irrigation scheme, or at least the start of it, where the River Min is divided into two, one arm taking water to irrigate the Sichuan Basin – just it has done for 2,250 years (Photograph 25).

Here, in the south-west of China, I was looking at a water management system that in its scale and impact was more impressive than anything I had seen in Petra, Rome and Istanbul. There I had witnessed small flows of water along reconstructed systems of questionable authenticity. Here at Dujiangyan the hydraulic engineering designed and built by Li Bing in 256 BC – almost a century prior to the reign of Aretas I, the first King of Petra, and more than 500 years before the Aqueduct of Valens was constructed – remained in full flow. It has never missed a single day. Moreover, unlike the monumental aqueducts and reservoirs I had seen elsewhere, at Dujiangyan it was entirely unclear where nature stopped and hydraulic engineering began. Here, I was ready to believe the claim that the scale and technological sophistication of water management in Ancient China – for irrigation, supplying cities, canal transport and a source of water-power for industrial production – surpassed that of all other ancient civilisations.[1]

The Dujiangyan irrigation scheme is arguably the greatest of the hydraulic-engineering projects that have pervaded and substantially structured the history of Ancient China. In its time, it was as monumental an achievement as the Three Gorges Dam is today. That dam was completed in 2009 to form a 2,300-metre-long barrier across the Yangtze River and contains the largest power station in the world; its construction displaced 1.3 million people, required more than 27 million cubic metres of concrete and is estimated to have cost 180 billion Yuan (£17.3bn). Mao Zedong had promoted the idea of the Three Gorges Dam in the 1950s, thirty years after the idea was originally conceived. He had also climbed the Qinyan Tower to view the Dujiangyan scheme, declaring that 'The irrigation project is the lifeline of agriculture.'[2] Two months later he launched 'The Great Leap Forward' in which water and natural resources were to be intensively exploited for agricultural and industrial expansion; thousands of dams, irrigation and water transport projects were built throughout China from the 1950s onwards, and continue to be built today. Mao Zedong must have been inspired by Dujiangyan – one cannot fail to be so. Perhaps he was also inspired by the statue of Li Bing. This sits within the adjacent Erwang Temple surrounded by incense candles and honoured as if Li Bing lives on as a God – a legacy to which Mao Zedong did not appear averse.

Two great historians

For this chapter I primarily draw on the work of one great historian, who in turn drew on the work of another. I need to do so because the archaeology of Ancient China, and especially of its water management, is so little known in the West. A suite of archaeological field projects are needed to relate the historical accounts of dyke construction, canal building and irrigation schemes to the evidence on the ground and to trace how far back into prehistory these may go. We need the Chinese equivalents of Sumio Fujii, John Oleson and Jim Crow – who exposed the ancient hydraulic engineering in the Jafr Basin, at Nabataean Humayma and of Constantinople respectively – to get to work. Fortunately, my recent discussions with Chinese colleagues suggest that they are indeed doing so. English-language publications providing non-Chinese readers access to the achievements of ancient Chinese water management – and the impact of its absence – are increasing in number.[3] I eagerly await further publications and new discoveries, but for now I must primarily rely on the work of Joseph

Needham, writing in the 1950s and 1960s, who in turn primarily relied on the writings of Sima Qian, who wrote between 109 and 91 BC.

Joseph Needham, 1900–1995, had already achieved academic eminence in the 1930s as a Cambridge biochemist specialising in embryology and morphogenesis. When three Chinese scientists came to work with him in 1937 a fascination with China, and especially with the history of its science and technology, gradually became overwhelming and would dominate the rest of his career.[4] Having mastered Classical Chinese, he became director of the Sino-British Cooperation Office in Chongqing between 1942 and 1946. This enabled him to travel widely in China and to begin accumulating a vast library of books about Chinese history and science which were shipped back to Cambridge. Needham moved to Paris in 1946 for a two-year post as the first head of the Science Division of UNESCO, before returning to his Cambridge college, where the University Press agreed to publish a series of books on 'Science and Civilisation in China'. This soon grew to seven volumes which were partly prepared by his friend and colleague Wang Ling, for whom Needham found a position in Cambridge. Needham devoted himself to the history of Chinese science until his retirement in 1990, producing no fewer than 15 out of the 24 volumes that have so far appeared – this having become an on-going project of the Needham Research Institute.

Needham's achievement was truly monumental, and this was recognised by his election as a Fellow of both the British Academy and the Royal Society, and by his appointment as a Companion of Honour by the Queen. Some believe that he exaggerated Chinese technological achievements, while many historians remain perplexed by what became known as the Needham Question: in light of its early scientific success, why was China overtaken by the West in science and technology? That, however, is a question that will soon become redundant because of the extraordinarily rapid growth of Chinese science in the 21st century that many expect will soon push the West into a back-seat role.

Volume 4 of 'Science and Civilization in China', published in 1971, covers *Physics and Physical Technology*;[5] Part III of this volume – no less than 937 pages – is concerned with 'Civil Engineering and Nautics', 167 pages of which address 'Hydraulic Engineering', and is sandwiched between sections on 'Bridges' and 'Nautical Technology'. Within these pages Needham reviews and interprets the history of hydraulic engineering from the first references in historical records of the eighth century BC up to his own day – between the 1950s and 1970s when the volume was prepared. He draws on some personal visits to sites; I was delighted to read that his visit to the Erwang Temple in 1943 to see the Dujiangyan irrigation scheme sounded much like mine in 2011. But Needham principally relies on the works of Chinese historians, and especially the monumental work of Sima Qian, known as the *Shih Chi*.

The *Shih Chi* is best described in English as the *Records of the Grand Historian*. It was supposedly written by Sima Qian between 109 and 91 BC as an account of Chinese history from the time of the Yellow Emperor, *c*.2696 BC, to the time of Sima Qian. Rather like Needham's own work, this was a multi-volume history, in Qian's case 130 volumes or scrolls that classified information into various categories. Just like any document, the *Shih Chi* must be interpreted rather than read as a literal fact; Needham does so brilliantly, combing its information, together with a multitude of other records, to write the story of hydraulic engineering in Ancient China, a story which he concludes is nothing short of an epic. Here I can only point to a few highlights that epitomise the Chinese achievement – and which perhaps help us understand why Chinese leaders today feel so compelled to construct monumental projects to control the flow of water.

Mountains and rivers, monsoons and silt

Controlling the flow of water has been a persistent need in China ever since the time of the earliest settled communities: prior to that, hunter-gatherers could simply have moved away from inundated areas. China is a vast country and hence generalising about how its topography and climate make it prone to flooding is necessarily rather simplistic (Figure 7.1). For our concerns the key feature is the four great river basins within which the ancient states emerged and flourished, and the mountain ranges that divide them: the Yellow River in the north, the Huai and the Yangtze Rivers in the centre and east, and the Pearl River in the south. All of these river basins were – and still are – prone to flooding, having massive water catchments which receive abundant quantities of rain brought by monsoon winds during the summer and spring.

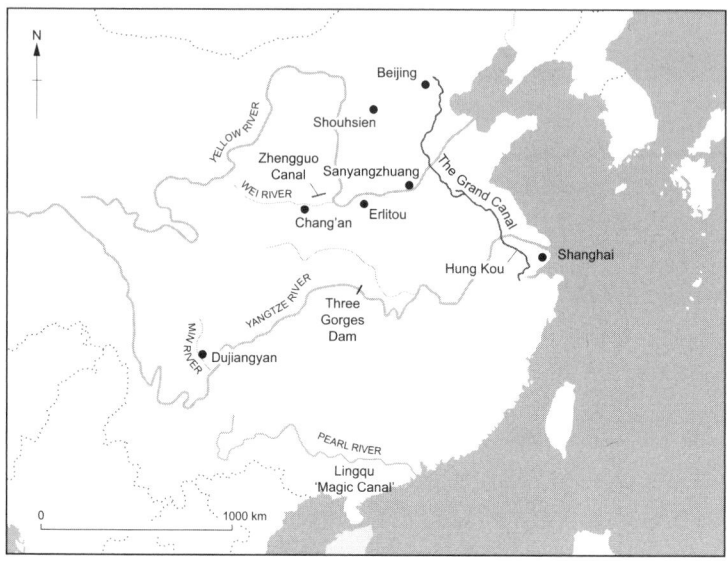

*7.1 Archaeological sites and hydraulic engineering projects
referred to in Chapter 7*

The Yellow River Basin has always been especially prone to flood-ing because for much of its distance it flows across a low-lying plain. It rises in the plateau of north-eastern Tibet and descends quickly eastwards through landscapes with soils formed by fine, windblown silt (loess) which are prone to erosion by deforestation and rainfall. It then has a long, slow flow across the North China Plain. The river is susceptible to enormous seasonal and annual fluctuations in rainfall, causing frequent and potentially devastating floods. With flooding comes the deposition of silt in astonishing quantities – the river is supposed to have once carried more than 1,000 million tons of it each year. Its source is the loess soil that covers the whole of the basin and beyond and is deposited as sand and silt on the plains. As such, for at least 2,000 years the river has been bordered by dykes seeking to pre-vent the flooding of agricultural land and settlements. Over the years, the river frequently overran the dykes or simply washed them away, devastating nearby settlements. Gradually they have been made higher and in turn the bed of the river has risen from the ongoing deposition of silt, much like the Tigris and Euphrates in Southern Mesopotamia.

The southern boundary of the North China Plain is marked by the

Qinling Mountains, beyond which is the Yangtze River Valley. This originates in the Tibetan mountains and then flows along the southern margins of the Sichuan Basin. This basin is encircled by high mountains with a mild, humid climate that allows a long growing season. As we will see, once irrigated it became critical to the success of the Qin kingdom. The Yangtze River carries far less silt than the Yellow River and famously flows through precipitous gorges, before entering its own plains and eventually reaching the sea. Those plains are also liable to flooding, although the total area at risk is far less than that of the Yellow River.

Warlords and emperors

Just as a thumbnail sketch of China's geography has been necessary for us to appreciate the hydraulic-engineering achievement, so too is one of its ancient history. The starting date is around 7000 BC when millet was domesticated in the Yellow River Valley and rice in that of the Yangtze. This led to permanent Neolithic villages and a complex array of cultures throughout China into the Bronze Age and beyond. The *Shih Chi* and other historical texts refer to two early dynasties, the Xia (2100–1600 BC) and the Shang (*c*.1600–1046 BC), but trying to relate these to specific archaeological sites is questionable – as is their actual existence. The most important site is Erlitou located in the Yellow River Basin. This has been excavated since the 1950s and provides evidence for urbanisation and early state formation in China but cannot be connected to any of the historical accounts. It seems most likely that a multitude of chiefdoms and small states proliferated throughout China during this time, rather than unified dynasties.[6]

These are first recognised in the historical accounts by reference to the Zhou Dynasty (1046–256 BC), which developed in the Yellow River Valley and is reported to have overthrown the Shang and expanded into the Yangtze River Valley. The historical accounts are likely to be more accurate when they describe hundreds of states during the eighth century BC and after, a period known as the 'Spring and Autumn Period' (722–476 BC). By the fifth century BC seven prominent states had emerged. They were in continual conflict and hence this is known as the 'Warring States period' (476–221 BC). One of these states, the Qin, rose to prominence and subdued the others, thereby extending its territory. Ying Zheng, the king of Qin, proclaimed himself emperor in 221 BC.

The Qin Dynasty lasted a mere twelve years. During that time, however, it standardised systems of writing, weights, measures and currencies; it imposed centralised government, began the Great Wall of China and achieved many engineering, scientific and cultural accomplishments, including the famous terracotta army. It was, however, soon overthrown by the joint forces of a peasant uprising and an aristocratic general, resulting in the much longer-lived Han Dynasty (202 BC–AD 220).

The Han Dynasty was one of the greatest empires of the ancient world. It rivalled the Roman Empire, its Western contemporary, in its military might and cultural achievement.[7] The Han undertook military campaigns to the north into Mongolia and to the west to the shores of the Caspian Sea; it prospered from economic expansion with the opening of trade with the West via the Silk Road. Following this period of relative unity and centralisation, China reverted again to a number of smaller, relatively independent states and warlords, initially referred to as the period of the Three Kingdoms – although there must have been many more. There appears to have been a great number of invasions and migrations, involving ethnic Turks, Mongols and Tibetans.

It wasn't until AD 589 that some degree of unity returned to China. This arose from the military success of Sui, ushering in the Sui Dynasty (AD 589–618), followed by the Tang Dynasty (AD 618–907). This is as far as we need to go because by that time a number of monumental works of hydraulic engineering had been undertaken, demonstrating the key role that water management played throughout the history of China.

Yu the Great and ideologies of water management

Chinese water management is supposed to have begun at the start of Chinese history in the Xia Dynasty. Stories about great floods pervade the myths and legends of many, if not all, ancient civilisations. China is no exception but its story has a fascinating twist. Rather than just being about the inundation of land and the survival of a few gallant people, China's story includes a reflection on the merits of differing approaches to hydraulic engineering.

The story is set in the time of the legendary emperor Yao, whom some would place at around 2300 BC in what is today the Sichuan province in south-west China. The countryside was being inundated with great floods and so Yao placed a man known as Kun in charge of devising a means of protection and control. For nine years Kun tried to do so by building earthen dykes, but these kept collapsing or being washed away. In light of his failure, Yao decided to first exile and then to kill Kun, whose body was cut into little pieces.

Yu, Kun's son, was appointed by Shun, the successor of Yao, to continue with the task – hardly an appealing prospect in light of his father's fate. Yu adopted a different approach. Rather than building dykes to alter the course of the rivers, he dredged their channels to enable them to contain the flood waters until they flowed into the sea. This was no mean task: it is said to have taken Yu thirteen years during which he passed by his house on only three occasions and never chose to call in to see his wife and children because of his devotion to his task.

Yu's body suffered for his exertions: his hands became calloused and his toenails fell off from standing in water for too long. But he succeeded in controlling the floods and was accorded the title of 'Yu the Great'. He took the place of Shun's son to become the next emperor and is said to have founded the Xia Dynasty.

The story of Yu is recounted in numerous different narratives and indeed media: at the Forbidden Palace in Beijing one can see a massive boulder of jadite carved several hundred years ago with the most intricate tiny figures removing silt from river bottoms to recount the story of Yu. Needham suggested that the story is of particular interest because it contrasts two approaches to water management: the Confucian and the Taoist way. These have, he argued, been in conflict throughout Chinese history and may indeed lie at the root of the dispute about the Three Gorges Dam and other mega-hydraulic works in modern-day China. And they are not, of course, merely about different approaches to water management: they are differing systems of morality.

Kun had adopted a Confucian approach: he attempted to build high dykes with the aim of confining, repressing and controlling the natural flow of water – and hence of nature in general. This failed and Kun was punished. Yu, on the other hand, sought to enhance the existing flows of the rivers by deepening their channels; he worked with, rather than against nature. This was a Taoist approach, one that Needham believes was fundamental to the early development of Chinese science and technology.

From legend to history, with evidence for a catastrophic flood

The prehistory of water management in China is yet to be explored; it is likely to reach back to the origins of agrarian society. Millet was domesticated in the Yellow River Valley and rice in the Yangtze River Valley, both by around 7000 BC. Neither could have been cultivated successfully without some form of water management to ensure sufficient water reached the crops in the dry season and to protect them from floods in the wet. It is inconceivable that the Neolithic and Bronze Age landscapes of China were not pervaded with dykes, ditches, wells and reservoirs. Some of the earliest historic texts include references to irrigation: one of the songs in the *Shih Ching* – the earliest known collection of Chinese poems and songs – refers to how water flows from a pool to flood the rice fields and is thought to date to the eighth century BC.

The oldest irrigation reservoir known to Needham was built between 606 and 586 BC to the south of the city of Shouhsien. He describes it as a great tank, 63 miles in circumference, which served to catch water draining from the mountains to the north of the Yangtze to then irrigate more than six million acres. It was around this time, or perhaps a century later, that large-scale water management projects appear to have begun in earnest. Needham attributes this to two factors. The first was the use of iron tools in the latter part of the fifth century BC, which significantly increased agricultural productivity and also, presumably, the capacity to undertake extensive engineering projects. The second may have been more important: the emergence of powerful feudal lords. These not only imposed taxes of grain on their peasant farmers but could also assemble vast labour forces to undertake engineering projects of a previously unimaginable scale. Needham asks us never to forget the role of sheer manpower in the engineering projects of ancient China, work that had to be undertaken with what he describes as a 'million men with teaspoons'.

The need for such engineering projects to not only irrigate land but protect settlements from flooding has long been appreciated by accounts of floods within the *Shih Chi* and other historical records, and from the deep expanses of silt in the river basins. But it was only in 2012 that the first archaeological evidence for the flooding of an ancient Chinese settlement became available – providing us with a Chinese Pompeii.[8] This is the settlement of Sanyangzhuang, located north-west of the modern course of the Yellow River and now buried below five metres of silt deposited by flood water. Ironically, the site was discovered during the digging of an irrigation canal in 2003. It dates to within the Wang Mang Period of the Han Dynasty (140 BC – AD 23), with the style of ceramic vessels suggesting 140–130 BC. It is a village consisting of at least four earthen-walled compounds containing courtyards surrounded by rectangular buildings, made from rammed-earth walls and with gable-ended roofs covered in tiles. There were at least two wells, lined with fired bricks, associated cultivated fields and a roadway. Ominously, the village had been built on top of thick sediment laid down by a previous flood of the Yellow River.

The archaeologists found that people had fled from Sanyangzhuang, dropping their tools and abandoning treasured possessions as flood water inundated their village. The flood immediately deposited a thick blanket of mud and then continued to entirely bury the village under three metres of sediment. The rammed-earth walls 'melted' into the watery slurry, and the roofs collapsed to seal the living floors below, which are yet to be excavated. Although it was a catastrophic flood, the water and silt were laid down gently, leaving the archaeological remains extraordinarily well preserved. A straight line of loom weights next to the remnants of the loom itself show that this had been taut with fibre when the weaver had fled; kitchen utensils and farming tools were left in their place of work; a neat pile of roofing tiles indicates an unfulfilled plan to repair a roof; valuable copper coins were scattered within a courtyard. The site was further buried below another two metres of sediment from a later flood after the Han Dynasty had come to an end; its future excavation will be an enormous task – another job for a 'million men (and women) with teaspooons'.

This remarkable discovery, one that will provide the first archaeological insights into rural life during the Han Dynasty, confirms the historical accounts of floods from the Yellow River beginning in AD 11 and continuing until AD 69/70 when the breaches in the levées were finally repaired. It is the most effective demonstration of why stories about great floods pervade the myths and legends of China and why hydraulic engineering had to be undertaken on such a grand scale. So now we must return to Dujiangyan.

Dig the channel deep, and keep the dykes low

The Dujiangyan irrigation scheme is the most famous, longest-lasting achievement of this period, having been started in around 270 BC, and is still quite breathtaking. The story of its construction is a fascinating blend of fact and fiction, with no clear indication of where the boundary lies between the two. It starts at around 300 BC, when the Shu Kingdom, now the region of Sichuan province, came under the control of the Qin state, this having subdued both of the previously warring tribes of the region, the Ba and the Shu.

At that time the extensive Chengdu plain was too dry to be productive, lacking any natural water supply. There was only one river, the River Min. This flowed from the mountains to the north and then hugged the foothills around the western edge of the plain, before emptying into the Yangtze. In the summer wet seasons and especially when swollen with melt waters from the mountain glaciers, the river would flood, devastating the villages and their fields; in the dry winter season the river would often be no more than a trickle, if that, and insufficient to water any crops. Moreover, each year its specific course varied, making it impossible to use for navigation.

Enter Li Bing. He is a figure who is accorded the same heroic status as Yu the Great, and whose story has striking similarities. There is, however, one key difference: the accomplishments of Li Bing are there to see for anyone who can travel to Dujiangyan for a day out at its truly lovely and inspiring world heritage site.

Li Bing was appointed by the Qin as the first governor of the Shu Kingdom. Some accounts say that he was given orders to make the region a strategic base for the Qin's future expansion. That required not only a more productive agricultural base but also a navigable river. Li Bing achieved both in one masterful feat of engineering.

That is a fact – the evidence is there to see – but first there is a story which may or may not be true. Upon his arrival Li Bing immediately appreciated the significance of the River Min and its potential importance for the success of his task. But he found the local people living in fear of the river, unable to contemplate an engineering solution to the recurrent problems of flooding, drought and overall uncertainty. The people were stymied from acting by their superstition and fear, worshipping the river as a god. Each year they would sacrifice two young girls, throwing them into the river to become its 'brides'. Nevertheless, the floods and droughts continued.

To gain their respect, and ultimately their labour, Li Bing announced that he would be honoured to sacrifice his own two daughters. He invited everyone to a grand feast and offered a toast to the future married couples – the wild, unpredictable water and his lovely daughters. Holding his glass aloft, he asked the River God to reciprocate the toast: nothing happened. Having been deeply insulted, Li Bing challenged the River God by plunging his sword into the water.

At this moment, and out of sight of the villagers, Li Bing's officials released two bulls, representing Li Bing and the River God. As the bulls fought, attracting the crowd's attention, Li Bing slipped away to then re-emerge suitably scratched and bruised. There were roars of support for him at which point his archers shot down the River God bull so that Li Bing was announced as the victor. The end of the feast also marked the end of the superstition and fear; as subjugator of the river, Li Bing now had a devoted labour force ready and willing to start work.

But what to do? It is said that he spent three years contemplating this question; he supposedly walked the entire length of the River Min before finding Dujiangyan and realising what had to be done: a subtle, Taoist, altering of the landscape that would have profound consequences for the environment and economy of the region.

First, he wished to divide the River Min into two by building an artificial island. Drawing on a labour force of 10,000 workers, he had boulders hauled into the middle of the river. But these were simply washed away, as were even larger boulders. So to make his island, Li Bing employed craftsmen to manufacture large bamboo cages that were filled with rocks and then placed into the water – replicas can be seen surrounding ornate fountains in the Dujiangyan Visitor Park today (Photograph 26). These were too massive to be washed away and so by piling them on top of each other the island was built – a task that supposedly took four years to complete.

Having made two channels separated by his island, shaped with a point facing towards the on-coming river referred to as the 'fish head', Li Bing now tackled an even greater challenge (Photograph 27). He wanted one channel – the so-called outer channel – to remain as the original course of the river and to direct the other channel – the inner channel – to the heart of the Chengdu Plain as a source of water for irrigation. Unfortunately, there was a mountain in the way, or at least the lower slopes of a mountain. Without gunpowder – not yet invented – it appeared that Li Bing's only option was to use hammers and chisels to slowly mine a channel through the solid rock, a task estimated as requiring thirty years to complete even with his devoted work force – an impossible timeframe for the needs of the Qin.

It is said that Li Bing found a solution by staring into the flames of a small fire. He ordered his work force to cover the rock face with grass and wood, stuffing this into whatever fissures they could find. He then set fire to it all, heating the rock as high as possible. Then he released the freezing cold water from the inner channel to gush over the super-hot rocks, causing them to crack and fragment; then in went the workers with their hammers and chisels to smash up the weakened rocks. Quite how many times Li Bing had to repeat this process is not recorded; but it is said that the task of mining a 20-metre-wide channel through the mountain slope still took eleven years to complete. The water could then flow freely through this narrow channel, designated as the 'precious bottleneck' towards the Chengdu plain. There, a web of channels and ditches took the water to the rice paddies, turning this into highly productive agricultural land.

Li Bing's work was still not complete. He had not yet solved the problem of the enormous fluctuations in the water supply from dry season to wet, and there was now a risk that the bottleneck would become blocked by silt and sand. To solve these problems, Li Bing cut two spillways through the island to connect the inner and outer channels just before the latter reached the bottleneck. Now, when the River Min was in torrent during the summer months, water would flow over the spillway from the inner to the outer channel from where it would flow to the Yangtze; the overflowing water would take with it much of the silt and sand that was being carried and consequently prevent this accumulating within the bottleneck. Conversely, during the winter, when water was becoming short, the level would remain below the spillway and hence the majority of the water would stay within the inner channel and be taken to the Chengdu plain. By this means the flow to the irrigation channels remained effectively constant all year round, with no risk of flood damage to crops and dwellings (Figure 7.2).

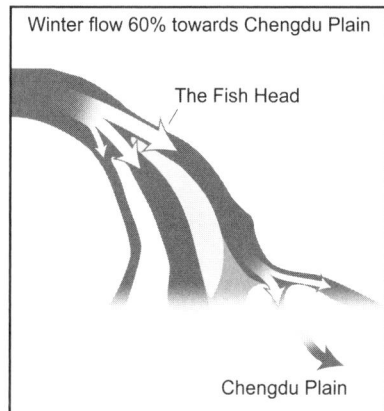 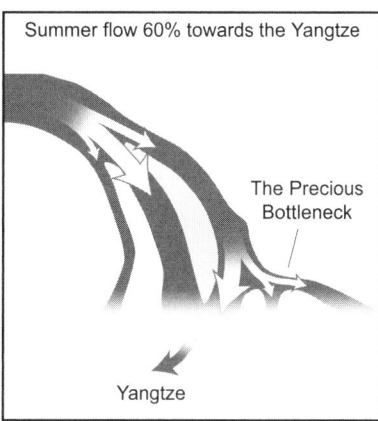

7.2 Changing seasonal water flow at Dujiangyan (after Gillet and Mowbray 2008)

While the risk of blockage of the bottleneck and deposition of silt in the new arable fields of the Chengdu Plain was solved, there remained a problem of silt accumulation in both the outer and the inner channel prior to the spillways being reached. There was no option but to undertake an annual removal of the silt – it simply had to be dug out. To ensure that sufficient would be removed, Li Bing sunk iron bars on to the riverbed when it was free of silt. These provided a marker: each year the workmen had to keep digging out the silt until the iron bars were reached – these too can be seen in the Dunjiangyan Visitor Park today (Photograph 26).

Needham describes how the annual removal of silt was undertaken – one must assume from personal observation. Every year in mid-October a coffer-dam was built that diverted all the water from the outer to the inner channel. This enabled the silt to be dug out from the outer channel to the predetermined depth. In mid-February, the coffer-dam was moved so that all the water was directed to the outer channel and silt excavated from the inner channel. On 5 April, the coffer-dam was entirely removed and the irrigation season began, marked by an annual festival.

Li Bing's achievement did a great deal more than simply turn the Chengdu Plain into the fertile rice-basket for China that it remains today. That in itself was extraordinary; in 1958 Needham describes how 930,000 acres were being irrigated and that when the ongoing canal building is completed, this would extend to 4.4 million acres (Figure 7.3). The flow of water was also used as a source of power: a stone tablet records how 'water-wheels for hulling and grinding rice, and for spinning and weaving machinery, to the number of tens of thousands, were established along the canals and operated through the four seasons'. The new waterways also became key for transport, of people and goods. Timber could now be floated downstream to dockyards in Chengdu to build warships. With a burgeoning population and a secure food supply, the Qin amassed an army that would be used to extend their authority and ultimately unify the country.

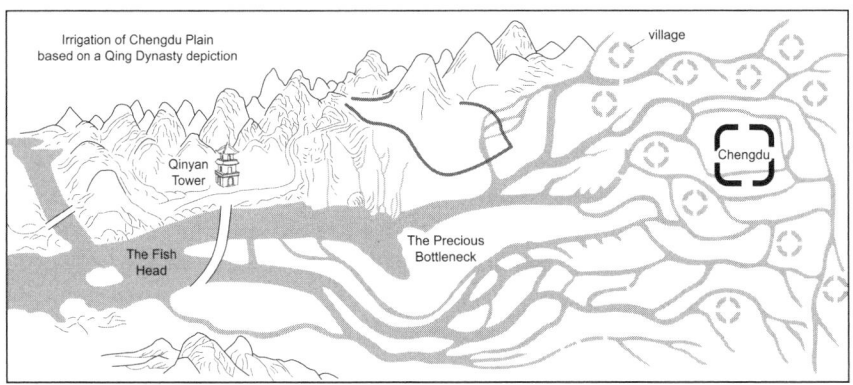

7.3 Irrigation of Chengdu Plain by the Dujiangyan scheme, based on a Qing Dynasty depiction

It should not be surprising, therefore, to find Li Bing's statue in the Erwang Temple, surrounded by burning incense and flowers, treated as if he were a god – the new River God (Photograph 28).

A cunning plan, backfired

Sima Qian, the Chinese historian who wrote the *Shih Chi*, recounts the origins of another great and long-lasting irrigation scheme of the third century BC. This supposedly arose from an attempt by the Han state to exhaust the resources of the Qin, their rivals, by persuading them into the construction of a canal 300 'li' in length, approximately 15 kilometres, thus preventing them from mounting an attack on the Han. A water engineer named Zheng Guo was sent with a plan for a canal from the Jing River west of Mount Zhong, to flow eastwards along the Northern Mountains and eventually into the Luo River, irrigating a vast swathe of land between the two. The Qin were persuaded and commissioned Zheng Guo. When the canal was more than half completed, the Qin realised that they had been tricked into the scheme. Zheng Guo was summoned for execution, but in his defence explained: 'It is true that at the beginning I deceived you, but nevertheless this canal, when it is completed, will be of great benefit to the Qin. I have by this ruse prolonged the life of the State of Han for a few years, but I am accomplishing a work which will sustain the State of Qin for ten thousand generations.' The king of Qin agreed with him, Zheng Guo was spared and the canal was not only completed in 246 BC but named after him in his honour – the Zhengguo Canal (Figure 7.1).

Zheng Guo's prediction was proved correct. The canal enabled 27,000 square kilometres of previously unproductive land to be irrigated and become fertile. The Qin reaped the reward in agricultural produce, enabling it to expand its armies and eventually conquer all other feudal states.

This canal was extended and developed throughout Chinese history; in 111 BC lateral branches were added to irrigate a more extensive tract of land; in 95 BC a parallel canal was added because silt accumulation in the Zhengguo Canal had reduced its effectiveness. Such modifications continued, always battling against the deposition of silt within the canal, with the Jing River eroding its bed and causing the intake to the canal to be moved to an ever higher level. By the time that Needham was writing, the intake to the canal had moved far up the river into its gorges and involved a 1,300-foot tunnel, a massive dam and 11 bridges to carry the canal across mountain streams.

Supplying the cities

I never know whether to be impressed or horrified at the scale of Chinese cities and their continuing rate of expansion. Beijing and Shanghai, each with a population of around 20 million people, always appear as gargantuan building sites with cranes looming in every direction. Chongqing, in the west of China, is the country's fastest-growing city, with 30 million people within its municipal area, while Chengdu with a population of a mere 14 million people is also rapidly expanding. London and New York, each with around 8 million, seem very modest in comparison. Six hundred of the 800 Chinese cities, and 30 of the 32 largest, are said to be suffering from a water short-age.[9] To provide its cities with a clean water supply and protect them from flooding, China is planning to increase its financial investment in urban water management from an astonishing US$77bn in 2011 to an astounding US$93bn by 2015.[10]

Although my visits to these cities have (so far) been for business other than archaeology, I always seem to stumble across the evidence for water management, getting glimpses of canals, weirs and water gates from the inside of taxis. These testify to water management having always been a concern for city planners. I once learned that the word *hutong*, which refers to the traditional alleys in Beijing, means 'well' in Mongolian; a great number of wells had been positioned in these alleys, with 1,245 being recorded in AD 1885.[11]

Supplying clean water, removing waste water and providing flood protection were challenges faced within the ancient cities of China, playing a key role in their planning and design. This was not only to support burgeoning populations but also to create the gardens and ornamental pools suitable for the upper classes and especially the emperor. This critical role of water first struck me during a visit to the Forbidden Palace in Beijing, the Imperial Palace from the 15th century until the abdication of the last emperor in AD 1912. The scale and gran-deur of its buildings – nearly 1,000 of them – is almost overwhelming and draws the visitor's attention away from a maze of water channels on and under the courtyards, and from massive cauldrons that had once held water to put out fires. When entering the Forbidden City, few visitors in their excitement to see the palaces and throne rooms notice they are crossing a surrounding moat. There was evidently a profound symbolism to water within the Imperial Palace, a symbolism that I have yet to grasp.

To explore urban water supply in Ancient China we can briefly visit Xi'an, the ancient capital located in the Wei River basin and surrounded by mountains in the centre of the country.[12] Prior to the Ming Dynasty of the 14th century AD, this city was known as Chang'an – which translates as 'Perpetual Peace'. It reached its peak around AD 750 (Tang Dynasty) with a population approaching one million people, vying with Constantinople to be the largest city in the world.

In 206 BC Liu Bang had made Chang'an the capital of the Han Dynasty and it soon became the political, economic and cultural centre of China. Located at the eastern terminus of the Silk Road, Chang'an was a great trading city with a cosmopolitan population drawing on the many ethnic groups of East Asia. By the seventh century BC Chang'an's population had become so large that a new supply of water was required.

A large reservoir, the Kunming, was created to receive water from the nearby Jiao River. This is estimated to have held around 35,000 cubic metres – about the same as a medium-sized reservoir today. Water flow into the reservoir from the river was controlled by a weir, ensuring that the reservoir and its outflow channels would not overflow in the wet season. Two channels left the Kunming Reservoir, one channel providing water for transport canals and the other water for drinking, washing and the ornamental parks. This second canal passed through a second reservoir and then divided into two further channels, one supplying the Imperial Palace and the other the residential areas of the city.

Chinese experts have described the Kunming Reservoir and its channels as an 'integrated hydraulic project which combined the function of water storage, diversion, drainage and flood control'. With this in place, by the end of the Han Dynasty (220 BC) the population of Chang'an had increased to around 300,000 people.

Failing to live up to its name, Chang'an became engulfed in the civil wars that followed the downfall of the Han Dynasty, with its water supply system and much of the city becoming severely damaged. This was a key reason for the relocation of the city three kilometres to the south-east once the Sui Dynasty became established in AD 589, it remaining as the imperial capital.

This new location provided access to water from the Chan, Jiao and Jue rivers. The new city was laid out as a rectangular grid within an exterior wall eight kilometres by ten kilometres in length. Eleven streets ran north to south and fourteen streets east to west, dividing the city into 108 rectangular wards, or *fangs*, each with its own enclosing walls. There was an East and a West Market. The streets were lined with fruit trees and were often wide, acting as fire breaks – devastating fires were a frequent occurrence in Chinese cities. The Imperial City was in the northern area of Chang'an, with two major palaces, the Taiji Palace and the Daming Palace, with the latter jutting out from the city wall. Beyond this palace, there were parks with orchards, vineyards and areas for sport such as horse polo (Figure 7.4).

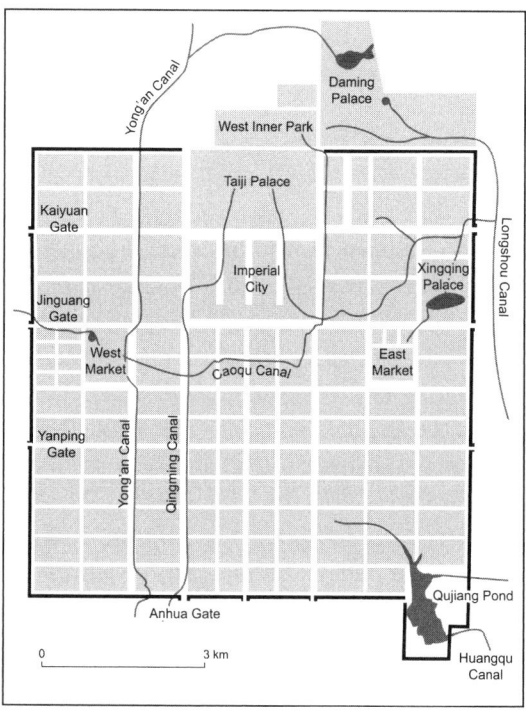

7.4 Sui-Tang Dynasty Chang'an, showing the fangs, *streets and canals (after Du and Koenig 2012)*

Water was supplied to the city via canals drawing from the rivers, and by wells within the *fangs*. The wealthy had their own private wells, while most city dwellers and workshops had access to public wells. Three canals provided water for washing, drainage and for Imperial use. The Longshou canal, built in AD 519, diverted water from the Chan River through the eastern city wall and into the Imperial City to the 'East Sea', a man-made pond in the Taiji Palace. A branch of this canal flowed northwards outside the city wall into the Forbidden Garden of the Daming Palace. A second canal – the Yong'an Canal – carried water from the Jiao River to the West Market, and a third canal – the Qingming Canal – brought water to the Imperial City from the Jue River. There has been some archaeological exploration of this canal, showing it to have been half-moon-shaped in cross section, 10 metres wide and three metres deep. It had been well-maintained, with its bottom covered in fine sand in some places to enhance the flow of water.

A fourth canal, the Caoqu Canal, was built in AD 742 and designed primarily for transportation. This connected the Jue River and the West Market, ending in a large pool. In AD 766 this was extended eastwards and then west to connect the Longshou, Yong'an and Qingming canals and to reach the Imperial City. A fifth canal, the Huangqu Canal, was built specifically to supply gardens and other scenic areas of the city, ending in the Qujiang Pond in the south-east corner.

There were many branch canals and underground drainage ditches. Altogether, the canals, wells and ditches constituted a sophisticated water network throughout the city, providing a water supply, transportation, drainage, a further means of fire control, and the ability to create parks and ornamental pools fit for a Chinese emperor.

Transportation canals, waterpower and the economic basis of Chinese civilisation

While enhanced urban transportation links via canals were a key element of city planning and were a welcome by-product of the Dujiangyan irrigation scheme, the provision of long-distance transporation was the prime motivation for a great number of water works as documented in the earliest historical records; it continued to be so throughout the history of China into the modern day. The desire for transportation came from the desire for power by the warlords and emperors: they secured this by transporting grain collected as taxation, by sending supplies to their campaigning armies and by bringing warships into battle. The presence of water transport did far more, however, than secure the powerbase of the Chinese warlords and emperors: it contributed to the astonishing economic growth that provided the basis for the cultural achievements of Chinese civilisation in the Sui and Tang Dynasties. As in third-millennium-BC Mesopotamia the combination of large grain surpluses from irrigated farming and canals facilitating trade was economically, culturally and politically electrifying.

The earliest recorded canal is the Hung Kou, otherwise known as the Canal of the Wild Geese or the Far-Flung Conduit. This connected the Yellow River with the Huai River, allowing barges to move between these two key economic areas of northern and east-central China. To call it a 'canal' is not entirely accurate because it was a complex of artificial waterways covering several hundred miles, the specific geography of which has been impossible to reconstruct.

The date of its initial construction remains unclear: some early writers attributed it to the Great Yu himself; others indicate that it may have been built between 360 and 330 BC, while Joseph Needham believes that a case can be made for a fifth- or early sixth-century-BC date. He suggests that its original role may have been to bring irrigation water to the Huai Basin, but this was soon usurped by improving transportation. By linking the Yellow and Huai Rivers, the Hung Kou connected several feudal states; Needham describes this as having been ominous for their survival because improved transport also significantly enhanced the capacity for warfare.

That was unquestionably the motivation behind the construction of another early canal, the Lingqu, otherwise known as the 'Magic Canal', and claimed by some to be the earliest known 'contour transport canal' in the world (Figure 7.1). As recorded in the *Shih Chi*, it was built in 219 BC on the orders of Emperor Qin Shih Huangdi to send supplies to his armies who had gone south to conquer the people of the Yue. It was a short canal, a mere three miles, but connected two extensive river networks. It ran across a saddle within the mountainous country that separated the centre of China from the south. On the northern side of the mountains, the Xiang River flowed northwards to the plains of Hunan and eventually to the Yangtze. On the southern side, the Li River flowed to the south, joining a tributary of the West

River, and then on to what is now Guangdong province and the South China Sea. The Magic Canal simply followed a slightly falling contour across the saddle to connect the Xiang and Li River systems.

It was not, of course, quite as simple as this. The headwaters of the Li River had to be canalised for almost twenty miles to enable barges to pass along it. At the other end the Xiang River had to be controlled, much as the Min River in Dujiangyan. A similar method was used – a canal parallel to the river with spillways to stabilise the water flow and allow the passage of barges.

This was an impressive achievement of hydraulic engineering, one that was continually developed through history such as by the addition of lock gates in the ninth century AD – previously barges had to be partly hauled to enter the canal. Although it has required repairs, just like the Dujiangyan system, the Magic Canal continues to function today after more than 2,185 years. But its significance goes far beyond the engineering work itself: the short three-mile section of canal was the final link in a great chain of waterways and canals that now stretched for more than 2,000 kilometres in a direct line from northern China, from the latitude of Beijing today, all the way to the South China Sea. Barges and other vessels are likely to have actually covered twice that distance if they had followed the entire length of this single waterway.

Canal building reached its apogee in the seventh century AD during the Sui Dynasty when a continuous canal route was completed running from the lower Yangtze River in the south, from what we now call Hangzhou, to what was then the capital in the north, the city of Luoyang. That was a distance of about 1,600 kilometres, and was further extended by another 160 kilometres when Peking became the capital in the 13th century AD. This remains the longest artificial waterway ever made, covering a distance equal to that between New York and Florida.[13] Quite appropriately it is known as the 'Grand Canal' (Figure 7.1). Some would argue that this is as significant a technical, economic and cultural achievement as the Great Wall of China, built between the seventh century BC and the 17th century AD; it has been described as one of humankind's 'stellar engineering achievements'.[14]

The key motivation for the Grand Canal was to allow the movement of grain from the agriculturally productive Yangtze Basin area to the armies stationed in the north that were defending the empire from the nomadic horsemen of the Asian steppe. It involved connecting stretches of much earlier canals, such as the Hung Kou – although this was entirely remodelled, creating a parallel canal to the ancient route, which had become clogged with silt. Historical records – the accuracy of which might be questioned – state that a labour force of no fewer than five million men and women was deployed to complete the first section of the Grand Canal in AD 605. It had a channel up to 10 metres deep and 30 metres wide, with 60 bridges and 24 locks. It is said that when it was completed in AD 609, the then Emperor Yang led a 65-mile-long flotilla of boats from the north to his southern capital at Yangzhou.

If supplying the northern armies with grain had been the original motivation of the canal, its consequences for China were far greater because it created a single economic market in eastern China, which in the seventh century AD was one of the world's most densely populated areas. The historian Steven Solomon describes how 'its waters teemed with commercial vessels of all shapes and sizes, powered by sails, oars, and paddle wheels … Rice-laden barges came and went from huge granaries that the government maintained at key junctures along its route to furnish the food lifeline for China's national security.'[15] One of its great advantages was that water transport became many times cheaper and quicker than its overland equivalent.

Canals may have played an equivalent role in Ancient China to their role in the Sumerian civilisation in the third millennium BC (Chapter 3). But the Chinese made an additional use of water to transform their economy: as a source of power. Steven Solomon has been a champion for this Chinese achievement arguing that 'for well over a millennium, China was human civilisation's leader in harnessing water as energy to do useful work.'[16] He describes how by AD 300 powerful vertical waterwheels were being used to operate trip hammers to pound iron into tools, hull rice and crush metallic ore. This was many centuries before they appeared in the West – a contributing factor to the 'Needham question' of why China was then overtaken by the West in its technological development. Solomon also describes how by AD 530 the Chinese were operating waterwheel-powered flour-sifting and flour-shaking machines based on the same essential design as the steam engine that powered the Industrial Revolution in the West. Waterpower was also employed in the art of silk-making that provided China with one of the great exotic goods valued throughout the ancient Old World and which enriched its empire for centuries.

At the Three Gorges Dam

On my way to visit the Three Gorges Dam my Chinese companion remarked that he thought there were missile bunkers in its vicinity and that the Chinese Government was committed to making a nuclear strike against anyone who threatens to attack the dam. That reminded me of Dujiangyan. In AD 207, 1,200 soldiers were dispatched to guard and maintain the scheme against any threat, it being as critical to the Chinese economy of the time as the Three Gorges Dam is today. But when I saw the Three Gorges Dam itself, this comparison seemed misplaced. In the Second World War Japanese bombers are supposed to have searched for the Dujiangyan scheme, having been informed that this was boosting China's war effort because it secured the food supply. They couldn't find it. Li Bing's engineering works had blended seamlessly with the natural world. That could never happen to the Three Gorges Dam – indeed I am told that the dam and especially the vast lake it creates can easily be seen from outer space.

One cannot avoid being impressed with the Three Gorges Dam. It is a testament to human culture, the capacity of humankind to design and construct works that fundamentally modify the natural world. The electricity it generates is claimed to prevent an annual emission of 120 million tons of CO_2 into the atmosphere that would otherwise be coming from coal-fired plants. The dam protects the middle and lower reaches of the Yangtze from flooding, while navigation on the river has been enhanced. At the date of my visit a system of locks for the ships was still being used, this having required a channel to be cut through a granite mountainside – another curious resonance with Dujiangyan. The locks will soon be replaced by a vertical lift, the final element of the project under construction (Photograph 29).

Reducing carbon emissions, protecting against floods, enhancing navigation are all laudable undertakings – but at what cost? 1.3 million people had to be resettled, their villages, towns and cities now flooded by the lake behind the dam; an unknown number of key archaeological sites were also inundated, removing the possibility of ever learning about critical events in the Chinese past. There were many objections to the dam on environmental grounds, and what some dismissed as doom-laden predictions are now being realised. The dam, or more specifically the lake it creates, has already destabilised surrounding cliffs and caused many landslides, resulting in a substantial death toll and the clogging of the river with sediment. Because it was built across two major faults, there has been a significant increase in earth tremors caused by the pressure from the additional weight of water on the land; there is a continuing risk of a massive dam-induced earthquake that would quite simply be a human and environmental catastrophe of an unparalleled scale. Loss of biodiversity, an increase in waterborne disease and pollution, and induced drought in eastern and central China are further reasons why the Chinese leaders have been distancing themselves from the project. Indeed, both President Hi Jintao and the Premier, Wen Jiabao, were absent from the inauguration of the project in 2008, while in 2011 the leadership made their first admission of the ecological and social challenges caused by the dam. There are no statues of them to be seen at the Three Gorges Dam.[17]

How different to Dujiangyan. Across the entrance to the World Heritage Site there are billboards surrounding a busy central square. These record a succession of Chinese leaders, from Mao Zedong to Hu Jintao, visiting Li Bing's irrigation scheme and photographed with the fish head or the precious bottleneck as a backdrop. They may well have been inspired by Li Bing's achievement; they may well have been trapped by the historical legacy of hydraulic-engineering works in China, prompting them to make ever greater and grander works. One can only wish that it had been possible for them to have followed the Taoist way – making the dams low and channels deep. Working with nature rather than against it, Yu rather than Kun.

Digging the channel deep again: the Yellow River goes digital

We can finish this chapter on a more positive note. Although China appears determined to build an even larger and ecologically more destructive project than the Three Gorges Dam, that known as the South–North Water Diversion Project, it is also embracing high-tech solutions that appear to be having some success. In July 2001 the Yellow River Conservancy Committee (YRCC) launched a project known as the 'Digital Yellow River'.[18] This aimed to use computers in the analysis and management of the river employing remote-sensing and global-positioning systems. At that time the Yellow River was in a dire state: 40 per cent of its water was highly polluted, it was choked with sediment, and extraction of water was so excessive that the river failed to reach the sea for 226 days of the year.[19]

The first digital hydrological station on the Yellow River began operation in June 2002. On 3 March 2010 it was announced that the YRCC had been awarded the Lee Kuan Yew Water Prize, an award that 'recognises outstanding contributions towards solving global water problems by either applying technologies or implementing policies and programmes that will benefit humanity'.[20] Cynics will always say that such awards have a political edge, but the YRCC Digital Yellow River project really does appear to have been a success: the river now flows continuously to the sea again, bringing a host of ecological and social benefits. The key to its success is the remote sensing to acquire real-time information about the state of the river and an automated system of opening and closing reservoirs and dams. This has resulted in the riverbed being deepened by an average of 1.5 metres for some 900 kilometres, more than doubling its rate of flow. Both Li Bing, who built Dujiangyan, and Yu the Great before him would have been proud: both had promoted the idea of digging the channel deep.

8

THE HYDRAULIC CITY

Water management by the kings of Angkor, AD 802–1327

I caught my first sight of the West Baray as my plane banked to land at the tiny airport at Siem Reap: it shone like a shimmering blue mirror amid the tall-trunked sugar palms, rice paddies and scattered wooden houses of Cambodia.

Siem Reap is surrounded by the temples of Angkor, the centre of the Khmer civilisation that reached its apogee between AD 802 and 1327 (Figure 8.1).[1] Millions of people come each year to visit the veritable man-made mountains of stone; many are decorated with the most exquisite bas-relief carvings of Hindu deities, dancers and scenes of ancient warfare, both real and imagined. Angkor Wat is the greatest of these temples: indeed, it is the largest religious building ever constructed on the planet (Photograph 30). But the Angkor kings had created other types of monument that are (for me at least) even more enigmatic: massive reservoirs and moats. The largest is the West Baray, a rectangular inland sea, 8 × 2 kilometres that had once held no less than 48 million cubic metres of water. When built during the first half of the 11th century AD, it had claims to being the largest structure yet made by humankind; it is one of a select few from the ancient world that can be seen from outer space.

For this chapter we have arrived in South-East Asia to visit the first of two civilisations that flourished within the tropics – the second will be the ancient Maya of Central America that we will reach in Chapter 10. Cambodia lies within the Lower Mekong Basin, the Mekong being one of the world's great rivers, rising in the mountains of Tibet and emptying into the South China Sea. It is the seventh-longest river in the world with the world's largest waterfalls, these marking today's boundary between Laos and Cambodia. The ancient city of Angkor is located in the middle of the basin, about 500 kilometres from the delta and between two geographical features that played a key role in its history. The first, 20 kilometres north-east of Angkor, is the mountain plateau of Phnom Kulen; it was the rivers from this plateau that ultimately fed the West Baray and other artificial reservoirs (barays) of Angkor. Moreover, all of the city's temples were built from red sandstone quarried from Phnom Kulen, it being a deeply sacred place for the Angkor kings and people.

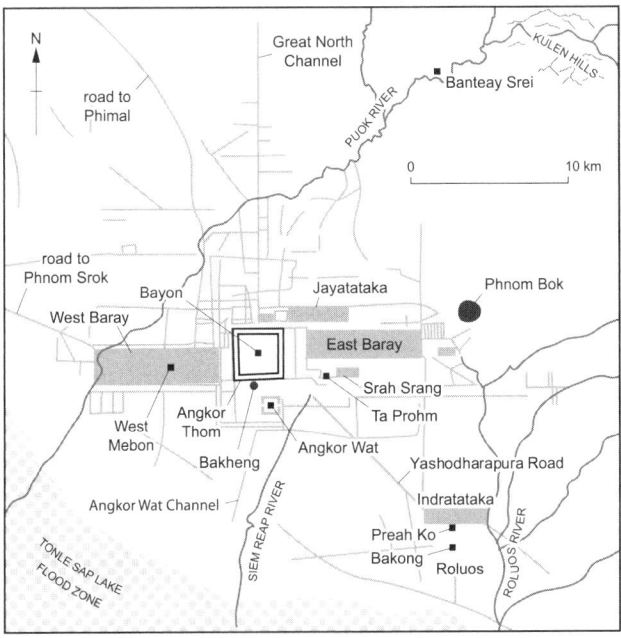

8.1 Barays (reservoirs), canals, rivers and a selection of temples at Angkor, referred to in Chapter 8 (after Fletcher et al. 2008)

The second geographical feature is found 10 kilometres south of Angkor: Tonle Sap, the 'Great Lake'.[2] This is appropriately named, as it is the largest freshwater lake in South-East Asia: 2,700 square kilometres in extent during the dry season and an awesome 10,400 during the wet. It expands because the Mekong Delta simply cannot cope with the monsoon rains and so water flows back along the Tonle Sap River and into the lake, submerging its vast flood plain. The Tonle Sap is renowned today for its abundance of fish, as was also the case in ancient times with fish providing a key source of food for the people of Angkor. The lake also provided something else: the setting for many battles during Angkor's turbulent history of warfare with neighbouring states and foreign enemies.

When flying into Siem Reap I had an outstanding view across the Tonle Sap and decided to visit its modern-day floating villages of Vietnamese refugees after seeing the temples and reservoirs of Angkor. When the plane turned and descended I saw the other inland sea, the one that is perfectly rectangular and man-made: the West Baray. With that image in mind, we can now explore what I discovered to be one of the most extraordinary feats of water management of the ancient world, that of the Khmer civilisation at Angkor. We will need to follow a succession of kings and archaeological discoveries that lead into an intriguing academic debate about why such massive reservoirs were ever built.

The debate revolves around the reservoirs. Were these built to store water for irrigation during the dry season or for flood control during the monsoon? Either would be a remarkable feat of engineering and fully justify the term for Angkor coined by the archaeologist Bernard Philippe Groslier in the 1970s: the 'hydraulic city'.[3] But others have rejected such functional interpretations proposing that the reservoirs were no less than an attempt to recreate heaven on earth: kings driven by their need to please the Hindu Gods rather than to quench their people's thirst. From the window of my plane, the massive West Baray certainly had the impact of a great cathedral built to glorify a revered god rather than a mere water-work to meet the needs of humankind.

We have two starting points. One is Jayavarman II, who was anointed as the first *Chakravartin*, 'king of kings', of the Khmer civilisation by a Brahmin priest in AD 802 on the sacred mountain of Phnom Kulen. He was followed by 14 other kings before the empire fell into disarray. Our other starting point is the year 1863. That was the publication date for the notes of Henri Mouhot, the French naturalist and explorer who 'discovered' Angkor.

Discovering Angkor

Henri Mouhot (1826–61) undertook expeditions into the interior of Siam, Cambodia and Laos in the middle of the 19th century, funded by the Royal Geographical Society and Royal Zoological Society of London. In January 1860 he reached Angkor and spent three weeks making detailed notes of the palaces and temples, the moats, pools and terraces. He died the following year of malaria in the jungles of Laos but his notes were published within his posthumous book of 1863 *Voyages dans les royaumes de Siam, de Cambodge, de Laos et autres parties centrales de l'Indochine,* and then within its 1868 English edition.[4] Mouhot's descriptions, sketches and claims that the Angkor temples were grander than anything left to us by Greece or Rome, along with his speculations that they had been built by some extinct but civilised and enlightened race, caught the imagination of the public. Mouhot became the Indiana Jones of his day, the man who had discovered a lost civilisation deep within the jungle.

He hadn't, of course. Angkor was never lost because the local people, the Khmers, knew all about the existence of the monuments; Angkor Wat itself was still home to Buddhist monks, as it has been ever since the end of the Khmer civilisation. Moreover, there had already been a succession of European visitors to the ruins, notably Portuguese traders and French missionaries, who had published their own accounts. But these simply hadn't been as eloquent and didn't reach the same readership as Mouhot's writings. It was his work that sparked the European imagination and expeditions followed to document and reconstruct the ruins. Such work continues today by multinational teams from around the world.

While Henri Mouhot may not have 'discovered' Angkor, he would certainly have discovered it for himself, just as any visitor is able to do today. I was one of thousands visiting the ruins as the tourist season was coming to its end. Despite the frequent crowds of camera-clicking tourists, I still found it possible to escape into empty shrines and gaze in silence upon the most exquisite, exotic and sometimes erotic carvings; I felt that my eyes were the first to witness such art since the time of the Angkor kings. Visitors have been helped to do this at Ta Prohm, a temple of the late 12th century, which has been left in its state of discovery. Massive tree trunks, branches and roots entwine themselves around heaps of collapsed masonry, sprawl along walls, twist through doorways and claim the ruins as their own (Photograph 31).

More than 150 years may have separated my visit from Henri

Mouhot's, but my personal sense of discovery was certainly profound because this tropical South-East Asian culture seemed so alien to my experience of ancient civilisations elsewhere in the world. And yet it wasn't the vast pillars, walls, sculptures and carvings that posed the greatest questions for me. On my visit I had left the tumbling tree-clad buildings of Ta Prohm and followed a tiny path through the surrounding woods, where I found the encircling moat: it was several metres wide and contained some glistening water, reflecting the peak of the temple that rose above the trees. All of the Angkor temples are surrounded by such moats, although few contain water today; many temples also have rectangular pools nearby, of which the West Baray is simply the largest. Why was there such a need for water?

The kings of Angkor

A definite answer will probably never be found because it is lost in the minds of the kings of Angkor who commissioned such works. The population of Angkor had numbered hundreds of thousands: courtiers, aristocrats, priests, soldiers, craftsmen, farmers and slaves. All we see today are the stone-built temples but there was once a vast array of wooden buildings housing these peoples: indeed, Angkor has been proposed as the most extensive urban complex of the pre-industrial world. When travelling in the back of a tut-tut (rickshaw pulled by a motorbike) between the temples, this being the cheapest and best form of transport, one is compelled to think that the present-day houses on stilts that line the roads, with their open-air fires, cooking pots and water buffalo, cannot be very different from those of 1,000 years ago in Ancient Angkor. Much of the agricultural production would also have been similar with rice providing the staple food supplemented by the cultivation of tree crops such as mangoes, tamarind and sugar palm, and a range of vegetables including millet, sorghum, mung beans and aubergine.

The majority of people within Ancient Angkor would have been peasant rice farmers, subject to military service, corvée labour and most likely heavily taxed to support the king and his vast bureaucracy, priests and courtiers. Other than from a few depictions of daily life on temple walls and a unique written account from a Chinese emissary to Angkor at the end of the 13th century, we know nothing about these peoples' lives: until we have extensive excavation of the domestic dwellings of Angkor, the history of this city will remain an account of its kings, their battles and their feats of hydraulic engineering.

The first reservoir builders

Jayavarman II was the first to be declared as the King of Kings. His name means 'protected by victory' and he was indeed a successful warlord. It was his military conquests that brought numerous kingdoms under his sole control to constitute Angkor's empire. Just like all of his successors, we know little of Jayavarman II as a person: our information comes from inscriptions that predominantly list kings and events but give limited insight into the daily life let alone the personalities of the kings.

We know that Jayavarman II made his capital at Hariharalaya near the present-day village of Roluos about 30 kilometres north of the edge of the Great Lake. This was positioned on its extensive flood plain, where rice farming would have been particularly productive. It is likely that the farmers of ancient Angkor used the same method as today within this region, one known as receding-flood agriculture – planting rice on recently inundated ground. Jayavarman II died at Hariharalaya in AD 835, as did his son and successor Jayavarman III in AD 877.

It was the next king, Indravarman I, who built the first of the massive reservoirs, the barays (Figure 8.1). An inscription tells us that on the occasion of his accession, he made the promise that 'Five days hence, I will begin digging.'[5]

This was, of course, a royal 'I' and he must have employed an army of diggers, presumably slaves, to construct a baray 3,800 metres long and 800 metres wide, holding 7.5 million cubic metres of water at the height of the monsoon. To do so, embankments were built as the walls of the reservoir, trapping the water flowing from the Kulen hills and along the Roluos River that was partly canalised to augment its flow. The baray was known as the Indratataka – the 'sea of Indra'; an inscription states that Indravarman I 'made the Indratataka, mirror of his glory, like the ocean'.[6]

This wasn't all that Indravarman built: two temples, Preah Ko and the Bakong, were constructed at his capital each with its own moat holding yet more shimmering water. Although modest in scale when compared to the temples built by later kings, these are nevertheless awesome monuments – mountains of stone covered in the most ornate and intricate carvings. They were built for Shiva, the Hindu god appointed as the supreme protector of the empire. The Indian Hindu gods had been adopted in the region long before the empire was founded, most likely introduced by Indian traders. After Shiva, Vishnu was the most venerated in a vast multitude of structures ranging from the magnificent stone temples to tiny wooden shrines that have left no trace today. It was not only the Hindu gods that had been adopted: Buddhism had also flourished in Angkor, while worship of the indigenous Khmer deities continued. All in all, Angkor had a remarkably rich and complex mix of religious activity that must have pervaded every aspect of state and daily life. In this regard it was not unlike Ancient Rome with its blend of religious traditions as exemplified by the mythraeum within the Baths of Caracalla.

Indravarman's successor, his son, built on a yet more massive scale. His name was Yashovarman, 'protected by glory', and his first priority appears to have been to build a larger reservoir than his father had done – a larger temple was also to follow. Yashovarman ordered the digging of the so-called Yashodharatataka, now known as the East Baray. This was eight times larger than the Indratataka, being 7.5 kilometres long, 1.8 kilometres wide and able to contain 50 million cubic metres of water: six million man-days would have been required to erect its embankments.

Yashovarman moved the capital of the empire to Angkor itself and built a five-stepped pyramid known as the Bakheng. That alone required 8.5 million cubic metres of sandstone hewn from the bedrock of Phnom Kulen and 4.5 million mud-clay bricks. It had, of course, its own surrounding moat – the control of water appearing to be as critical to these kings as the control of the vast labour force required to build the temples.

Yashovarman died in AD 900. He was followed by a succession of *Chakravartin* who continuously outdid each other in the ambition of their architecture. With the hindsight of history we should perhaps be most grateful to Rajendravarman II, under whose patronage the construction of a temple known as Banteay Srei was begun towards the end of his reign and was dedicated in AD 968. During my visit to Angkor this relatively small temple, 20 kilometres north of the city of Angkor, made a far greater impression on me than Angkor Wat itself and brought into sharp focus the question about the relationship between religion and water.

Banteay Srei, the 'Citadel of Women', is the temple's modern name: in the ancient inscriptions it is known as Tribhuvanamahesvara which translates as 'Great Lord of the Threefold World'. That is taken from the linga that sits inside one of its shrines – a linga being a short stone pillar derived from a phallus and representing the essence of the god Shiva to the people of Angkor. The temple has three concentric rectangular enclosures, the central one being reached by a causeway across a surrounding moat and containing ornate shrines which are completely covered in the most intricate carvings (Photograph 32).

The striking difference between Banteay Srei and all other temples at Ankgor is its size: this is a miniature temple appearing more suitable for children than adults. As such, one can see the intricacies of the carved lintels at eye level rather than by having to crane one's neck. The hard red sandstone has hardly weathered in the last 1,000 years, leaving the deep reliefs with sharp edges, some emerging as almost complete three-dimensional sculptures. Many of the Hindu gods and their stories are there to see: multi-armed Shiva engaging in the (ultimately destructive) dance of life; Hayagriva clutching the heads of daemons he has slain; and Vishnu represented as a horse. Hundreds of human figures, animals and combinations of both, along with lace-like foliage, are exquisitely carved into the stone, including the most beautiful bare-breasted dancing girls.

On the day of my visit, the temple was made all the more stunning by the brilliant blue sky: the sun was shining off the sugar palm trees that surround the site. Crossing the causeway, the red stone, silver-grey bark, green leaves and blue sky came together as one within the reflections of the moat. Once again I left the beaten tourist track and headed off along a meandering path through woodland that took me to yet another rectangular expanse of water, one unseen by the majority of tourists. This too was shimmering; it seemed to me to be as spiritually charged as the carvings themselves on the temple walls.

Building heaven on earth

It was during the reign of Suryavarman I ('protected by the sun', AD 1011–49) that the most outrageously ambitious reservoir was constructed – the West Baray. He also built a huge royal palace and was one of the greatest military leaders, subduing revolts throughout the empire and securing oaths of allegiance. The West Baray was continued by his immediate successor, Udayadityavarman II (AD 1050–66). He raised the height of the dykes that contained the water and built an island temple, the West Mebon, within which a colossal reclining statue of Vishnu was housed.

The succession of *Chakravartin* continued, as did the relentless warfare and increasing extravagance of the building programmes. Suryavarman II (AD 1113–50) extended the empire into what is now Vietnam and then built Angkor Wat. This is the largest and most magnificent of all Angkor temples, indeed some would argue of all religious buildings ever made, with a grandeur that overwhelms even that of St Peter's in Rome. Angkor Wat was, in effect, a self-contained city as well as the state temple built for Vishnu. It covers 200 hectares, with the temple located on a 1.5-kilometre-square island surrounded by a 200-metre-wide moat. It consists of multiple enclosures, terraces and galleries; as one climbs to reach the highest level with its five towers, the claim that Angkor Wat was built as a representation of the Hindu universe on earth becomes compelling: the towers being the five peaks of Mount Meru, the home of the Hindu gods, the moat representing the surrounding mythical oceans.

The bas-reliefs of Angkor Wat are breathtaking in their extent and exquisite detail, depicting great battles, both real and imagined. With water constantly on my mind, I was drawn to a scene that contained swirling marine life – fish, crocodiles, serpents and turtles. This was one tiny part of a 49-metre-long relief depicting the 'Churning of the Sea of Milk', one of the Hindu creation myths. A massive snake has its body coiled around a mountain; 88 devas (gods) hold on to its head while 99 asuras (daemons) line up holding its tail. Vishnu watches over them as they take their turns in pulling to rotate the mountain and churn the cosmic ocean, the Sea of Milk. This was supposed to generate *amrita* – the elixir of immortality. Alas, their cooperation eventually failed: the gods went back on their promise to share the *amrita* with the daemons and chaos prevailed.

The Bayon, built by Jayavarman VII (AD 1181–1215), is only just behind Angkor Wat in its magnificence, and somewhat ahead in its enigmatic nature. This is another man-made mountain, one created by a multitude of towers carved with massive faces that stare solemnly and silently in all directions across the landscape. Inside there is a labyrinth of gloomy passages and staircases. Winding my way around I suddenly found myself face to face with Shiva, or at least his essence in the form of a linga – an erect stone pillar within an alcove; further on, I came across water once again, now at the bottom of a well deep in the bowels of the temple; I climbed winding staircases until I was on the roof and met another enormous stone face gazing serenely across the landscape of Angkor. Just as at Angkor Wat, many of the walls are covered with stunning bas-reliefs; but here they include scenes of daily life such as market trading and cooking food.

Jayavarman VII also built reservoirs, of course. His largest baray was the Jayatataka ('Sea of Victory'), although this was a modest 3.5 × 0.9 kilometres in size. He also built the so-called 'Royal Bath', the Srah Srang, which remains full of sparkling water today (Photograph 33). This has a lovely sandstone terrace with huge lion sculptures flanking steps down to the water. I sat on these, tired after my temple visits in the hot sun, exhilarated by what I had seen, but also mentally exhausted and confused.

It was during the reign of one of the final *Chakravartin*, Indravarman III, that a particularly important event for later historians of the Angkor civilisation occurred: the arrival on 12 August 1296 of Zhou Daguan. He was an emissary from the Yuan Dynasty Court in Beijing and spent almost a whole year at Angkor. On his return he wrote his *Memoirs of the Customs of Cambodia*, an invaluable source of every-day life of the king, his courtiers and his people.[7] Zhou described an opulent state; but the empire was now being challenged by Vietnamese and Thai armies and the civilisation was in decline.

One sign of the empire under stress may have been a sudden and short-lived intolerance of Buddhism. For centuries Buddhism and Hinduism had existed side by side and often in a rich religious mix, but sometime during the 13th century every Buddhist image in Angkor

– and there were many thousands – was broken or defaced. Which of the last kings of Angkor ordered such vandalism is unknown and by the time of Zhou's visit in 1296 Buddhism was flourishing again, in light of his accounts of monks in saffron robes.

The last king of the Angkor civilisation was Shrindrajayavarman who reigned between 1308 and 1327 as Thai armies advanced and the artistic endeavours declined. No stone temples or inscriptions are known following his death in 1327. The invading Siamese were certainly one of the causes of Angkor's decline; another is likely to have been the growing significance of international commerce for which other cities, notably Phnom Penh positioned in the Mekong Delta, were better located. And as we will shortly see, a failure of the hydraulic-engineering system may have also paid a vital blow to the power of the kings and the vast urban sprawl of Angkor.

It had been a remarkable half-millennium of cultural achievement: neither before nor since in human history has there been such a massive programme of temple construction. It is the ruins of those temples and their art that draws millions of tourists to Angkor each year today; few of those visitors are aware of the equally astonishing and in many ways far more mysterious feats of hydraulic engineering that were undertaken by the Angkor kings. The West Baray may be quite evident to see from the plane, but on the ground an expanse of water doesn't have the same draw as temples and their carvings. The other barays are now empty of water and effectively invisible in the landscape except for those with an archaeologist's eye: those able to detect subtle topographic undulations marking the previous presence of a dyke or changes in vegetation that reflect variations in the moisture content of the buried soil below.

It is not only the barays that are now lost from sight: there had once been a vast and complex network of ponds, channels and canals that had linked the barays and the temple moats together. Indeed, this hydraulic spider's web had been invisible even to the archaeologists who worked at Angkor throughout the 19th and 20th centuries. It was only with the use of sophisticated aircraft-borne radar in 2002 that the true hydraulic-engineering achievement at Angkor became known to the modern world. So here we must turn to another remarkable history: that of mapping Angkor.

Finding a hydraulic city amid the temple ruins

John Thompson, an English explorer, followed in Henri Mouhot's footsteps and took the first photographs of Angkor, an exquisite collection that was published in his 1867 volume *Antiquities of Cambodia*. By that date Cambodia had become a French protectorate and so the next century of work at Angkor was undertaken by French antiquarians, art historians and archaeologists, right up until the Cambodian civil war of 1970 that resulted in the tyranny of the Khmer Rouge.

Following further expeditions and travellers' accounts, a more formal approach to Angkor arose with the establishment in 1899 of the École Française d'Extrême-Orient (EFEO). Initially led by Jean Commaille, this colonial organisation assumed responsibility for the conservation of monuments. Commaille removed tons of earth and debris to expose the full extent of Angkor Wat and the Bayon, before being murdered by thieves in 1916; he was buried at Angkor within a tomb near the Bayon. The next Head of Conservation of the EFEO, Henri Marchel, was also to die at Angkor – although in his case it was out of choice because he loved the place, passing away peacefully in 1970 at the age of 101.

Marchel had become curator at Angkor in 1919 and did not retire until 1957. During the 1930s he pioneered the application of 'anastylosis', the dismantling of a monument stone by stone, numbering each stone in turn, rebuilding the core using modern materials, notably concrete, and then placing each of the stones back into their correct position. Anastylosis replaced the use of unsightly concrete beams to prevent the collapse of buildings and was first applied at Banteay Srei. That was not long after an attempt by André Malraux, a French traveller in the region, to steal carvings from that temple for which he was prosecuted – although such cultural vandalism did not prevent him from becoming the first French Minister of Cultural Affairs, serving between 1959 and 1969.

Two archaeologists in the middle of the 20th century transformed the study of Angkor away from being entirely concerned with reconstructing temples, recording artworks and deciphering inscriptions to a concern with the social and economic organisation of the city. The first was the Russian-born Victor Goloubew (1873–1945), who pioneered the use of aerial photography. By revealing extensive canals and reservoirs, this began to expose what had been a hidden secret of the Angkor ruins: a large number of complex hydraulic works. Goloubew suggested that these had served both a ritual and a practical purpose, the latter being the irrigation of rice fields during the dry season.[8] Bernard Philippe Groslier (1926–86), appointed as Director of Archaeological Research of the EFEO in 1959, ran with this theme and characterised Angkor as the 'hydraulic city'.

Groslier transformed the EFEO into a modern, professional organisation and led the first systematic survey and mapping of the city and its environs. He wrote extensively about Angkor, developing his hydraulic-city hypothesis in a series of publications between 1960 and 1980. Drawing on his own surveys and those of Goloubew, Groslier proposed in 1966 that 'Angkor is primarily an immense system of hydraulic works: main supply channels, dikes, reservoirs, field runnels, all designed to store up freshwater, collect rainwater, and redistribute it to the meticulously planned network of rice fields below.'[9] He went further within a seminal article in 1979 by proposing that it was a failure of the hydraulic system that caused the collapse of the Angkor state.

Uncertain rainfall and feeding the masses

Groslier proposed that the great barays had been constructed to accumulate water during the monsoon to be used for irrigating rice fields during the dry seasons, or even in the wet season whenever the monsoon rains were insufficient. The soils were, he believed, highly fragile and the rainfall equally uncertain. The scale of the temples suggested that there must have been a vast population to feed. As such, Groslier believed that irrigation must have been essential. To anyone visiting Angkor today this is not an unreasonable hypothesis: thousands of labourers must have been required to construct the stone temples, let alone all of the wooden buildings that have no longer survived. The sheer size of the barays and the complex canal network that Goloubew's and Groslier's surveys revealed imply that these must have served some functional purpose, one going beyond any ideological motivation for their construction.

According to Groslier, water simply seeped through the southern embankment of the barays into channels that directed it downhill, running into one field and then the next, heading towards the northern edge of the Great Lake. By this means, the whole area between the barays and the Great Lake was irrigated, an area that Groslier called the 'hydraulic zone'. He calculated this to contain 86,000 hectares for the cultivation of rice. By estimating that one hectare of irrigated land can produce 1.46 tonnes of rice, and that one tonne can feed 4.8 persons, Groslier proposed that water from the barays would have supported 600,000 people.[10] He argued that a further 429,000 could have been dependent on the margins of the Great Lake itself, irrigated by the annual rise of the lake waters (and producing 1.1 tonnes of rice per hectare), and a further 872,000 by a dry rice culture in the area to the north of the barays (producing 0.63 tonnes of rice per hectare). In total, Groslier estimated the population of ancient Angkor to have been 1.9 million.

Influenced by Wittfogel's theories, Groslier believed that the scale of the irrigation system required rigid centralised control for its construction and management – and hence the power of the kings. Likewise, when the barays became full of silt causing the irrigation system to fail, the kings lost their power base and the city and empire collapsed.

Bernard Philippe Groslier was still undertaking fieldwork and writing at the outbreak of the civil war in Cambodia in 1970 which engulfed Angkor in hostilities. He only left in 1973 after being severely injured by a Khmer Rouge bullet and after many of his notes had been lost. Groslier continued research in the region, undertaking surveys in Thailand, Malaysia and Burma, before finally returning to Paris.

Challenging the hydraulic-city hypothesis

Was Bernard Philippe Groslier correct that Angkor had been a hydraulic city? In one regard no one has deferred from this view since his publications of the 1970s: how could they possibly have done so with so many reservoirs, canals, moats and ponds? There has, however, been a debate as to whether the hydraulic engineering had been for god or man. Groslier certainly did not reject the idea that the barays and moats were serving an ideological role in representing the heavenly oceans on earth while also functioning as a means of irrigation. The anthropologist W. J. van Liere, however, writing in 1980 rejected their role in irrigation and/or flood control. He argued that religious considerations alone had dictated the layout of the reservoirs, canals, ponds and moats, all being oriented east–west and north–south, irrespective of any undulations of the topography. He could find no evidence that water from any of these constructions had ever been taken to the surrounding fields.[11]

Van Liere went further than simply rejecting any irrigation function for the Angkor hydraulics: he suggested that the reservoirs, ponds and moats would have impeded agriculture by holding water for sacred purposes when it would have been more usefully left to drain on to the rice fields. Moreover, he questioned the overall technical ability of the Angkor builders, doubting whether they could have constructed an effective irrigation system even had that been their desire. The anthropologist Philip Stott developed van Liere's arguments in 1992, arguing that:

The hydraulics that did exist, such as the great baray at Angkor itself, were just like the temple mountains, essentially a part of the urban scene, providing religious symbolism, beauty, water for bathing and drinking, a means of transport, and perhaps a supply of fish. Yet, not one drop of their water is likely to have fed the rice fields of Angkor.[12]

The next main challenge to Groslier's hydraulic thesis came in 1998 from Robert Acker of the School of Oriental and African Studies, London.[13] He had been trained as a geographer and addressed two questions: first, would the additional amount of rice arising from the irrigation have made a significant difference to the size of the Angkor population? Second, are the barays appropriately positioned to provide water for irrigation?

For the first of these Acker recalculated the population estimates for Angkor. He began by using the same yield estimates for rice as Groslier, 1.46 tonnes/hectare from irrigated land and 1.1 tonnes from rain-fed agriculture. This was after having considered whether ancient yields would have been higher, because of higher levels of rainfall, or lower because of less productive strains of rice. Overall, Acker felt that these opposing trends would have cancelled each other out and hence yields would have been much the same as today. Consequently, Acker continued to use Groslier's estimates.

The Hoover Dam, seen from the Mike O'Callaghan–Pat Tillman Memorial Bridge

The 19,600 years old site of Ohalo, located on the shore of Lake Tiberias, undergoing excavation in September 1999. The remnants of a circular hut can be seen in the lower left of the picture

Tell es-Sultan, Jericho, surrounded by the modern-day town

Semi-subterranean dwellings with pisé (mud-based) walls and floors at the Neolithic site of WF16, Jordan

The Neolithic settlement of Beidha, southern Jordan

The Neolithic 'barrage' at Wadi Abu Tulayha, Jafr Basin, Jordan – the earliest known dam in the Levant at 7500 BC

BELOW: Excavation of the Neolithic well at Sha'ar Hagolan, Israel

ABOVE: View towards the reconstructed north-west entrance of the Minoan palace of Knossos, Crete

LEFT: Drains at the Minoan palace of Knossos, Crete

LEFT: Courtyard at the Minoan palace of Phaistos which most likely functioned as a rainfall-harvesting device, with a raised processional walkway. Heather, Nick and Sue Mithen demonstrating the Phaistos walk

BELOW, LEFT: Granary or water cistern? At the Minoan palace of Phaistos, Greece

BELOW, MIDDLE: The city wall of Tiryns, Greece, showing the 'cyclopean' Mycenaean masonry of the west side

RIGHT: Approaching the Treasury at the Nabataean city of Petra

BELOW, RIGHT: Remains of the terracotta pipe and channel in the siq at the Nabataean city of Petra

ABOVE: Section of the Nabataean 'Ain Jammam aqueduct channel north of Humayma, Jordan

LEFT, TOP TO BOTTOM:

Humayma, Jordan, showing the rectangular Roman fort towards the top of the picture with the Nabataean aqueduct running in front of it, and a range of structures from Nabataean and later periods

Walking through the Nabataean city of Petra, Jordan, looking east towards rock cut tombs, with the Great Temple on the right of the picture

The Great Temple at Petra. The ornamental pool and garden had been located on the far side of the pillars

Arches carrying the Aqua Claudia in the Parco degli acquedotti, Rome

Aqua Claudia in the centre of Rome showing the double tier of channels

Caldarium at the Baths of Caracalla

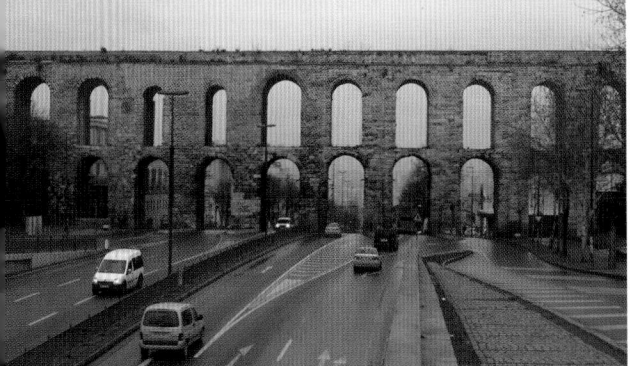

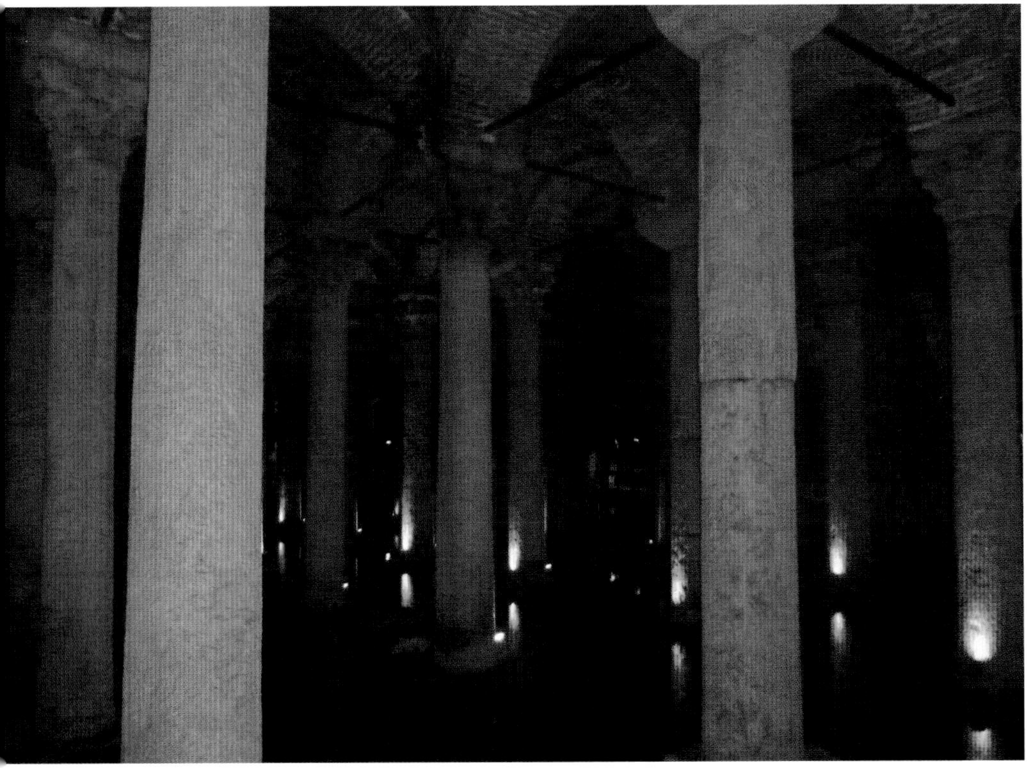

TOP TO BOTTOM:

The Byzantine Aqueduct of Valens, now known as the Bozdoğan Kemeri, Constantinople/Istanbul

The Kurşunlugerme, close to Gümüşpınar, carrying the long Byzantine aqueduct from Vize to Constantinople

Inside the Byzantine Yerebatan Saraya, or the Basilica Cistern, Istanbul

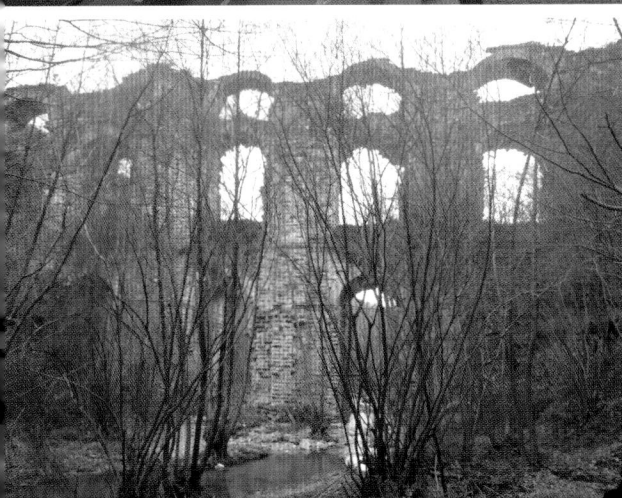

In contrast, Acker thought that Groslier had seriously overestimated the amount of land that lay to the south of the barays and was available for irrigation, calculating this to be 25,500 hectares rather Groslier's estimate of 86,000. Without irrigation, this land would have been able to produce rice in the wet season via rain-fed agriculture; so by taking the difference in yield from such rain-fed agriculture (1.1 tonnes/hectare) and what would have been produced by irrigation (1.46), and then allowing for the production of rice by irrigation in the dry season, Acker calculated the barays would have enabled an additional 29,600 tonnes of rice to be produced over what would have otherwise been the case. By using a slighter lower level of consumption than did Groslier, one tonne for five rather than 4.8 people per year, Acker calculated that the additional rice from irrigation would have fed only 148,000 people. He thought that this was an insignificant number when compared to those that could have been supported around the margins of the Great Lake and by dry-land farming. As such, he dismissed the idea that barays had been constructed to meet the subsistence needs of the population.

Acker went on to also dismiss the idea that irrigation was needed to mitigate the variations in rainfall from year to year. One of his key arguments was that in recent times Cambodia has supported a population at least three times the maximum estimate for Angkor – the country had a population of around six million in the middle of the twentieth century. That population was sustained by rice farming with no recourse to hydraulic engineering – neither massive reservoirs nor irrigation canals. So why would hydraulic engineering for irrigation have been necessary when the Angkor kings had ruled?

Acker further argued that the placement of the barays in the landscape also undermined Groslier's hypothesis of the hydraulic city: they are simply not positioned to make effective use of their water for irrigation. The Indratataka, for instance, which had held 7.2 million cubic metres of water, sufficient to irrigate 3,000 hectares during the wet season, was placed just five kilometres from the edge of the Great Lake, providing a mere 2,400 hectares of land for irrigation. But it also lay directly north of the city of Hariharalaya, which would have impeded the flow of water and used up a significant amount of land that could feasibly have been irrigated. The East Baray (54.2 million cubic metres of water) was positioned north of the Indratataka and hence made that baray redundant, while having much of its potentially irrigable land covered by the Indratataka itself and Hariharalaya.

With regard to the West Baray, Acker proposed that this had a capacity far beyond what could have possibly been needed to irrigate the available land. He calculated this as 156 million cubic metres of water, sufficient to irrigate 10,416 hectares during the dry season and 62,500 during the wet. Yet there were only 14,000 hectares available in its hydraulic zone. This is so limited because the West Baray is on the right bank of the Siem Reep River into which all its waters would have drained before reaching the expanse of rice fields to the north of the Great Lake. Acker looked at each of the other barays, the Jayatataka, Preah Kahn Baray and South-East Baray and found the same type of problems: none of these were positioned to make effective use of the water they contained.

Acker concluded that 'Angkor had at its disposal far more arable land than its population could have ever farmed' and 'water in practically unlimited quantities'.[14] If the rice fields did need artificial watering, this was provided by the many hundreds, perhaps thousands, of small and medium-sized tanks dug into the ground to a depth below the water table. So what did Acker think was the purpose of the barays and the great canals?

One possibility is that they were for flood control. If the annual monsoon was particularly heavy, water would have gushed off the highlands of the Phnom Kulen and could have flooded through the city, potentially destabilising the stone buildings and washing away the soils of the rice fields into the Great Lake. So the canals and barays might have functioned to channel and then contain such flood water, also explaining why they had such high banks around them.

An alternative is that the barays were built and used for entirely religious reasons, having no functional value at all: they were no more and no less than a re-creation on earth of the lakes that surrounded Mount Meru, home of the gods. According to Groslier, Angkor was meant to be a microcosmic re-creation of the universe, with the primordial ocean shaped and ordered by the king in the same way that the universe was shaped and ordered by the gods.[15] As one commentator has written: the baray was a 'receptacle for the glory and power of the king', the way in which he 'manifest[ed] his divinity'.[16] As such, and just like a cathedral, they had to have an immense size to serve that purpose.

The hydraulic city maintains its flow

In 1975 Cambodia fell under the control of the Khmer Rouge, resulting in four years of arguably the most horrific atrocities ever undertaken on planet earth. With liberation in 1979, it was found that the monuments of Angkor remained unscathed, largely because the Khmer Rouge adopted the city as the ideal to which the enslaved population should strive. But the landscape was left peppered with landmines which had to be cleared before restoration and archaeological research could resume.

In 1993 Angkor became a World Heritage Site. It is now the focus for research by teams from throughout the world – France, Germany, Japan, New Zealand, Hungary, Australia, China and even Hawaii. The École Française d'Extrême-Orient still exists, but all work at Angkor is now coordinated by the Cambodian Government agency called APSARA.

The international teams collaborate, combining their particular areas of expertise. The Greater Angkor Project (GAP) is one such collaboration between ASPARA, the EFEO and the University of Sydney. This has recently completed a remarkable new mapping of Angkor, one that compels us to return to the notion of a hydraulic city, partly designed for extensive irrigation of rice.[17]

Bernard Philippe Groslier commissioned maps of Angkor but these were crude by modern standards and were left unfinished, constraining our understanding of Angkor as a settlement rather than a series of relatively isolated temples and barays. It was another Frenchman, Christophe Pottier of the EFEO, who produced the first detailed maps, although these were restricted to the southern area of Angkor. He plotted the occupation mounds, local temples, household ponds, canals and channels, recording several thousand features and demonstrating that sets of fields were linked to specific temples. His fieldwork also showed that, contra Groslier, van Liere and Acker, the barays did indeed have inlets and outlets and were fully connected to a series of channels.

While Pottier was undertaking his mapping on the ground, new forms of survey were taking place from the air – in fact from space. In 1994 the space shuttle Endeavour created a radar image of Angkor, covering the northern region, beyond the extent of Pottier's maps and showing the complexity of the archaeological landscape. This included the presence of a 25-kilometre-long channel running from Angkor Thom to the northern hills, appropriately named 'The Great North Channel'. In 2000 the University of Sydney commissioned an airborne radar survey by the Jet Propulsion Laboratory of NASA. This used a method known as AIRSAR that could penetrate cloud cover and create detailed and high-resolution images of Angkor, being able to detect the most subtle undulations in the landscape and vegetation cover. Both of these methods indicated where rice fields, channels, ponds and so forth may have once been located.

The key achievement of the Greater Angkor Project between 2003 and 2007 was to create one single map of Angkor and its environs, covering 3,000 square kilometres. This used a Geographic Information System to integrate all existing sources of information: Groslier and Pottier's surveys, the space shuttle and airborne radar surveys and a wide range of other archaeological maps that had been generated during the past century.[18] A key part of this mapping work was field-work on the ground: going out to investigate the various humps and bumps identified by the radar by looking for scatters of tile, brick and pottery. The work was completed in 2007 and its publication has transformed our knowledge of Angkor: we can now see a city with a far more complex, and indeed convoluted, hydraulic management than even Groslier could have ever imagined (Figure 8.1). To quote the project team, the mapping revealed 'Angkor as an extensive settlement landscape inextricably linked to the water resources that it increasingly exploited over the first half of its existence' and 'Even on a quite conservative estimate, Greater Angkor, at its peak, was therefore the world's most extensive pre-industrial low-density urban complex.'[19]

Water management at Angkor

The new mapping of Angkor showed that the city had an elaborate configuration of channels and embankments built from massive quantities of clayey-sand, the most readily available building material.

These served to reduce and control the water that flowed into the city area from the hills to the north; they acted to contain the water and then to disperse it, either for irrigation or simply to avoid its deleterious effects.

Roland Fletcher, Professor of Archaeology at the University of Sydney and one of the leaders of the Greater Angkor Project, has led the interpretation of the hydraulic system, proposing that it constituted a tripartite system.[20] There was a northern zone between the hills and the barays that functioned to collect water within a series of massive channels and to direct it to the barays and moats. This zone has a series of east–west embankments, some running for as much as 40 kilometres, behind which water would have accumulated. They were connected to north–south channels that took the water into the barays, being designed to work with the rivers that ran through the landscape and were diverted where necessary. Fletcher and his colleagues emphasised the technological sophistication of the hydraulic engineering, especially with regard to a masonry spillway to cope with excess water during the monsoon: hard, irrefutable archaeological evidence now exists to show that van Liere and Acker were entirely wrong to doubt the technical abilities of the Angkor engineers.

The central zone of the hydraulic system was formed by the massive barays and temple moats that were fed by the northern collector system; these contained the water prior to it being distributed to the south. Again, a quite unexpected degree of complexity is found with a system of channels that directed water in, out and between these reservoirs.

The southern zone was a suite of channels that dispersed the water from the barays and moats to the south and east of the city. The most massive of these left from the south-west corner of the West Baray and ran for 40 kilometres to the south-east; another exit channel drained to the south-west, the most direct route into the Great Lake. A major dispersal channel has also been detected from the Angkor Wat moat, one that joins a complex network of embankments and channels that dispersed water across the landscape south of the city.

There was unquestionably a significant degree of careful planning and hydraulic engineering required to create this tripartite water management system. It was one that evolved over the 500 years of Angkor's growth, under a process of additions and cumulative modifications. That it was not perfectly designed as a whole, as Robert Acker has argued, should come as no surprise – the nature of complex systems is that they are constantly modified and tinkered with. They also frequently serve multiple functions. This certainly appears to have been the case for the hydraulic system at Angkor: it controlled the monsoon flood waters when these were excessive, irrigated the rice fields during the dry season and captured sufficient water for the creation of the heavenly lakes of the gods on earth.

If only Bernard Philippe Groslier had lived to see how his idea of a hydraulic city has been so compellingly vindicated. But what about his other idea, that the failure of the water system caused a collapse of the Angkor empire?

Climate change and hydraulic-management failure

Groslier had proposed that the water management system at Angkor had been devised to remedy the uncertainties of the monsoon and particularly the length of the dry season. He lacked, however, any measure for quite how 'uncertain' the monsoon would have been during the 500 years of Angkor's rise and fall; presumably he drew on his own experience of SE Asia with regard to how the intensity and duration of rainfall in one year can be quite different to that of the next. It was not until 2009 that the first measure of monsoon uncertainty for the Angkor period was devised, this coming from a study of tree rings from 1,000-year-old cypress trees.[21] This showed not only that Groslier was correct to suppose a considerable degree of climate uncertainty throughout the Angkor period, but also that this became particularly marked in the middle to late 14th and early 15th centuries: the dramatic swings between intense drought and heavy rainfall appears to have caused an Angkor demise.

Growing in the highlands of Vietnam are some rare and especially old cypress trees, *Fokienia hodginsii*. Professor Brendan Buckley from Columbia University and his colleagues searched these out and took cores through their trunks to get a sample of the tree rings from throughout the period of their growth. Each tree ring records one year of growth and in general terms its width records how much water was available. Buckley and his colleagues were able to produce a 759-year-long sequence for between 1250 and 2008, recording the variation in the intensity of the monsoon – just the type of data that Groslier would have loved to have seen.

This record showed that the monsoon got weaker – that is, there was less rain – in the middle to late 14th century, with some periods of sustained drought. Shorter periods of more severe drought occurred in the early 15th century, with the single driest year of the whole record being in 1403. But what is especially important is that during this period of the middle to late 14th and early 15th centuries, the tree rings also record some of the wettest years, the most intense monsoon. So in the context of an overall decline in rainfall, the period during which Angkor is known to have declined and then become effectively abandoned was one with the highest degree of rainfall variation – years of drought interspersed with years of intense rain.

Roland Fletcher and his archaeology colleagues have found the impact of such climatic variation in the ruins of Angkor. One of the major channels that runs from Angkor Wat to the Great Lake was found to be packed full with coarse sands and gravels of the type that would result from a sudden flash flood.[22] Such sediment would have been derived from the northern zone, indicating the soils in that region had become susceptible to erosion, probably from a combination of deforestation and drought. A thick mat of foliage was found below the sediments. This was radiocarbon-dated and indicated that the flood had occurred sometime between 1220 and 1430 – just the period when the tree ring sequence was indicating such climatic fluctuation.

Some of the engineering work at Angkor indicates that the hydraulic system had to be modified to cope with water shortages. There were, for instance, a whole series of alterations to the input and output channels of the East Baray, some of which appear to be designed to cope with severely reduced water flow. These included the main intake channel being moved from the north-east corner of the reservoir to halfway down its eastern bank, suggesting that there was a shortage of water to fill the baray.

Brendan Buckley, Roland Fletcher and their colleagues emphasise how the whole water management system at Angkor had become too complex and convoluted to cope with a changing environment. This may not have directly caused the collapse of the Angkor state but it certainly made it far more vulnerable to other external influences, such as conflict with the Siamese armies and the growing significance of trade rather than agricultural production as a source of economic and political power. Whether or not the periodic failures and ultimate demise of the hydraulic system at Angkor directly undermined the authority of the kings, as Groslier had surmised, remains unknown. But it is now unquestionable that Angkor had indeed been a hydraulic city: with that debate finally over a new phase of archaeological work can begin on understanding not only how the hydraulic system developed over time, but about daily life within the sprawling urban landscape of Angkor.

At the West Baray

I had one late afternoon left at Angkor. The next morning I had a taxi booked to take me to the Tonle Sap Lake so that I could take a boat ride to the floating village of Vietnamese refugees before my midday flight back to London – to see their floating houses, shops, schools and workshops. I decided to spend my spare time taking a closer look at the West Baray, having so far only seen it from my plane on arrival. I had spent all morning and early afternoon at Angkor Wat, lingering over the bas-reliefs and climbing the steps to the highest towers, where I found orange-robed Buddhist monks, and did not know whether they were residents or tourists like myself.

There was a bustling mass of taxis and tut-tuts outside the entrance to the Angkor Wat causeway. I asked a tut-tut driver to take me to the south-west corner of the West Baray, knowing that this was where Christophe Pottier had found the traces of an outlet channel signifying the role of the reservoir in irrigation or flood control. The tut-tut driver looked sceptical, explaining in his broken English that tourists didn't go to that far corner of the baray, it being at least 40 minutes away and with 'nothing to see'.

Within ten minutes or so we had left all the other taxis and tut-tuts behind and were driving along a trackway between rice paddies and wooden houses, and then along the bank of the West Baray. The expanse of water was indeed truly vast, although for much of the way I had to sneak glimpses through the thick vegetation that lined the bank. The track had deteriorated and the tut-tut struggled through deep muddy ruts, which looked quite fresh, suggesting some recent traffic. Indeed, as we approached the corner of the baray, parked cars appeared, as did motorcyclists and groups of walkers, all heading my way. When my tut-tut had to pause, I asked my driver what was happening and he explained that the day was a public holiday and families were gathering for picnics and swimming.

There was indeed a party atmosphere at the corner of the baray; not a tourist in sight and no opportunity to go poking around for archaeology. Cars and motorcycles were 'parked' in every nook and cranny off the track and a line of stalls were selling all sorts of food – exotic vegetables, juices, fried fish, barbecued fish, baked fish and kebabed frogs. Behind the stalls, people were lounging on rush mats and in hammocks, sleeping, eating and talking; beyond was the water itself, in which hundreds of people were swimming and bobbing about in enormous black inner tubes (Photograph 34). It was a fantastic sight. Having spent the last few days wondering whether the barays had been made for flood control, irrigation or to worship a Hindu god, this reminded me of one other key function of water: it can be such a source of fun and enjoyment. Well done, Indravarman, Yashovarman, Suryavarman and the rest of the water kings of Angkor.

9

ALMOST A CIVILISATION

Hohokam irrigation in the American South-West, AD 1–1450

When touching down at Sky Harbor International airport at Phoenix, Arizona, one isn't so much in the vicinity of the most extensive prehistoric irrigation system in North America as quite literally right on top of it. Buried below the tarmac of the runways, the terminals and the fire station are the in-filled ditches and canals that had once irrigated the desert.[1] These had been made by the Hohokam people, constructing their first canals close to the start of the first millennium AD, when they lived in small villages adjacent to the Salt and Gila rivers in what would become Arizona. Within a few hundred years they were living in settlements that housed 1,000 people; they had monumental architecture, elaborate material culture, and far-reaching trade networks. Just like the early settlers in Mesopotamia, Egypt and other regions of the Old World who had also irrigated their fields, the Hohokam appeared well on course towards civilisation.[2]

But something went wrong. Within a thousand years the Hohokam culture had collapsed: their multi-storey buildings and walled settlements were abandoned. Their canals were dry and filling with silt, their fields returning to desert scrub. They left behind no recorded language; they neither invented nor adopted the wheel and very little is known about their belief system or hierarchy. The people themselves were soon forgotten. Their name today, the Hohokam, means 'those who have gone' or 'all used up'. This comes from the language of the Akimel O'odham people, the Native Americans who were encountered by the Spanish missionaries who arrived at the end of the 17th century and whose descendants now live within reservations adjacent to the Gila River.[3]

Here we have a story of how irrigation enabled a society to flourish but was unable to take it across the cultural threshold into the domain of civilisation. And so whereas the names of the Sumerians, Egyptians and the Incas are widely known, who other than academics have heard of the Hohokam? Like my own journey to visit their archaeological remains, it is a story that both begins and ends at, or at least near, Sky Harbor airport.

Not entirely lost

As my plane came in to land I caught a glimpse of the dry bed of the Salt River running through the middle of Phoenix: an arid brown scar cutting through the grey concrete, tarmac and glass-windowed world of the sixth most populated city in the United States. Water no longer reaches this far from its source in the mountains of eastern Arizona, it being drawn away for irrigation before it reaches the city. The modern irrigation is essential for sustaining the population – almost 1.5 million in the city alone. Essential too is the power generated from hydroelectric schemes. This makes possible the air conditioning that makes life bearable when summer temperatures average over 100°F (38°C).

The present-day irrigation system is, in fact, a legacy of the past because the Arizona Canal follows precisely the same route as a canal first excavated by the Hohokam as much as 2,000 years ago. Indeed, it has only been by making use of the dry ditches of the Hohokam canals that Phoenix has developed at all. Other than the Spanish missionaries, Europeans did not settle in the region until the 1860s. The legend is that Jack Swilling, an ex-Confederate from the American Civil War, came 'out west' in the 1850s seeking fame and fortune by prospecting for gold. One day he was struck by a view across the Salt River Valley, recognising its agricultural potential and observing the ruined settlements and long dry ditches of the Hohokam. He set to work rebuilding the canals calling his business venture the 'Swilling Irrigating and Canal Company'. He had numerous other ventures, building mills, mines, saloons and dance halls, and becoming a farmer, rancher and politician. Soon there was a thriving settlement in the Salt River Valley known as Pumpkinville on account of the large pumpkins that grew beside the canals. After several name changes, this became Phoenix, recognising that it was a city born out of the ruins – the canals – of a former culture.

Excavations at Hohokam settlements began in the 1880s and were reported in the *Boston Globe*, bringing the former culture to the attention of the Eastern establishment.[4] In the 1890s work began at what had become known as Casa Grande, a four-storey timber and mud-walled Hohokam building, located adjacent to the Gila River. That became the first designated national monument in the United States, following a decree by President Woodrow Wilson in 1918.[5] By then surveys of the Hohokam canals were underway, followed by an aerial survey in the 1930s. Excavations were also taking place at Pueblo Grande, adjacent to the Salt River. This is the largest known of all Hohokam settlements and is likely to have had a population in excess of 1,000. It had a platform mound on top of which the elite rulers had resided, two ballcourts, many adobe houses, plazas and a 'big house' similar to that at Casa Grande. Pueblo Grande was located at the head of a canal system drawing water from the Salt River to irrigate surrounding fields, and consequently was in a position to control the water flow to settlements further down the system. The part-excavated and part-reconstructed ruins of Pueblo Grande are now located immediately adjacent to Sky Harbor airport, the course of its canals running below the runways.[6]

Visiting the Pueblo Grande ruins today is essential to understanding the Hohokam achievement. The ruins and a museum are within an archaeological park; one can stand on the remains of its platform mound and see a modern-day flowing canal just a few metres away, following the same course as when the Hohokam had stood on the mound and looked across to their fields. Although the mound is crumbling today, it once stood eight metres above the surrounding flat land and incorporated a maze of rooms, courtyards and other buildings. There are no clear entrances or steps and it had been surrounded at ground level by a two-metre-high wall: access to the mound and especially its summit appears to have been highly restricted.

The function of the rooms on the top of the mound remains unknown, but several were of two storeys and many had high ceilings. Excavated remnants of minerals and pigments within the rooms suggest ceremonial activities. The most striking aspect of the mound is the panoramic view it gives of the very flat surrounding landscape, even with its diminished size today. One looks down on to reconstructed pit houses and adobe houses within compound walls, a

well-preserved ballcourt and in the distance, the low-rise office buildings of modern Phoenix and the airport. The Hohokam view would
have been of fields, canals and the surrounding village with at least
one Hohokam big house of three or four storeys 800 metres to the
north. The ballcourt is one of two in the area, an oval-shaped depression with a raised viewing ramp all round for spectators to watch the
game.

The settlement of Pueblo Grande was scattered with cemetery areas
in and among the adobe houses. There was little consistency in the
burials: most were cremations, others were buried either supine or
in crouched positions; some burials were within the houses while ten
have been excavated from the platform mound itself. Grave goods
were frequently placed in the burials but there is little evidence of
wealth or status differences between individuals. That is perhaps surprising given the clear signs of differences in social status coming
from the settlement layout and architecture.

Origins: hunter-gatherers in the Sonoran desert

Visiting Pueblo Grande provides the first lesson about the Hohokam;
the second comes from taking a walk into the surrounding desert. This
is part of the vast Sonoran Desert of the south-west United States and
provides essentially the same landscape as that in which the Hohokam
first began to dig their canals.

The desert initially appears incredibly harsh, especially if like me
you visit in late July when the temperature was approaching 48°C. It
receives less than 20 centimetres of rainfall per annum, while winter
temperatures can regularly dip below freezing. The small amount of
rainfall is erratic and uneven, coming in both the summer and the
winter. This has knock-on effects, causing uncertainty about the
seasonal flowering of the plants and hence the activity of birds and
animals. It also affects the river flow and the capacity for irrigation.
But despite the heat, visiting the desert is an absolute delight: it teems
with life and is stunningly beautiful.

The vegetation is dominated by the tall saguaro cacti, into which
birds have pecked out holes for nests; surrounding these is a rich array
of desert plants, some familiar from our houseplants at home, such as
the yucca and the smaller cacti that come in so many different prickly

forms; other plants are exotic, such as the creosote, agave, jojoba, cholla and ocotillo. In the higher elevations of the mountains, there are juniper, oak and pine trees. Insects are prolific on the ground and on the plants; so too are the insect- and seed-eating birds, and small rodents – mice, rats, ground squirrels and rabbits. There are lizards, snakes and tortoises and once there had been wild deer, antelope, lion and coyote – maybe there still are. Hawks are frequently seen swooping, while eagles glide high in the perfectly clear blue sky.

The desert was once cut by rivers which created lush green corridors, housing a further range of plants and animals. A few glimpses of such corridors can be gained today when sufficiently upstream from the modern drainage schemes (Photograph 35). The rivers had provided an abundant supply of fish until they became dammed in the 20th century. For the Hohokam the two key river systems were that of the Gila River and its principal tributary, the Salt River. The Gila River originates in the mountains of western New Mexico and the Salt River in those of east-central Arizona; they both have substantial watersheds, draining a total area of 186,000 kilometres.

For those who know what to look for and when, the desert landscape can provide a remarkable range of food and raw materials – there are at least 200 species of edible plants.[7] Some of the required knowledge still exists amongst the Akimel O'odham, although they are now as dependent on the supermarkets and stores as the rest of us. The first Native Americans in the region had been the so-called Palaeo-Indians of the last ice age, dating to at least 8500 BC.[8] They shared a grass- and herb-covered landscape with mammoth, giant sloth and other mega-fauna. All of those beasts became extinct at the end of the ice age probably due to climate change but possibly helped along the way by the spears and traps of the Palaeo-Indians.

The flora and fauna of today's Sonoran desert developed within the warmer climates after 7000 BC. They were exploited by desert-adapted hunter-gatherers during what is known as the Archaic Period. Key wild foods were the bean pods from the mesquite trees, fruits and buds from the saguaro, hedgehog, and prickly pear cacti, seeds from goosefoot and tansy mustard, along with a multitude of roots and bulbs. Rabbits are likely to have been the main source of protein, supplemented by occasional kills of larger game. The desert plants also provided materials for construction, notably the internal 'skeletons' of the large saguaro cacti and reeds from the river margins.

At around 1000 BC, these Archaic hunter-gatherers began small-scale farming, cultivating the domestic varieties of maize, squash and beans that had originated in Mexico. Farming was undoubtedly a challenging task in the desert environment because of the low quantity and high unpredictability of rainfall. Dry farming would have relied on that rainfall, probably diverting run-off short distances on to fields. Flood water farming may have also been undertaken. This required a river with sufficient flow to have a flood plain that could be cultivated with confidence that the crops would be watered by the next flood. Some archaic settlement sites are located in just the type of riverine marsh locations that would have lent themselves to such flood water farming.

Farming developed to supplement rather than to replace the wild plant foods. The mix of the wild and domestic provided not only a healthy diversity in the diet but also spread the risk of food shortage. There was always the chance that a potential source of food would not be available: the wild harvest might fail because of drought or seedlings washed away within a flash flood. It would be unlikely, however, that all sources of food would be devastated within a single year.

The Hohokam culture appears to have emerged from these so-called Archaic desert-adapted hunters/gatherers/farmers – although some archaeologists believe they were immigrants into the region from Mesoamerica.[9] Archaeologically the Hohokam culture is formally defined by a shared type of pottery, one with a distinctive red colour. It was by mapping the scatters of this pottery that the focus of this culture was identified in the Phoenix Basin, an area of just under 6,400 square kilometres of Sonoran desert vegetation, encompassing the Salt, Gila, San Pedro and Santa Cruz rivers. While this appears to have been the core area of the Hohokam, their distinctive pottery is found throughout an area of 80,000 square kilometres, covering most of south-central and southern Arizona. The sharing of such material culture is often assumed by archaeologists to reflect the sharing of a language and even a religious view.

In this regard, the earliest Hohokam communities were not dissimilar to prehistoric Native Americans living throughout North America: they had small villages with pit houses and an economy based on a mix of hunting, gathering and farming. They used bows and arrows, snares, traps, nets and clubs, primarily taking rabbits, while throughout the desert they had a scatter of wild plant processing sites with roasting pits, mortars and metates. They had domesticated dogs. Pottery making and a range of other craft activities were pursued and through a shared style of artefacts they expressed their common culture. Some trade was undertaken with neighbouring groups. Overall, there was nothing special about the early Hohokam. But then, at around AD 50, the Hohokam started doing something quite different to communities elsewhere: they began to build canals.

Canals and irrigation

Whether the Hohokam independently invented the idea of canals and irrigation or whether they learned about why and how to build canals from groups further south, with whom they were trading, remains unknown. It was once fashionable to look all the way south to Mesoamerica to find the original stimulus for the Hohokam canal developments, where there is substantial evidence for water management prior to AD 50 as we will see in the next chapter. But a 1998 archaeological survey in the Tucson Basin, a mere 150 kilometres south of Phoenix, found that archaic farmers adjacent to the San Pedro River had begun making irrigation canals as far back as 1250 BC.[10] The inhabitants of a settlement now known as Las Capas were found to have been cultivating maize, squash and beans in the same Sonoran Desert environment as the Hohokam, using irrigation canals considerably more advanced than those in use within Mesoamerica at the same time period. Las Capas and its irrigation system were abandoned around 800 BC, never having developed to the scale that would be achieved by the Hohokam in the Phoenix Basin during the first millennium AD. Whether the Hohokam were descendants of the Tucson Basin farmers, learned canal irrigation techniques from them, or made their own independent innovations remains unknown – and one of the most interesting archaeological questions to explore.

Whatever the case, the Hohokam are likely to have started with little more than extended ditches diverting water from springs or rivers on to nearby farming plots. Within a few hundred years, however, lengthy networks of canals had been created: hundreds, maybe thousands, of miles of these sprawled out from the Gila and Salt rivers, and on a smaller scale throughout the Hohokam region (Figure 9.1).[11] With these, the Hohokam irrigated fields of maize, squash, beans, pumpkins and cotton, transforming themselves into fully fledged farmers and sparking a course of cultural development that so very nearly became a civilisation.

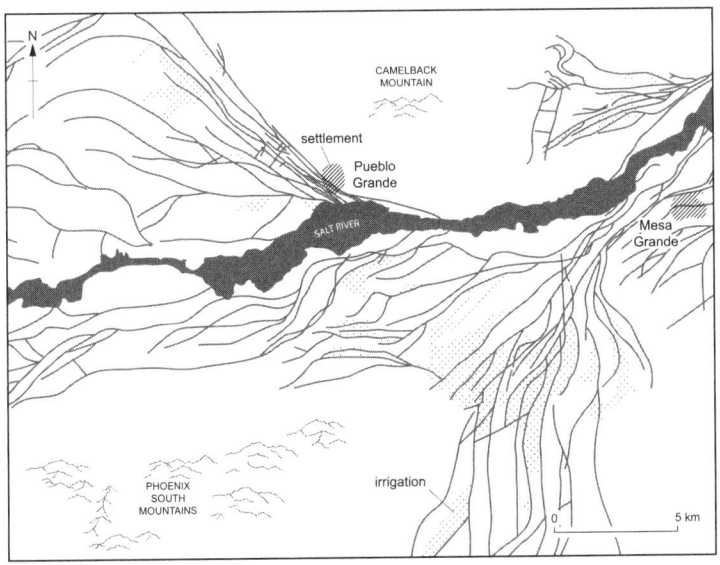

*9.1 Irrigation canals of the Hohokam in the Salt River Basin
(based on maps within Pueblo Grande Museum, Phoenix)*

The canal system was built in a piecemeal fashion and so archaeologists struggle to know how much of it was in operation at any one time. There were a number of main canals drawing water from either the Salt or Gila rivers, or one of their major tributaries. Although not surviving, there is likely to have been a series of weirs across the river channels that raised the water level sufficiently for it to flow into the canals, reaching head-gates made from logs and brushwood. From here the canals would split into a number of smaller distribution canals, the flow of water into these being controlled by opening or closing sections of the head-gate (Figure 9.2). These distribution canals took water towards the fields, releasing it into the soil via an array of even smaller lateral canals.

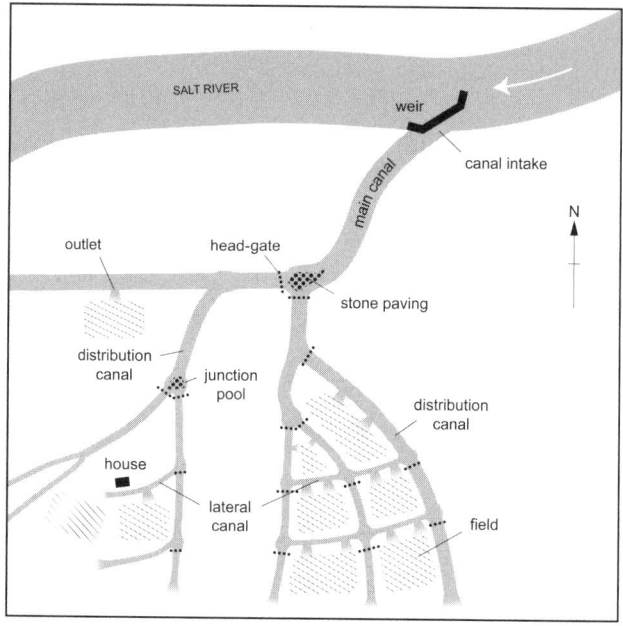

9.2 Hohokam irrigation system components (after Masse 1991)

Constructing this canal system was harder work than in other parts of the world, such as Mesopotamia (Chapter 3). There the canals were dug into relatively soft silt and the main river had already been naturally raised above the surrounding land by levées, which simply had to be sectioned for the water to flow into the canals. In contrast, the Hohokam had to dig through sunbaked soil using digging sticks, stone picks, axes and hoes. Unlike the Sumerians, the Hohokam lacked any draft animals to help with either cutting the soil or carrying the spoil away, and had a greater challenge of where to put the canal intakes and head-gates to ensure an appropriate gradient could be found for the water to flow neither too fast (and hence eroding the banks) nor too slow (and hence dropping its sediment and silting up the canal).

Impermeable bedrock was often close to the surface, resulting in more reliable surface flow than where the water could sink through porous sediment, sometimes disappearing from the surface entirely. Such 'bedrock reefs' created natural hydraulic heads in the rivers, the ideal place to divert water into a canal or even directly on to adjacent fields. The Hohokam made good use of such bedrock in this manner, which then influenced where their settlements were located and which ones experienced the most substantial growth. Indeed, the distribution of settlements and the canal network were very closely related to the variations in the local topography.

Some of the canals were especially large – 15 metres wide, three metres deep and with banks three metres high creating a potential six-metre depth of water. Seeing is believing, and I was struck by the magnitude of the now dry canal paths that have been preserved at the so-called 'Park of the Canals' within Mesa, a suburb of Phoenix – even though these would have been only medium-sized for the Hohokam (Photograph 36). Walking along these canals today, over a landscape that appears to be completely flat, it is easy to imagine how the early nineteenth-century settlers were impressed and then sought to replicate the engineering feats of the Hohokam.

Settlement growth and excavations at Snaketown

With this Hohokam irrigation system underway, there was a substantial expansion of the number and size of settlements which must reflect both indigenous population growth and migration into the Hohokam region from elsewhere – people attracted to its enhanced food security and trading opportunities. A distinctive settlement pattern can be detected of villages, hamlets, farmsteads and field-houses, all broadly following linear arrangements along the edges of canals. Villages were permanently occupied by at least 100 people, with a few growing to a substantially greater size towards the end of the Hohokam period.[12] None, however, ever gained the status of a town and urbanism is central to any definition of 'civilisation'.

The villages had some common features and layout. They had pit houses arranged in residential groups around a central plaza, shared areas for rubbish disposal, cemeteries, outdoor work areas and large earth ovens. As well as the central plazas, most villages had two other forms of public structures: ballcourts and platform mounds.

More than 200 ballcourts have been recorded, all constructed between AD 750 and 1200.[13] The ballcourts are oval depressions about 30 metres in length and 15 metres wide, with the excavated earth heaped up to provide an encircling embankment, creating a viewing platform for the playing of the game inside. We know that ball games were at least one function of these structures because ceramic figures and rock art depictions have been found, showing the players equipped with pads on their arms and hips.

Ballcourts are a pervasive element of Mesoamerican culture, being found in Aztec and Mayan cultures (Chapter 10). The Mesoamerican ballcourts had a slightly different form, having straight sides like a tunnel, a flat floor and sometimes stone rings in their side walls; the Hohokam ballcourts may be the most northern expression of a cultural phenomenon widespread and probably originating in Mesoamerica or an entirely independent entity. We have documentary accounts of the games played in the Mesoamerican ballcourts, showing that they were symbolic of the passageway between the upper and lower symbolic spiritual worlds. Playing the ball game was a means of communicating with the gods and sometimes ended in the ultimate sacrifice by the winner. We have no idea whether that was also the case for the Hohokam, but the highly consistent form of the ballcourts suggests that some shared elements of Hohokam ideology were being expressed and reconfirmed by whatever activities took place within them.

The first platform mounds appeared a couple of centuries after the first ballcourts, at around AD 900. These were originally quite small – 10 metres in diameter, less than a metre high. Some were enclosed within palisades of posts indicating that access was restricted. Some had structures built on top of them while others had hearths. But there are no signs of residential structures on the earliest mounds, these appearing to have been locations for the performance of ritual.

We have limited knowledge about Hohokam ritual and cosmology. In light of their desert environment, the unpredictability of rain and their dependence on irrigation, it would not be surprising if water had been a key theme in their religion. Hence ritual activities on the mounds might have been to placate the gods, seeking to avoid both floods and droughts. Water-related imagery was a theme of their pottery with painted decorations of water birds, fish, snakes, turtles and tadpole-like shapes[14]. Such forms are also seen carved into rock faces, engraved on to shell artefacts and abstracted into their wavy geometric designs. We must assume that these had symbolic significance, suggesting that water had an important role in Hohokam ideology.

Some archaeologists suggest that the glittery surface of Hohokam undecorated pots, created by using schist as a temper in the clay, was intended to mimic the sparkle of sunlight on water.[15] The largest source of such schist was from a prominent stone peak called Gila Butte, rising 500 feet above the Gila River just a few kilometres from Snaketown, one of the largest Hohokam villages.

Initial excavations at Snaketown took place in 1934–35 and helped define the key characteristics of Hohokam culture; further excavations were made in 1964–65 by Emil 'Doc' Haury, the pre-eminent Hohokam scholar of his day, who exposed an extensive area of the village.[16] A re-analysis of Snaketown's archaeology in the 1980s by David Wilcox of the Museum of Northern Arizona revealed a substantial degree of spatial order within the settlement.[17] At its centre there was 'nothing', or rather a central plaza around which all of the key structures were built. The plaza was ringed by eight artificial mounds, one having a palisade and at least two being capped with caliche – a sedimentary rock hardened with calcium carbonate. Within the ring of eight mounds were found the largest houses, cemeteries and individual burials; one relatively large ballcourt was located outside the ring of mounds to the east and another smaller ballcourt to the west. One of the mounds seemed to be a focus for pottery manufacture, possibly a centralised production for the community and/or for trade. There appear to have been up to 1,000 individual houses at Snaketown over the course of its 500-year history. Wilcox assumed a house life of 15 years to conclude that the maximum population at any one time would have been in the order of 300.

The Snaketown canal system had made use of the hard bedrock at the base of Gila Butte. Some think that this was more than good engineering design: might there not have been an idea that the water literally flowed from the schist-rich mountain, some of which was embedded into each and every pottery vessel?

Coping with droughts and floods

I have already emphasised that rainfall is highly unpredictable and patchy within the Hohokam region. Irrigation systems can mitigate the risk by drawing on water from extensive catchments, hence exploiting not only rainfall but also ground water sources. While this was certainly the case for the Hohokam irrigation systems, they were nevertheless susceptible to dramatic changes in the extent of flow within the system.

We have a fantastic record for the specific challenges faced by the Hohokam from tree rings that cover the period between AD 740 and 1370 – precisely the period of Hohokam expansion and demise.[18] The widths of tree rings provide a measure of annual rainfall, which can be used to calculate the flow rate of rivers within their vicinity. During the 1980s Donald Graybill from the University of Arizona's Tree Ring Laboratory analysed tree rings from ancient trees in northern Arizona and used these to reconstruct the flow of the Salt River, which, he argued, would also provide a measure for that of the Gila River.

Graybill found that there was constant fluctuation in the flow rates, as would be expected. He also found evidence for periodic massive flows – floods – that would have had devastating consequences for people living within mud-walled pit dwellings adjacent to rivers. The precise timing of these floods is significant for the development and ultimate demise of the Hohokam. There were four years of quite exceptional floods between AD 798 and 802. These would have inundated the settlements and washed away all but the most robust weirs and head-gates; sediment would have been eroded from the river bottoms and banks and then deposited within the distribution and lateral canals.

Such floods would have changed the whole micro-topography of the river system, requiring the irrigation system to be redesigned and rebuilt. Graybill and his colleagues suggest that this would have cost the Hohokam several seasons of cultivation, causing them to temporarily increase their reliance on wild foods. This probably had to be repeated on an even larger scale just under a century later, when in AD 899 there was the largest river flow of the whole recorded 640-year-long period. This most likely created extensive floods throughout the Salt–Gila Basin, again leading to the loss of several seasons of cultivation.

Floods of less severity are documented throughout the period but the next most striking events are between 1322 and 1359: these were years of dramatically low flow rate, suggesting that there may have been insufficient water within the irrigation system especially for those at the end of the distribution canals. Within this 38-year period of effective drought, however, came another year of flood, in 1356. Further tree ring evidence suggests another sequence of massive floods in the years 1380, 1381 and 1382.

All irrigation systems are challenged by floods and droughts. The Hohokam appears to have been especially sensitive because in spite of so much design and effort in construction, some key elements were simply missing. There is no evidence, for instance, that secure dams were ever constructed that could have held back a large river flow or stored water for time of drought. Nor do we have evidence for drainage channels from fields. Without being able to remove flood water from inundated fields, the Hohokam were not only losing access to the crops but also risking the build-up of salts within the soil that reduce fertility – the fate of the Sumerians (Chapter 3).

We will explore the consequences of the fourteenth-century droughts and floods for the Hohokam shortly. But here we need to consider how within the context of both the routine level of fluctuation and the floods of the ninth century, the irrigation systems had been designed, constructed, maintained and frequently repaired.

Was there central planning or informal cooperation?

As I explained in my introductory chapter, one of the ongoing debates is whether irrigation systems had required centralised control for their construction – as Wittfogel had proposed in his great 1957 work, *Oriental Despotism*. There is no direct evidence to support this proposition for the Hohokam: we have no trace of a ruling elite when canal systems were first constructed and then continually repaired. The case for an elite can only be made after 1200, which marks the start of what archaeologists call the Classic Hohokam period, characterised by the appearance of the 'big houses'.

Prior to this date there were no houses strikingly larger or more prominent than others, no burials with especially rich grave goods, no rock carvings depicting figures of seeming authority. In short, nothing to suggest there were formal positions of leadership. That is not to say that there were no 'natural-born' leaders or those who gained high status by virtue of their achievements – all societies have these, even so-called egalitarian hunter-gatherers. But these positions were not inherited, having to be earned on a daily basis.

Just as there appear to have been no formal chiefs within each community, there is no evidence that any one settlement had authority over others. With settlements evenly spaced along the main and distribution canals at about five-kilometre intervals, one has the impression that each small community constructed and managed its own stretch of canals and fields.

Even if there had been no overall control there must have been widespread cooperation between communities and between families within each community about the design, construction and maintenance of the canal system. Building the main canals would have required a substantial labour force, more than a single village could have provided. Agreements would have been required about the control of water through the head-gate to make sure that those towards the end of the distribution channel would get their share. The term 'irrigation communities' has been used by archaeologists to capture the need for cooperation between a set of communities sharing a main canal. So how was this cooperation achieved?

Ballcourts, trade and feasting

Here we might return to the ballcourts. We can be confident that the playing of games – or whatever other group activities occurred within these structures – provided occasions when people from neighbouring communities came together. It is attractive to imagine inter-village competitions, the Hohokam equivalent of the Premier League or Major League baseball, but we have no evidence that this was the case. The greatest similarity with today is likely to be that the occasion of a ball game was also one for business. That might have meant trade and exchange or discussion and planning about the design, construction or maintenance of a canal system.

I suspect that the degree of cooperation required for the Hohokam irrigation could have been undertaken at such aggregations of people for ball games, or at least during the feasting events before or after the games. This might explain why the spread of the ballcourts to their widest distribution occurred in the period immediately after the destructive floods of AD 798–800, requiring enhanced levels of cooperation to redesign or repair the irrigation systems.

The ball games may have been occasions for trade. We know that extensive trade and exchange was being undertaken, not only within the Hohokam communities but extending much further afield.[19] Copper bells, of a size to be stitched on to clothing or hung on a thread, are found at a few Hohokam sites and could only have come from Mexico – the Hohokam themselves neither invented nor adopted metalworking. Around 200 copper bells have been recovered, with 28 coming from a single burial at Snaketown, suggesting a necklace.

Copper bells may have been matched in their status value by pyrite-encrusted, sandstone-backed mirrors, also originating from Mesoamerica, as is likely for the parrot and macaw bones found in Hohokam settlements, although these birds may have been indigenous to the region. If these birds came from afar, we don't know whether they arrived as live or dead specimens, but we assume they were desired for their colourful feathers for ceremonial and ritual activities. Such items are likely to have been circulated between Hohokam groups, and with that circulation went discussion about irrigation systems.

The Hohokam traded shells coming from the Gulf of California or the Pacific coast, either as unworked raw material or as shell jewellery, such as bracelets and armbands. A variety of precious stones were also traded – obsidian, pyrite, steatite – while the presence of many foreign pottery types at Hohokam sites suggests either the pottery itself was being exchanged or more likely whatever foodstuffs and liquids it contained. Such materials testify to long-distance trade, but there must have been a substantial flow of goods within and between the Hohokam villages, with much of the exchange taking place at the ballcourt events.

A great deal of the exchange would have been in perishable goods. The historically documented Native Americans from the Sonoran Desert traded an immense diversity of these, including cactus seeds and syrup; agave cakes and fibre; wild gourds; peppers; acorns; dried meat; skins; pigment; and salt. We must assume that the Hohokam did the same. Some of the non-perishable and utilitarian items appear to have had specialised production centres, such as for pottery and stone plant-processing tools, manos and metates.

Trade is, of course, a catchall term and is likely to have included several types of exchange processes, especially when so many diverse materials had been involved. One way of characterising this is by thinking of three 'spheres of exchange': a 'prestige sphere' focusing on high-status items for village leaders probably exchanged at feasts; a 'social sphere' including shell ornaments, baskets, textiles and pottery, which would have been about cementing social relations, perhaps in the context of marriage and other social occasions; and a 'utilitarian sphere' for food, raw materials and tools, which might have been undertaken at markets and trade fairs. Circulating within each sphere, riding on the backs of the traded items, would have been ideas, proposals, plans and agreements about the construction and maintenance of the irrigation systems.

The prestige goods exchange sphere probably involved feasting. At the settlement of Grewe close to the Gila River, a large number of earth ovens have been found close to the ballcourt. While one might think of these as the Hohokam equivalent of the fast-food venues at football or baseball games, a more likely parallel is the ostentatious lunch in the chairman's box for distinguished guests. At Grewe the largest and the wealthiest houses are close to the fire-pits and the ballcourt.

So the feasting was most likely an occasion for the wealthy individuals to impress the guests to their village, and by so doing build up reciprocal obligations which might be called in at a time of need. My guess is that it was in the context of such feasting that items were exchanged and agreements made about the new design or the construction or the repair of a canal system that would be of benefit to all. Indeed, it might be telling that trade, in all of its forms, appeared to reach its maximum extent between AD 800 and 1100, the period immediately after the devastating floods when the irrigation systems needed rebuilding and community cooperation was at a premium.

Cultural transformation of the Classic period

A transformation occurs within the Hohokam world around 1200, most strikingly within the villages of the Salt–Gila Basin irrigation communities. The ballcourts go out of use and there is a dramatic reduction in the extent and amount of trade – not surprising if that primarily occurred in the context of the ball games.[20] There are also significant changes in the settlement pattern: Snaketown is abandoned and never occupied again, even though it remained exceptionally well located for irrigation. Grewe, on the other side of the Gila River, is also gradually abandoned, the population appearing to shift to a new settlement a few kilometres away that we know today as Casa Grande. As the name indicates, that is where a new form of building is made – a big house – while its ballcourt is the last known to be built by the Hohokam.

After almost 1,000 years below the scorching sun of the South-West, the Big House at Casa Grande still looms high above the Sonoran Desert (Photograph 37).[21] Made only of multiple layers of adobe mud and wooden supports, the four-storey building was divided into several rooms, connected by access routes and roof hatches. The lower storey was packed with mud while the fourth storey appears to have been a single room.

The Casa Grande settlement as a whole, and the Big House in particular, provide excellent views along the Gila River, far better than those gained from either Snaketown or Grewe. We must assume that people from those settlements, or at least their most powerful families, relocated to Casa Grande because of the enhanced opportunity to monitor, and hence control, events along the Gila River.

Casa Grande was first 'discovered' in 1694 by an Italian Jesuit missionary, Father Eusebio Francisco Kino, who used the Big House as a landmark for himself and early travellers crossing the Sonoran Desert. He conducted several masses within its mud walls, thus consecrating it as part of Western culture. By the late 19th century the newly built Pacific railroad was bringing many visitors to the Big House, causing it to suffer wear and tear from tourists and treasure hunters. In 1892 Casa Grande became the very first 'National Monument' in America, but this led to more tourists and even a series of historical pageants within the Big House from 1926 to 1930 with up to 13,000 visitors gathered among the ruins. Today the Casa Grande is clearly visible from Highway 87, largely due to a now iconic 1932 cover, protecting the building and making up part of an impressive museum and archaeological park.

Big houses were also constructed elsewhere after 1200, notably at

Pueblo Grande. Another change in monumental architecture was the quite literal rise of the platform mounds. Small mounds had long been a key feature of the Hohokam villages, appearing to have been locales for ritual activity and accessible to all members of the community for communication with the Hohokam divinities. As from around 1200 the existing mounds were substantially enlarged into rectangular structures with retaining stone walls, and then placed within compounds by the construction of surrounding walls. Structures, probably domestic dwellings, were built on top of the mounds, which also became the site of inhumations. Houses were also built within the compound walls, supporting the notion that these had become residential complexes.

The construction of compounds enclosing domestic buildings was a general feature of the Classic-period Hohokam settlements, these replacing the clusters of pit houses of the earlier settlements. The new compound groups were larger with more distance placed between the compounds than there had been between the pit house clusters, perhaps reflecting greater social distance. Only one compound within each settlement had a platform mound, a clear sign that its residential groups had secured higher status than others and chose to both express and maintain this by monumental architecture. Like the ballcourts, the mounds would have required a large labour force for their construction, one far larger than could have been provided by the group occupying the mound itself.

Quite how this labour was secured remains unclear, but there are two points to bear in mind. First, whereas the ballcourts appear to have been explicitly for the community as a whole, the ball games most likely appealing to all groups of people who had open access to the activity, the platform mounds and big houses seem to be for the immediate benefit of just one social group – those living within that compound. It seems unlikely that other people could get access to the mound or even observe what activities were occurring. Second, the monumental mounds of the Classic period evolved from the smaller pre-existing mounds that appear to have been the location for ritualistic activity. One interpretation is that the high-status social groups were now claiming direct access to the Hohokam divinities by placing themselves on the platform mounds and hence using religion to legitimise their position –

claiming a sacred authority to mediate for the larger community with the gods.

This may have been through enacting rituals at times of astronomical events which only they could predict, experience or announce. Casa Grande has several small holes cut into its walls which align with the equinoxial sunrises, summer solstice sunsets and some lunar observations. At Pueblo Grande on the Salt River, one room in particular on the platform mound has been noted for its astronomical significance. The 'Solstice Room' has two doorways which allow the sun's rays to pass straight through only on the summer solstice, the longest day of the year. At the platform mound of Mesa Grande to the south of the Salt River, another room marks the winter solstice in the same way.

To summarise, as from 1200 we see a number of inter-related changes in Hohokam settlement and society: the abandonment of some settlements and the growth of others; the replacement of clusters of pit houses by larger residential compounds placed greater distances apart; the construction of 'big houses'; the transformation of platform mounds from being publicly accessible ritual locales to being monumental structures controlled by a single social group within the community.

Trying to manage or grasping power?

It is difficult to imagine that these changes were not associated in some manner with the challenges being faced by the irrigation systems. Although the tree ring record does not record especially dramatic events – either floods or droughts – around the time of the transformation, there were always ongoing fluctuations. What had changed was the population level: substantially more people needed feeding either by intensifying production or by expanding the extent of cultivated land and hence the reach of the irrigation system. Moreover, expanding trade networks required the sustained production of craft items and perhaps a surplus of food to trade or to provide for the ostentatious feasts.

It was in this context of an irrigation system under pressure that some individuals or families appear to have taken the opportunity to gain control, residing on the platform mounds and hence legitimising their status by a connection to the gods. Here one can take a generous or cynical view of their motives.

On the one hand, one can imagine that managing the irrigation systems had become so demanding that the informal ballcourt forum for discussion was no longer viable: someone had to take overall control. It was quite easy to do so. The Salt–Gila Basin landscapes are extremely flat and low-lying. Simply by elevating oneself a few metres, either on to the top of a mound or the upper storey of a big house, one can gain a view across the key canal junctions of the system and observe activity within the fields. Decisions can then be taken swiftly about where irrigation water should be released or where water is needed – especially if one is claiming authority from the gods to do so. This arrangement might be to the benefit of all: tolerating the self-appointed leader on the platform mound or in the big house, and providing whatever tribute is required, may have enabled the irrigation systems to be maintained at the required full capacity throughout the fluctuations in flow.

One can take a different view. By 1200 the irrigation systems may have been working so effectively that food surpluses were being generated. Those individuals or families with the requisite inclination and abilities might then have exploited this situation to secure a power base for their own aggrandisement. By placing themselves safely within their big houses and platform mounds, they could control the flow of water to the other family groups and communities of their choice, demanding tribute in return. Whether or not this benefited their own village or the irrigation community as a whole was of no consequence: these individuals' desire for power could now be realised by exploiting the irrigation system and the food surplus they produced. Again, one might wonder why the irrigation farmers agreed to pay rent or tribute to these leaders. The simple fact is that they may have had no choice: they were tied to their canals and fields and so unable to move away, becoming an easy target for the new elites.

Two pieces of archaeological evidence incline me towards the second view. First, the platform mounds had a very large number of storage rooms compared to those found within the compounds that lacked such mounds. This suggests that food surpluses were being stored and were under the control of whoever resided on the mounds. Second, the Classic period sees the appearance of numerous fortified or defended hilltop sites around the Salt–Gila Basin,[22] indicating a heightened level of conflict and control within Hohokam society.

'Those who have gone'?

Whatever the motives of those who occupied the platform mounds and big houses, the Classic Hohokam period had a relatively short life. Within 100 years of Pueblo Grande, Casa Grande and the other centres reaching their peak of architectural and cultural complexity, they were abandoned. The Hohokam effectively disappeared.

The most likely cause is the dramatic environmental fluctuations of the 14th century: the drought years of 1322–55, the massive flood of 1356, and then further floods between 1380 and 1382. Whatever new levels of management had been put into place, the irrigation system simply did not have resilience against these fluctuations when having to support such a substantial population. So these droughts and floods are likely to have once again washed away the head-gates and weirs, eroded the canal beds, deposited sediment and fundamentally changed the micro-topography of the river valleys.

Why did the Hohokam just not start again, as they must have done after the floods of AD 798–802 and 899? One reason might be that if the new leaders had been claiming divine authority to rule, this too would have been washed away when they were unable to prevent both droughts and floods. That would have left the Hohokam without leaders to plan and manage the system; they were also without the ball-court system and trading networks that had previously enabled the requisite planning to piggyback on social activities.

It may also have been the case that there were only so many times that the Hohokam could start again in the Salt and Gila rivers systems – there may simply have been no more places available for yet another head-gate to be constructed. A millennium of intensive farming may have taken its toll in other ways: repeated irrigation without adequate drainage may have led to the accumulation of salts within the soils and hence a loss of fertility. Or perhaps there was simply cultural exhaustion.

What happened to the Hohokam? The best guess is that they simply downsized their culture, returning to a dry-farming, hunting and gathering lifestyle much like the one that they had started with more than 1,000 years prior to their demise. Indeed, the Native Americans encountered by the Spanish missionaries in the 17th century and then by the first European settlers in the 19th century were most likely the direct descendants of those who built the big houses, the platform mounds, ballcourts and the largest prehistoric irrigation system in North America. The Hohokam had never gone away at all.

Starting over

As my air-conditioned flight took off from the sweltering heat of Sky Harbor Airport and I climbed above Phoenix and the Sonoran Desert, it wasn't the dry Salt River bed that caught my eye. Instead, I was struck by a glistening rectangle of water that is now the Town Lake of Tempe, a Phoenix suburb. This two-mile-long artificial lake, proudly completed by the citizens of Tempe in 1999, is part of a flood control project in the otherwise dry riverbed. It has returned at least a portion of the Salt River to its former glory, re-creating the original natural habitat and providing a marina, paths for biking and hiking, and a splash playground for children.

But this was a tiny sparkle within a vast urban sprawl. The population of Phoenix grows at an inexorable rate, as does demand for electricity for its air-conditioned homes and workplaces, and water for irrigation and to quench its people's thirst. The Hohokam had survived for 1,000 years in this arid desert region; it is a mere 200 years since Jack Swilling and nineteenth-century settlers found their abandoned canals. As I headed to the next destination on my journey through the ancient world it seemed unlikely that the present-day population of the Salt–Gila river basins would match the Hohokam achievement.

Dujiangyan irrigation system, China, constructed as from 270 BC, from the Qinyan Tower looking east towards the Sichuan Basin along the divided River Min joined by a spillway

Fountain in Dujiangyan Visitor Park, China, containing the iron bars that Li Bing used to mark the level to which silt had to be annually cleared from the river, and pebble-filled bamboo cages as used by Li Bing to construct the island that divided the river course

The 'fish head' at Dujiangyan, China, built originally by Li Bing to divide the River Min into two channels

TOP: Statue of Li Bing in the Erwang Temple at Dujiangyan, China

BELOW: The Three Gorges Dam, China

ABOVE: Angkor Wat, Cambodia

RIGHT: Ta Prohm, Angkor, Cambodia

View across the moat to the inner enclosure at Banteay Srei, Angkor, Cambodia

The south-west corner of the Royal Bath, or Srah Srang, Angkor

Swimming in the West Baray, Angkor, Cambodia

ABOVE: Salt River, Arizona, creating a green corridor through the Sonoran Desert, not dissimilar to that enjoyed by the Hohokam in the first century AD

RIGHT: Hohokam canal ditch, located on the Park of the Canals, Mesa, Arizona with Steven Mithen providing scale

BELOW RIGHT: The Big House at Casa Grande, Coolidge, Gila River Basin, Arizona

The Maya site of Edzná Mexico, looking east from the steps of the House of the Moon temple. The Great Acropolis supporting the Building of the Five Storeys is to the right

The Maya site of Edzná Mexico, looking west through the middle of the ballcourt. The remains of a stone ring through which the ball was passed can be seen midway along the right-hand slope

Water lily, symbol of the Mayan Holy Lords

CLOCKWISE FROM TOP LEFT:

Machu Picchu, Peru, soon after sunrise on 6 August 2011, looking north across the Central Plaza towards the peak of Huayna Picchu, with the roaring Urubamba River in the valley below

Inca Pachacuti statue in the centre of Cusco, Peru

Terraces on the east side of Machu Picchu, looking towards the Central Plaza

Aqueduct bringing drinking water 749 metres from the spring on the eastern side of Machu Picchu peak to the citadel

Fountain 10 at Machu Picchu, illustrating both the functional and artful provision of water. One of 16 fountains aligned along the Central Staircase

CLOCKWISE FROM TOP LEFT:

View looking east across the 13 terraces at Tipón, Peru, towards the royal residences

Ornamental issue of the spring on to the third terrace at Tipón, Peru

Flying steps within terrace walls, staircase and water channel with cascades at Tipón, Peru

Terraces at Pisac, Peru, looking north-west along the Sacred Valley of the Urubamba River

10

LIFE AND DEATH OF THE
WATER LILY MONSTER

Water and the rise and fall of Mayan civilisation, 2000 BC–AD 1000

Losing one's toenails in the quest to understand ancient water management is a great deal farther than I have been prepared to go in writing this book. But such was the sacrifice made by the archaeologist Ray T. Matheny, following in the footsteps of Yu the Great of China (Chapter 7). Matheny devoted much of his life to the study of the Mayan city of Edzná in the Yucatán peninsula of Mexico. This had reached its cultural apogee at around AD 750 with magnificent temples surrounding a central plaza, and a population of several thousand. Matheny had begun work at Edzná in the early 1960s and began to suspect that it had relied on canals for transport, irrigation and drinking water. He might be considered as being to the Mayan civilisation what Bernard Philippe Groslier had been to that of Angkor: contemporaries but unknown to each other, they were the first to appreciate the significance of water management to the rise and fall of their respective rainforest civilisations.[1]

Excavating in the dry season – by far the most pleasant time to work in the highly seasonal rainforest of the Yucatán – Matheny found that he was unable to understand how water flowed around the site. His curiosity led him to build a makeshift airport in order to fly his own plane over the dense rainforest and map the differences in vegetation, which he believed to indicate the presence of water features on the ground. Based on his findings he decided that he had to work in the wet season – a previously unthinkable idea for Mayan archaeologists – to further understand the system.[2]

For much of his field seasons between 1971 and 1974 Matheny stood in waterlogged ditches to map what turned out to be the most elaborate system of canals and reservoirs of the Mayan civilisation. Losing his toenails may have been the least of his worries: he was surrounded by mosquito-infested swamps, shared ditches with poisonous snakes and had to be alert for swarming bees. But Matheny, now 86 years old, is made of unflinching stuff: before he began his relatively tame archaeological exploits he had parachuted out of a burning B-17 bomber in 1944, and before that – when only five years old – had clung on to his grandmother's legs as she leant from the open cockpit of a biplane to repair the engine.

All of this puts me to shame during my own visit to Edzná. Wearing my great big boots I had merely tiptoed into the edge of the forest surrounding the temples to find the overgrown ditches of the canals that Matheny had discovered. I left rather quickly, irritated by a few mosquitos and frightened by something moving in the undergrowth. Looking back it was nothing more than a lazy iguana having a scratch.

I hadn't needed to see the ditches to appreciate the remarkable cultural achievement that is Edzná.[3] Having inspected the engraved images of the rulers of the city, known as the K'uhul Ajaw or 'Holy Lords', climbed the steps of its Great Acropolis, and visited the ball-court, I stood transfixed within the central plaza. Just as at Angkor in Cambodia, I was surrounded by the man-made mountains of a rain-forest civilisation (Photographs 38 and 39).

Rise and fall of the Mayan civilisation

At its peak between AD 250 and 900 – the Classic period – the Mayan civilisation consisted of a network of kingdoms within both the rainforest-covered lowlands of the Yucatán peninsula, southern Mexico and Belize, and the highlands to its south, those along the Pacific coast of today's Guatemala, Honduras and El Salvador. The civilisation was at its most grandiose in the Yucatán central lowlands (Figure 10.1).[4]

The emergence of the civilisation remains obscure. The earliest settlements with distinctive Mayan characteristics are known from both the pacific highland region and the lowlands. The most notable highland site is Kaminalijuyú, dating to 800 BC, which gained its wealth and power by controlling the trade of jade and obsidian into the lowlands.[5] In that region we know of the small rural site of Cuello, dating to 1200 BC. Here there were several houses made from wooden frameworks and thatched roofs that clustered around courtyards. The burial of what appear to be selected individuals with ceramic vessels at around 900 BC suggests that a social elite had become established.[6] Cuello was once claimed to date as far back as 2000 BC. Although those very early dates are now discredited, a distinctive Mayan culture in the lowlands may well stretch back that far in time. What we do know is that by 600 BC Mayan centres such as Nakbé and El Mirador had been established in the lowlands with monumental architecture including pyramids, temples and plazas. By AD 250 a large number of kingdoms had been established throughout the lowlands, with the emergence of massive political and ceremonial royal centres such as at Tikal, Caracol, Calakmul and Edzná.

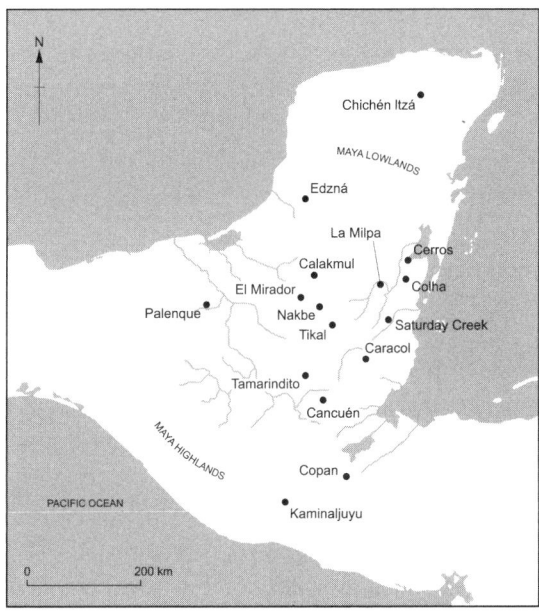

10.1 Maya sites referred to in Chapter 10

These centres had dense clusters of monuments of awesome proportions. Temples were erected on the apex of stepped pyramids, rising above the canopy of the rainforest, with windows aligned to celestial events. The palaces had rooms and courtyards which provided residence for the Holy Lords, the rulers whose position combined both political and ideological authority. Some palaces were so large and multi-levelled that we call them acropoli. Another important class of monument was the ballcourt, built on a larger scale and with a different layout from those of the Hohokam. The Mayan games involved moving a rubber or stone ball, with the ballcourts of the Postclassic period having hoops positioned on the sloping walls of their courts through which the balls had to be passed (Photograph 38). A stone relief from a ballcourt of the Postclassic site of Chichén Itzá appears to show a player of the game – presumably the loser – being ritually sacrificed: his head is held in the hand of another player, while blood spurting from his severed neck appears as serpents. The presumed victor's other hand holds a knife.[7]

Pyramids, palaces and ballcourts provided both the expression and the source of power for the Holy Lords who ruled each kingdom. They claimed their power by divine authority, and used this to muster vast labour forces to build temples, palaces and pyramids. Engravings on stone stelae – vertical standing monoliths – depict the Holy Lords in elaborate paraphernalia, with hieroglyphic inscriptions telling us their names, duration of rule and military conquests – the kingdoms were frequently at war with each other.

The temples, plazas and ballcourts were the scene for public ceremony and ritual. Thousands would gather as the Holy Lords enacted rituals to please the gods and venerate the ancestors, ensuring that the rains would come and the crops would grow. They smashed elaborate vessels, pierced their own bodies to let blood, and enacted human sacrifice. The people gave allegiance to their Holy Lords by providing tribute: food, raw materials and labour.

What, one might ask, did they get in return? It wasn't food: unlike within other ancient civilisations or even the Classic period Hohokam, there were no storerooms at the royal centres. The types of crops grown within the rainforest – maize, beans, squash – along with the dispersed farmsteads and tropical climate were not conducive to centralised storage, making the accumulation and redistribution of food impossible. Nor was it protection: if threatened, the people could disperse into the rainforest, and we have no evidence that their farmsteads had been defended. A clue as to the Holy Lords' hold over their people is provided in the titles that some chose to adopt: the Lord Water of Caracol, the Water Lily Lord of Tikal.[8]

We know about such names from the deciphering of the Mayan script. This was based on characters known as glyphs with as many as 500 in use at any one time and more than 1,000 different glyphs known throughout the Mayan period. This was not simply a pictogram script, as with Egyptian hieroglyphics; a significant number of the glyphs had phonetic meanings directly related to the sounds of human speech. The Maya had not only written on stone stelae but also produced long texts written on paper prepared from tree bark. Regrettably, thousands of these were destroyed by Spanish priests, leaving us with no more than three relatively intact scripts.

Just as impressive as their writing was their mathematics, primarily deployed in calculating calendars. The Maya – or at least some of them – used a base 20 and a base 5 within their numerical system, employed the concept of zero as a place marker and manipulated figures of many millions in their calculations.

Mayan religion was extremely complex, and inextricably linked to their natural surroundings. They worshipped a multitude of gods, but particularly the sun, the moon, rain, animals and planets. The Mayan concept of the world was pictorially linked to the branches, trunk and roots of a tree, the 'Tzuk-te'. It had three levels: the Heavens, the Earth and the Underworld, otherwise known as Xibalba and divided into nine levels. Entry to Xibalba was through a cave or opening in the ground, similar to the geological sinkholes found throughout the lowlands and known as cenotes.

Sacrifices were made to the gods of Xibalba, including the payment of a blood debt, through self-piercing and blood-letting, and, as considered above, the possible ritual sacrifice of ball-game losers and other victims. They believed that violent deaths were a pathway to heaven. Temples were built in astrological alignment, particularly to Venus, with their steps and tiers often designed to relate to the 365 days of the year and the nine levels of Xibalba.

Nothing in the Maya belief system was left to chance. Travel and trade around the Mayan empire was, and still is, difficult in the dense rainforest jungle. A network of sacbe, or causeways, linked some Mayan centres but they had neither wheeled vehicles nor beasts of burden. Trade was essential: the mosaic environment required foodstuffs to be exchanged while salt had to be acquired from coastal communities. The elite Maya needed to trade further afield, to the Pacific coast and the Caribbean, to secure prestige goods such as fine ceramics, jade, obsidian, quetzal feathers and spondyllus shells. As we have seen with the Sumerians, Nabataeans and Hohokam, such trade was essential to sustain both the working population and the elite.

In summary, no one is in any doubt that the Mayan civilisation with its temples, mathematics and material culture was one of the greatest achievements of the ancient world. So why did it collapse?

We should rephrase that question to refer to multiple collapses because throughout its long history kingdoms and royal centres rose to prominence and then declined, some being entirely abandoned for ever, while others gained a new lease of life.[9] 'Peaks and troughs' is how some archaeologists prefer to describe Mayan history. Several cities went into terminal demise around AD 300, which marks the end of the Preclassic period, notably the great city of El Mirador. Later on there was a widespread abandonment of the royal centres around AD 900, marking the end of the Classic period. Those in the northern lowlands continued to flourish in some form, but were also abandoned by AD 1450, less than a century before the Spanish conquistadors arrived in the Yucatán.

The abandonment of the royal centres marked the collapse of political authority, the rule of the Holy Lords, rather than of the population. It appears that on each occasion people migrated to new areas or became more permanently dispersed in the rainforest. In some areas local communities continued largely unaffected by the so-called 'Mayan collapse', such as at Saturday Creek, occupation of which was continuous up to AD 1500.[10] Indeed, we must note that the Mayan culture has continued to thrive into the present day, their religious beliefs now being a blend of the ancient Mayan and Catholicism.

Spanish conquistadors of the 16th century were the first Europeans to encounter Mayan farming communities, followed by European explorers in the middle of the 19th century. They found ruins of the great Mayan centres, engulfed by the rainforest, and immediately began asking questions about what caused the rise and fall of Mayan civilisation. There are as many theories as there are Mayan archaeologists, but water – its availability and management – has emerged as being an essential ingredient of any feasible scenario.

Bajos, cenotes *and* aguadas: *an unlikely landscape for a civilisation*

The cultural achievement of the Maya is made all the more astonishing because their lowland centres were built in a landscape with hardly any permanent water supply. The majority of the Yucatán is what geologists call a karst plain – one whose landforms are dictated by underlying bedrock of highly soluble limestone. The limestone is often exposed to form small ridges, and is riddled with solution channels, making it function like a sponge.[11]

Because limestone dissolves so easily, it leaves hardly any residue. With the absence of any other types of rock to provide clay or sand, the soils of the Yucatán are often thin and sometimes non-existent, although where they do develop they can be extremely fertile. Rather than forming rivers, the rainfall rapidly infiltrates and drains away through limestone fissures into underground channels, with the majority being rapidly discharged into the sea. The water table is often deep below the surface, more than 30 metres and beyond the capability of the Maya to excavate wells.

The landscape is riddled with underground caves and cavities. The ceilings of many have collapsed to leave sinkholes known as *cenotes*, usually cylindrical with steep sides and often very small openings. In some cases they reach the water table, providing a permanent supply of water, the level of which rises and falls with the seasons.

The basic pattern of rainfall experienced by the Ancient Maya is essentially the same as that of today: highly seasonal with an unpredictable start to a wet season and a long dry season. Between May and October rainfall can be heavy, as much as 1,500 millimetres, often coming in torrential thunderstorms. Some of this accumulates for short periods into large seasonal swamps known as *bajos* and some into shallow clay-lined basins known as *aguadas*. Floods and hurricanes are not infrequent and can do untold damage to farmsteads, crops and small-scale water management systems such as dykes and canals. Very little rain falls during the rest of the year, the long dry season between November and April, when the Yucatán becomes a green desert.

Uncertainty about the precise timing of the rainfall provided a challenge to the Maya. The beginning of the rainy season was highly variable: if the rain came later than expected, the planted seed would fail to germinate; if the rain came early, the seeds would rot in the ground. A further problem was that the *bajo* water was susceptible to pollution. The margins of the swamps were attractive for settlement and farming but the effluent from household waste would have flowed directly into the *bajos*, without the benefit of passing through a thick soil horizon filter to remove the pollutants. Moreover, the rate of natural decomposition of the swampy vegetation in the hot and humid conditions made the *bajo* water prone to stagnation. These problems were compounded by the encroachment of salty marine water, an extensive body of which underlies much of the northern third of the peninsula, sitting below and often mixing with the freshwater table.

The challenging nature of this landscape for human settlement has long been recognised. In 1566, Diego de Landa, the Bishop of Yucatán, wrote how 'nature worked so differently in this country in the matter of rivers and springs, which in all the rest of the world run on top of the land, that here in this country all run and flow through secret passages under it'.[12]

More recently, Professor William Back of the US Geological Survey described how: 'the paucity of fertile soil, the scarcity of potable [drinkable] water, and the shortage of other natural resources makes this one of the most inhospitable regions in which a sophisticated civilisation has ever developed'.[13]

Making a living and providing tribute

It was, nevertheless, within this heavily rained-upon and water-scarce landscape that the Mayan civilisation flourished. The basis of this had to be a highly productive farming system generating sufficient surplus to support the Holy Lords, their retinues, administrators, scribes, craftsmen, armies and so forth. Quite how such surpluses were achieved still remains unclear. We lack any evidence for the large-scale, intensive farming that characterises the civilisations of the Old World and must assume that household-based cultivation of maize, beans, squash and tubers had sufficed.

The majority of farmers lived in farmsteads dispersed throughout the rainforest, both close to the royal centres and in their extensive hinterlands. Farmsteads typically had several houses, facing an open area of patio – their spatial arrangement being mirrored at the royal centres by the palaces and temples surrounding the plazas. These scattered farming settlements appear to have been self-sufficient but provided tribute in the form of food and labour to a royal centre.

Being able to relocate their settlements within the rainforest, the farmers always had the opportunity to switch their allegiance from one royal centre to another.[14] As such, the Holy Lords from Tikal, Calakmul, Caracol and elsewhere had to continually work and compete for their loyalty, and hence their tribute.

One means of doing so was by persuading the farmers that they had the most effective communication with the supernatural world by the use of ever more elaborate ceremonies, rituals, feasts and ball games. Another was through military conflict, a frequent topic within the iconography. We know, for instance, that the kings of Caracol defeated those of Tikal twice in the latter half of the sixth century, resulting in shifts of population within their hinterlands. Such warfare was primarily about status rivalry rather than securing resources through coercion: a Holy Lord's success in combat was another means to persuade the populace that he carried the greatest favours of the gods and ancestors.

Farmsteads and communities outside of the central lowlands were able to locate themselves next to permanent rivers and could remain beyond the reach or need of the royal centres. At the settlement of Saturday Creek, for instance, located in the flood plain of the Belize River, the Mayan community appears to have lived in relative comfort from 900 BC to AD 1500, with no sign of outside interference or payment of tribute to a Holy Lord.[15] Neither did they build themselves temples or ballcourts, or exhibit any other signs of a dominant elite. It would appear that with no constraints on the availability of water and fertile soil, it was impossible for one family to secure a power base. Nevertheless, the Saturday Creek people did engage in many ritual activities, involving ceremonial deposition of obsidian blades and ceramic vessels. These appear as local, small-scale versions of the same rituals being enacted at the royal centres.

'Slash and burn' was once thought to have been the method of ancient Mayan farming, this being used by the mid-nineteenth-century Maya when first encountered by Europeans. Slash and burn, or swidden agriculture, involves felling the trees and other vegetation from a plot of land, burning these when dry, using the charcoal to bulk up the soil, and then planting crops on an annual basis. Plantings and harvesting can be repeated for a short succession of years but the soil soon becomes exhausted and the plot must be left fallow.

Those nineteenth-century Maya had planted maize, beans, squash and tubers within their plots. They, however, were living at a far lower population level than their ancient forebears, whose numbers must have exceeded several million. Slash and burn alone would have been insufficient, a significant proportion of the landscape having to be left uncultivated. If the ancient Maya had used slash and burn at all, it must have been just one element of a portfolio of farming methods. Others would have included garden plots between households, raised fields made by cutting drainage ditches round the margins of the *bajos*, and the use of terrace walls to retain soil and moisture. Arthur Demarest, Professor of Anthropology at Vanderbilt University, Tennessee, and one of the foremost Mayan scholars, has characterised Mayan agriculture as being diverse in its practices, dispersed throughout the landscape and sensitive to the micro-environmental variations.[16]

Many foodstuffs would be been acquired from within the rainforest. These included exotic items such as chocolate beans from cacao trees, vanilla and allspice. The forest was also a source of raw materials, including saps and resins, bark and wood for construction and fuel. Domestic turkeys were kept in pens and bees kept in hives for honey. Many wild animals were hunted for their meat, fat and skins, notably deer, agoutis, tapirs and turtles.

However effectively the Maya made use of their soils, seasonal lakes, *aguadas* and the products from their forest, they would only have been able to sustain their royal centres by managing the water supply. The earliest settlers relied on the natural accumulations of water within depressions, but hydraulic engineering had begun within the lowlands by at least 1000 BC, when shallow ditches draining the margins of swamps are known from northern Belize. During the Preclassic period, major centres such as El Mirador and Cerros were positioned to receive run-off from extensive watersheds. Professor Vernon Scarborough of the University of Cincinnati, the leading authority on water use by the Maya, describes this as a form of 'passive water management' – a logical extension of the much earlier use of natural depressions.[17]

By the time of the Classic period, however, water management had been transformed and was now being undertaken on a monumental scale at the royal centres. There was no one set method: each centre appears to have devised its own particular means, appropriate for its own local environment. What they all shared was an imagination and a capacity to engage in the hydraulic engineering of an entire landscape. We will look at three royal centres: Tikal, Edzná and Calakmul.

Reservoirs at Tikal

Tikal was arguably the most magnificent of the royal centres and is certainly the most archaeologically studied.[18] Located within the central lowlands, now in modern-day Guatemala, it reached its peak of architectural achievement, and military and political power, during the Classic period, between AD 200 and 900. It had a long succession of Holy Lords and a history of absorbing neighbouring communities and entering into conflict with competing centres. It maintained some form of interaction with the city of Teotihuacan in the Valley of Mexico, 1,000 kilometres to the west, possibly involving intermarriage between the two royal families.

The archaeological site is vast and the temples awesome. They tower more than 70 metres high amid an array of palaces, pyramids, plazas, ballcourts, administrative buildings, inscribed stone monuments and huge reservoirs. There are more than 3,000 structures in all, covering 16 square kilometres, constructed on limestone ridges above the surrounding swampy lowlands. The buildings cluster into a number of architectural complexes linked by raised stone causeways (Figure 10.2).

They were not, of course, all built in one go. Excavations of Tikal's North Acropolis have revealed more than 1,000 years of construction within this single building alone, including twenty successive floors, several destructive events – possibly as part of ritual activity – and then rebuilding. Having begun with a floor plan of 6 × 6 metres, it finished with one of 100 × 80 metres, supporting four funerary temples. On its south side, the North Acropolis faced the Great Plaza, 125 × 100 metres, where an audience of many thousands would have watched and participated in ritual performances led by the Holy Lord – the same rituals as they conducted within their own households but undertaken on a magnificent scale. When the North Acropolis was excavated, a plethora of ritual deposits were found under its floors, on their surfaces and within niches, including obsidian artefacts, shells, stingray spines and sea-worms. The performances had been grand theatrical events – visually, acoustically and emotionally.

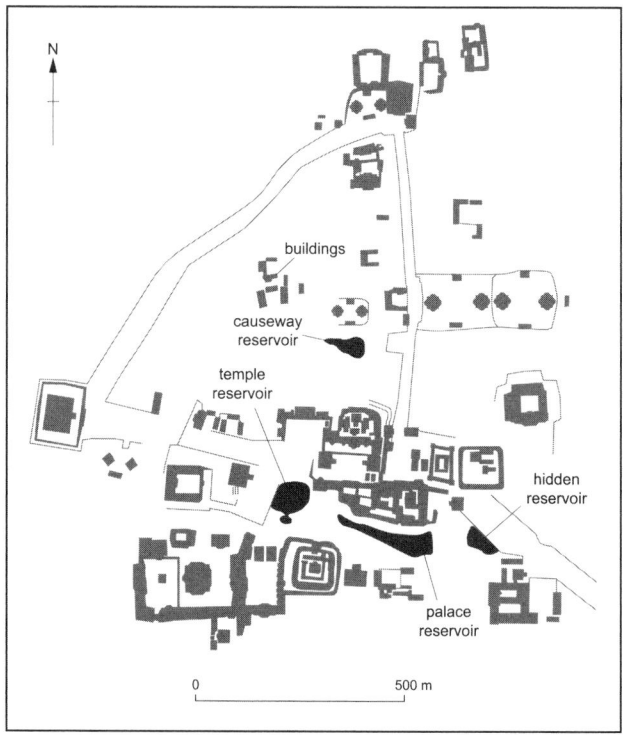

10.2 Schematic settlement plan of Tikal showing location of reservoirs (after Lucero 2006)

Another of Tikal's monumental buildings, Temple 1, had housed the burial of one of Tikal's most powerful Holy Lords, Hasaw Chan K'awil – 'Heavenly Standard Bearer' – who had ruled between AD 682 and 734. He was laid to rest on a jaguar pelt and entombed with over 20 vessels, slate plaques, alabaster dishes, carved and incised bone objects, and more than 16 pounds of jade artefacts. His tomb overlooks a plaza from where thousands of people would most likely have watched the internment of their dead king.

While impressive, one must recall that the monumental complex is simply the centre of Tikal: the bulk of the population would have lived in wooden structures, on the hillsides and in the lowlands. Dense clusters of structures have been found, between 187 and 307 structures in every kilometre square within the immediate vicinity of the centre and between 112 and 198 in each kilometre square for some distance beyond.

What is truly astonishing is that Tikal was located in the seasonal green desert and entirely lacked a natural supply of water: no springs, rivers or lakes in its immediate vicinity. At least six larger reservoirs were used to store as much of the wet-season rains as could be captured to see the population and their crops through the dry. These were constructed from either natural depressions in the limestone or within the quarries from where the stone had been mined for the construction of the plazas and temples, the bedrock being sealed by clay deposited by run-off water or deliberate plastering.[19] The reservoirs were situated within the monumental complex, as integral and as significant to Tikal as were the temples.

The reservoirs have been documented and interpreted by Vernon Scarborough and his colleagues. He characterises the citadel of Tikal as a 'human-modified watershed' or even a 'water-mountain'.[20] Such hydraulic engineering of an entire landscape could surely have only been achieved by centralised planning and construction under the authority of a political elite. How else could the massive labour force have been assembled and coordinated?

The extensive paved surfaces – plazas, courtyards and platforms – on which the monuments were constructed sealed the underlying limestone and acted as catchment surfaces for the rainfall. The causeways between the plazas functioned as both processional routes and walls to direct the run-off into the reservoirs. As such, the Holy Lords were directly associated with the control of water.

With an average rainfall of 150 centimetres per year, the central reservoirs would have captured more than 900,000 cubic metres of water, sufficient for the needs of approximately 60,000 people. Some would have been used for drinking, perhaps just for the elite and their retainers, while much was released – probably under the direction of the Holy Lord – down clay-lined ditches along the flanks of the slopes to reservoirs adjacent to the margins of the *bajo* where crop cultivation was undertaken. These reservoirs appear to have begun as natural *aguadas* in deep beds of impermeable clay adjacent to the *bajo* swamps and were then artificially extended. Here the water was stored, partly to provide further drinking water and partly for irrigation, released when required to flow along ditches and on to the crops during the dry season.

Another set of so-called residential reservoirs were located downhill from the central precinct within the densely populated zone of the city. These were then further supplemented by smaller reservoirs, less than a metre deep, for individual household use.

Despite its lack of natural water resources, Tikal was one of the most powerful and successful Mayan kingdoms. It grew steadily in strength from 200 BC to the ninth century AD, after which its power declined prior to abandonment, along with other sites in the region.

Canals at Edzná

Edzná is located 50 kilometres south-east of modern Campeche with its wonderful Spanish Colonial walled town. When Ray Matheny first visited Edzná in the 1960s it was still covered in dense rainforest, located within a valley close to hills and a seasonal lake – a *bajo* known as the Pik. Today much of the forest has been cleared for farmland, although the monuments themselves are still surrounded by dense trees which provide some impression of how it must have appeared to both Matheny and the Maya themselves.[21]

The Maya had first made a settlement at Edzná in 600 BC, probably being attracted by the relatively deep fertile soils, adjacent *aguadas* and a seasonal lake some distance away but still accessible. A permanent water supply was entirely lacking and consequently the initial community must have been either small or have engaged in some form of water management from its earliest times. The settlement grew rapidly between 250 BC and AD 150, with the construction of many ceremonial monuments. There followed a halt in its expansion, and possibly a major contraction in size, coinciding with the end of the Mayan Preclassic period. Unlike other centres that became abandoned at this time, Edzná, began to grow again; it survived another hiatus at AD 560, to reach its zenith at around AD 750. It then entered a period of slow decline through the ninth century, finally becoming abandoned around AD 950.

At its peak the ceremonial centre of Edzná consisted of a great multi-tiered acropolis, a ballcourt, several large pyramids and other buildings around a central plaza. The surrounding area was a mass of raised building platforms and small *aguadas* serving the local population and stretching out into the dense rainforest.

Just like at Tikal, this centre thrived without any source of permanent surface water: no rivers, lakes or springs. There were no *cenotes* (sinkholes), while the opportunity to dig wells was effectively denied since the water table was at least 15 metres below the surface in the wet seasons and more than 50 metres below in the dry. Unsurprisingly no wells have been found at Edzná and only 12 small chultuns – bell-shaped cisterns carved in the bedrock – to supplement the *aguadas*. It appears that small-scale water storage was eschewed in favour of something far grander: a network of massive canals converging on the central ceremonial complex along with a suite of reservoirs. Together, these had a capacity for more than two million cubic metres of water, all collected from run-off. The extent to which these were all entirely man-made canals, modifications of natural features in the limestone or even entirely geological in origin remains unclear. Many were only identified and mapped with the help of aerial photographs that Matheny took in 1971 from his own light aeroplane in both the wet and the dry seasons – excavation is required to establish their precise nature.

Whether cultural, natural or some combination of the two, this water system appears to have operated on a grand scale in conjunction with some of the monumental buildings. If the former, then thousands of labourers and decades of work must have been involved in its construction. As at Tikal, this suggests the control of a centralised authority. What Matheny called the 'Principal Canal', was the largest – 50 metres wide and running south for 12 kilometres from the centre of the complex towards the Pik, the seasonal swamp (Figure 10.3). As such, it seems likely that this was not only used for water storage but also for transportation; we know that a great deal of heavy stone was brought into the complex. When Matheny began mapping this canal in 1973 it was still collecting water to a depth of one metre. Water drained into it from hill slopes on its eastern side, the excavated soil having been heaped up on its west.

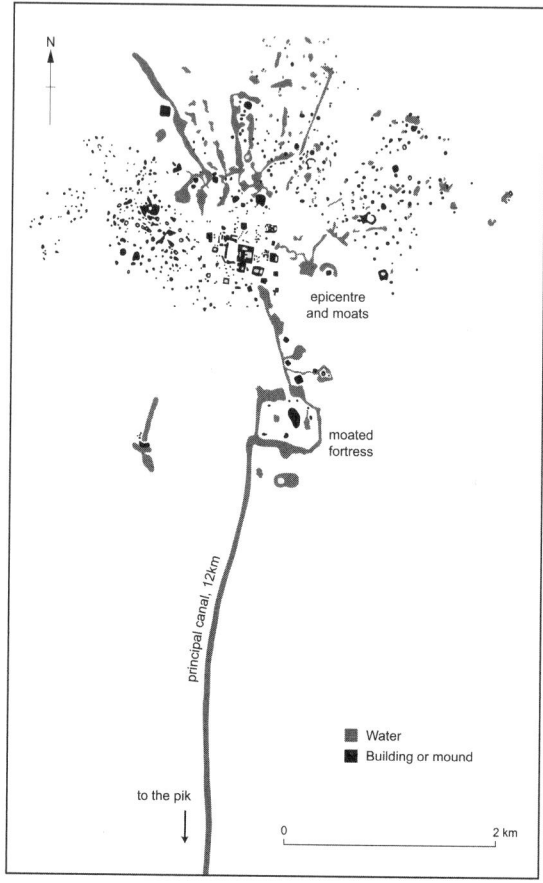

10.3 Schematic plan showing canals at Edzná (after Matheny et al. 1983)

Matheny discovered a moated fortress along its course. This was a ten-metre-high rectangular mound almost completely surrounded by water on which several Mayan buildings were constructed. He found no signs of warfare and some suggest the building had a religious significance in light of numerous iron oxide nodules found on the causeway across the moat – archaeologists frequently resort to 'ritual' when they lack any other explanation.

The canals and reservoirs on the northern side of the ceremonial complex converged on the complex of monumental buildings. The longest stretches of canals are only a maximum of three kilometres and hence these do not appear to be transport routes and blur the distinction between a reservoir and a canal. Although they probably extended into the next valley to reach the headwater springs of the Champotan River, their main source of water was run-off. The Maya took advantage of the natural slopes in the Edzná valley, building reservoirs in the higher ground to collect water from the hills and then using the canals, controlled by stone-built gates, to channel this to lower reservoirs close to the centres of population.

Despite the committed work by Matheny and his team, we remain unclear about the extent and nature of the hydraulic features to the south of the complex associated with the large canal. In 1972 a land reclamation project had begun with an army of chainsaws and bulldozers. The rapid removal of vegetation exposed archaeological features to the north of Edzná which required immediate attention by Matheny. As a result he lacked the time and resources to undertake the planned excavations to the south and this area remains to be surveyed.

Calakmul: encircled by water

For a third example of Mayan water management we can visit the city of Calakmul, located today in the central lowlands of Mexico, about 30 kilometres north of the Guatemalan border.[22] This city was discovered in 1931 and has two impressive pyramids, the largest reaching 45 metres above the tree canopy. It was only during the 1980s and 1990s that a team led by William Folan from the Universidad Autónoma de Campeche systematically mapped the city and its environs, recognising that it too had been dependent upon a sophisticated system of water management.

Calakmul had been occupied from Middle Preclassic to Postclassic times, a span of at least 1,200 years until around AD 810. It is situated on a dome of limestone 35 metres high surrounded by low-lying terrain. The centre is estimated to have covered 70 square kilometres with a population of 50,000 people at its peak in the Classic period, during which it controlled a region of at least 8,000 square kilometres.

The central complex covers 1.75 square kilometres and is delimited to the north by a massive wall, six metres high and almost two metres wide, extending for over a kilometre with several entrances. There are almost 1,000 structures within this central complex, dominated by a number of massive pyramids, temples, plazas and stelae. The design and layout of some of these buildings allowed astronomical observation: by standing on one of the buildings one could have used the alignments of others to determine the dates of solstices and equinoxes.

Excavations within one of the palaces in the 1980s discovered a tomb of what must have been one of the Holy Lords of Calakmul. The tomb contained a 30-year-old male lying fully extended on his back on a woven mat with his arm crossing over his chest. The body had been doused in red pigment, wrapped and then placed on top of five dishes and next to a cloth adorned with hundreds of shells. The tomb was full of burial goods: elaborate pottery vessels, 8,252 shell beads, 32 jade beads, a large shell, a stingray spine and a block of red pigment. Most impressive were three jadite masks. One had been for the man's face and had been made from 175 individual pieces of jade, with shell eyes, lips and teeth; another jade mask depicting a jaguar had been placed on the man's chest while a third on his belt had three stone pendants attached that were probably intended to make sounds as they clinked together.

It would have been under the authority of such rulers that landscape hydraulic engineering as seen at Tikal and Edzná could have been achieved. Something similar was undertaken at Calakmul. As well as surveying the archaeological monuments, Folan and his team examined the topography of the immediate region. They found that Calakmul, like so many other Mayan centres, had been situated next to a major *bajo*, known as El Labertino, one covering an area of 34 × 8 kilometres. This would have provided a seasonal supply of water and is likely to have been the location for the initial settlement. To ensure that a permanent water supply was available as the city grew in size, the Maya of Calakmul undertook a series of landscape modifications that resulted in the 22 square kilometres of the city, its inner zone, being entirely encircled by a series of interconnected *aguadas*, canals and dry stream beds (*arroyos*).

This hydraulic system was based upon thirteen *aguadas* that had a total capacity of almost two million cubic metres of water. Two especially large *aguadas* were located along the edge of the El Labertino *bajo* and were fed by rainfall and run-off; once the largest of these had filled up it overflowed into a canal that carried the water to the second largest of the *aguadas*. Two smaller *aguadas* elsewhere in the encircling ring of water were also connected by a 280-metre-long canal. Other sections of the ring were formed by *arroyos*. These small riverbeds may have been have been dry outside the wet season, although they also appear to have been modified perhaps to maximise the collection of run-off.

It remains unclear whether this water supply system was designed by the man found within the tomb or another of the Calakmul Holy Lords; we do not know whether it was a single enterprise or constructed in a piecemeal fashion over several centuries. Whichever is the case, it had been central to the growth and power of Calakmul as a royal centre.

Diverse and pervasive water management

The three case studies of Tikal, Edzná and Calakmul provide little more than an introduction to the diverse water management systems of the Maya, found in both urban and rural settings. They were diverse because the environment was diverse, especially when one also encompasses the southern highlands and the coastal wetlands: the Mayan civilisation developed and thrived within an ecological mosaic presenting each location with its own specific hydrological challenge to address.

At the Classic period city of Palenque in the southern lowlands of Chiapas, Mexico, water management using aqueducts, dams, channels, drains and a bridge was concerned with controlling flooding and erosion caused by nine streams that cut through the city.[23] Flood control was also the key driver for water management at the Classic-period city of Copán in western Honduras, undertaken by a complex system of overground and underground conduits, drains, plazas and causeways.[24] At La Milpa, Belize, reservoirs were created by blocking the upper portions of watersheds. A system of sluice gates and check dams controlled the release of water from the reservoirs. One purpose of this appears to have been to regulate soil moisture in pockets of

flatter upland soils, which could then become highly fertile and intensively cultivated.[25]

In the wetlands now located in northern Belize, the Maya dug
canals, diverting water and using the upthrow to create raised fields,
used for growing avocados and maize.[26] In other rural areas, such
as the uplands of the Petexbatún region of Guatemala, Late Classic
period water management was about trying to maintain rather than
remove water from the soil. Early Classic period forest clearance and
farming had led to soil erosion, population contraction and then
regeneration of the forest. When people returned in the Late Classic
period, their cultivation was supported by an elaborate system of terraced walls that checked soil erosion and helped maintain moisture.[27]
Within this region, a 60-metre-long dam and reservoir holding 2,000
cubic metres of water were constructed at Tamarindito, enabling the
adjacent terraces to be irrigated.[28]

Canals, dams and reservoirs are further examples of water management involving significant degrees of community endeavour. We
should also note household level water management, notably with
regard to the use of man-made and natural wells.[29] With the almost
constant flow of new discoveries about the extent of Mayan water
management – a press release on 26 August 2010 announced the discovery of two massive reservoirs at the Classic-period city of Uxul, the
floors of which had been sealed with fragments of broken pots[30] – it
is clear that water management pervaded the lives of Mayan people,
whether a rural farmer or a Holy Lord.

The iconography of water

It might be going too far to say that water was an obsession for the
Mayan people but it was certainly frequently both on their minds
and in their hands, even when facing drought. Pottery vessels may
not only contain water but also carry a memory of water within
their walls, it having been a critical constituent of the clay. So when
a Mayan peasant was handling the simplest of vessels or a Mayan
king was handling the most elaborate, they were both symbolically in
touch with water. Those who made the pottery vessels did not want
them to forget this, frequently painting the vessels with images of a
watery world.

Patrice Bonnafoux of the University of Paris has analysed such imagery and found it to be especially prominent during the few centuries after the collapse of many centres that marked the end of the Preclassic period at AD 250.[31] He described the Maya as having a fascination with aquatic iconography. The most pervasive image on the pottery vessels is what he calls the 'water band', a series of parallel lines in the middle of which runs a strip of dots and points. The band may be either straight or undulating, representing still and moving water respectively. The bands are frequently complemented with stylised depictions of water-lilies and shells; sometimes there are egg-shaped motifs that look like bubbles. Interspersed within this imagery are water creatures: aquatic birds hunting for fish, alligators and frogs swimming among the water lilies, turtles lying still, fish nibbling at the flowers (Figure 10.4).

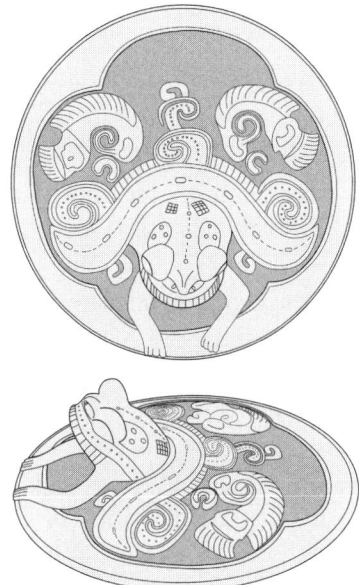

10.4 Lid of Maya vessel depicting frog emerging from water
(after Bonnafoux 2011)

Water lilies were an especially important plant for the Maya, so much so that some of the nobles referred to themselves as Ah Nab or the water lily people (Photograph 40). As well as on pots, water lilies are depicted on stelae and murals, and were part of the royal head-dresses. One of the Holy Lords of Tikal even took the title of the Water Lily Lord. We must assume that lilies were frequently present on the surface of reservoirs and *aguadas*. Although consuming significant quantities of water, they would have also conserved water by lowering its temperature and hence reducing evaporation. As well as being so attractive, the water lilies' key importance may have been as an indicator of clean water because they cannot tolerate high levels of acid, algae or calcium.[32]

The tranquil watery world depicted on Mayan ceramic vessels is often threatened by monster-like creatures. These come in various forms but share long curved snouts and typically have scrolls for eyes; some of them have feathery signs, which could be the depiction of fins and fish caught within their beaks. They are probably all representations of the same spiritual entity which Patrice Bonnafoux chooses to call the 'water monster'.

There appears to have been no distinction made between creatures of the freshwater and seawater worlds. This is perhaps not surprising when one recalls the particular geology of the region with seawater encroaching inland and sometimes forming a layer underneath the freshwater within the deepest *cenotes*. When the water bands, lilies and water creatures are all put together, the iconography of the pots appears to be depicting the luxuriant life within the *bajos* and *aguadas* at the height of the rainy season.

This imagery is found throughout the history of the Mayan civilisation but is especially prevalent in the first few centuries of the Classic period, between AD 200 and 560. During the Preclassic period the water theme has no particular prominence, with images of terrestrial plants and animals being just as prevalent. The same is found within the Late Classic period, during which water imagery may even take a secondary role. But it is the dominant theme in the Early Classic period, suggesting a particular mental preoccupation with water during this phase of the Mayan civilisation.

This appears to be mirrored in the items that were valued by the Holy Lords and were placed within their tombs: those of the Early Classic period have a much higher frequency of shells and other marine-derived items than is found in either Preclassic or Late Classic tombs. Such items include stingray spines, corals, sea urchins and water-worn pebbles.

Bonnafoux continues his interpretation of Mayan iconography by considering the physical form of the vessels. He finds the most distinctive feature of the Early Classic period to be the lid: never before or after was this so popular. He suggests that vessels with lids and feet were symbolic representations of the Mayan universe: the lid stands for the surface; the body of the vessel represents the aquatic underworld, it being so frequently decorated with water bands and aquatic creatures; the feet symbolise the mythological pillars that support the world. Some of the lids are decorated with three-dimensional models of aquatic creatures, such as frogs and water monsters, frequently represented by only their upper halves (Figure 10.4). They look as if they are climbing out of the water, leading Bonnafoux to suggest that the interior of the vessel itself is symbolic of a cave or a *cenote* leading to the underworld.

Into the watery underworld

That *cenotes* were conceived in this manner is demonstrated by the remarkable finds from the so-called 'sacred *cenote*' at Chichén Itzá, a centre within the northern lowlands.[33] This great Mayan centre rose to power in the Postclassic period, after the collapse of many other centres especially in the southern lowlands. The sacred *cenote* is one of two *cenotes* at the site. The other one has the purest water and was most likely the source of drinking water. The unusual blue waters of the sacred *cenote* are reached by a 300-metre white limestone sacbe or path. It is a circular hole, 50 metres in diameter, which opens into a 27-metre-deep bowl under the ground. According to rumours from early Spanish accounts, the Maya used to throw precious objects and human sacrifices into the *cenote* in worship of the rain god Chaac.

This seems to have been verified by the dredging of the *cenote*, which has produced over 30,000 items of gold, jade, copper, turquoise, obsidian, pottery and incense along with the bones of 200 adults and children whose bodies showed signs of injury before death. A layer of blue sediment was found in the mud at the bottom of the *cenote*. This indicated that the sacrificial objects – both artefacts and bodies – had been coated in the Mayan blue ceremonial pigment made from copal, indigo and palygorskite.

One of the earliest Spanish accounts about Chichén Itzá recounts how young women were apparently thrown from a height into the sacred *cenote*. If they survived, they were pulled up by rope and had to pass on received prophesies gained from within the *cenote* about the arrival of future rain. One does, of course, need to be cautious about European accounts of native peoples, so often riddled with racism about perceived savages. But this account does have a feel of authenticity in light of the excavated finds from the *cenote*.

The passage into a watery underworld was the principal metaphor for death for the Maya elite and was represented within the iconography of stone carvings and the elaborate burials of the Holy Lords.[34] This passage might occur by falling or sinking into the underworld or into the gaping mouth of a monster, which itself might represent the reservoirs within the royal centres. A canoe trip was another means by which the Holy Lord could pass into the underworld. Several depictions of this were carved on to the long bones found within the burial of Hasaw Chan K'awil at Tikal. These represented the Holy Lord within a canoe being escorted into the underworld by animal spirits and 'paddler gods'. In two of the carvings the canoe is dipping down into the underworld with the back of the Holy Lord's wrist pressed to his forehead.

Water and ritual at the royal centres

We have seen that the most extravagant royal centres were in the central lowlands, where rainfall was at its most seasonally variable and water at its most scarce. The Holy Lords were dependent upon tribute of food and labour to construct their temples where they enacted elaborate rituals and theatrical performances for massed onlookers. It was via such ritual that they communicated with the gods, ensuring that the rains would come, crops would grow, success would be gained in battle or whatever else was on the Mayan wishlist.

We have also seen that the Holy Lords had no control over the populace via the distribution of food: there was neither intensive farming nor centralised storage of food for redistribution. What they did have, however, was water – within their reservoirs, canals and *aguadas.* And that is what the people needed during the dry season. Vernon Scarborough succinctly summarised the resulting articulation between water and power: 'As a critical and scarce resource during the lengthy dry season, water was politically manipulated by a Maya elite to centralise and control power during the Classic period.' He notes that other means were also used to establish and maintain their authority, including warfare, conspicuous consumption of luxury goods and ritual. But the fundamental need for water made its control a powerful organising force within the Mayan Lowlands.[35]

Lisa Lucero, Professor of Anthropology at New Mexico State University, further developed Scarborough's arguments to explore how water, ritual and political power were intimately linked within the Mayan world.[36] She suggests that during the dry season the Holy Lords distributed water to the populace in return for their tributes of food and labour. Water was released from the reservoirs to irrigate the fields and access was allowed to those with containers to carry it away. Moreover, the Holy Lords provided the capital – building materials and labour – to repair the reservoir and canal systems following their destruction during floods and hurricanes. These transactions between the Holy Lord and the people were mediated by ritual: through theatrical performances the Holy Lords communicated with, pleased and appeased the gods, ensuring that rain would fall to fill the reservoirs once again. Those performances were enacted in the temples and plazas of the royal centres and were reflected in the mirror-surface of the adjacent reservoirs, magnifying the drama and reinforcing the association between the Holy Lord and water.[37]

This scenario readily explains the pervasiveness of water imagery within the Maya iconography. Moreover, the Holy Lords not only wished to associate themselves with water but especially with clean water. And hence we see the adoption of the delicate water lily as one of the key symbols of power.

This appears to be a system in which all parties benefited to varying extents – except, of course, those who had to be sacrificed in the rituals: the Holy Lords gained because they secured their power base and the people because they secured access to water during the dry season. With a dispersed farming system, they were able to move their allegiance to another royal centre if that seemed a better bet for a reliable water supply – either by having larger reservoirs or having a Holy Lord who appeared more favourable to the gods.

So how did it come to an end?

The Mayan collapse

What caused the demise of the Mayan civilisation, or to use a more dramatic phrase, the Mayan collapse? Before attempting any answers we must be clear that this is not a single event. There were, for instance, the decline and abandonment of numerous kingdoms around AD 200, the end of the Preclassic period. A widespread 'collapse' happened again at the end of the Classic period at AD 900, but this was not universal – kingdoms in the northern lowlands continued to flourish in the Postclassic period. Interspersed between these periods of prominent decline certain royal centres had dramatic reversals in their fortunes. Could there be a single explanation for all of these events?

Speculation has been rife ever since the 'lost cities' were first reported by Europeans in the middle of the 19th century: a peasant revolt; disease and epidemics; foreign invasion; the collapse of trade routes. In the more environmentally conscious 20th century the sustainability of the Mayan lifestyle came into question. Slash and burn, then assumed to be the predominant farming method, would have led to deforestation on a massive scale. This would have been exacerbated by the vast quantities of firewood required for the limestone-based cement and stucco used in the huge building works. Indeed, the cycle of competition between royal centres, with ever more elaborate buildings, rituals, trade goods and so forth, immediately smacks of an unsustainable system, one that could be pushed into a spiral of decline by any one of several factors. Warfare appears certain to have played a part, this being the principal theme of the last of the stelae erected in the royal centres prior to their abandonment.

Could there have been a prime mover for the Mayan collapse, some event that triggered all of the others into action? Most probably there was: drought.

Although long discussed, the drought theory has most recently been championed by Richardson Gill in his 2000 book *The Great Mayan Droughts: Water, Life and Death*. Gill is a retired banker who first became interested in the Maya in 1968 when he visited Chichén

Itzá on holiday from his work in the family bank. When the bank collapsed in the Texas economic crisis of the early 1980s, he took the opportunity to study anthropology and archaeology at university, focusing on the Maya. He wanted to follow a hunch and solve the mystery of the Mayan collapse. We should not be surprised, therefore, that his drought theory has been controversially received by much of the academic establishment, if not casually dismissed in some quarters. But hard evidence has been accumulating during the last decade. When this is seen in the context of the constant water challenge faced by the Mayan population and how, according to Lisa Lucero, the political system relied on the distribution of water, the drought theory becomes compelling.

Gill's first step was to refine the specific dates at which the centres were abandoned. For this, he carefully inspected the dates carved on the stelae to find the last dates at which they had been erected. This provided Gill with four distinct periods of abandonment, which he described as four successive collapses: the Preclassic collapse, AD 150–200; the Hiatus collapse, 530–90; the Classic collapse, 760–930; and the Postclassic collapse, 1450–54. Within the Classic collapse he also identified geographical variation in its dating: 810 in the west and south-west of the Maya region, 860 in the south-east and 910 in the centre and north.

Gill's next step was to undertake research on the historical accounts of the region to explore whether droughts have recently occurred and with what impact. He discovered that there had been a devastating three-year drought between 1902 and 1904; droughts of a similar intensity were also recorded for 1330–34, 1441–61 and in the 16th century, when an estimated half of the population of the Yucatán died from famine or disease. When combined with his own childhood memories of droughts in Texas during the 1950s, Gill became convinced that drought had caused the demise of the Mayan civilisation. And so he decided to search for scientific data to prove his theory – a personal mission as much as an academic research project.

This was during the 1990s, when scientists were gaining a much better understanding of how ancient climate and weather can be reconstructed from various types of evidence, notably sediment cores from lakes and marine deposits, tree rings and stalagmites. Gill was able to amass a remarkable collection of data that began to show that his four dates for Mayan collapse did indeed correlate with periods of drought.

Gill found that in 1984 Wibjörn Karlén from Stockholm University had identified multi-year periods of drought in tree rings around AD 800, 860 and 910.[38] Glacier advances in Arctic Scandinavia in 800, 860 and 910 also matched his dates for the Classic collapse. Although this evidence came from the northern hemisphere, climate scientists were beginning to understand the connections between patterns of weather around the world. Models of global climate change were demonstrating that periods of severe cold in the northern hemisphere cause a southward shift of the Inter-Tropical Convergence Zone – the area near the equator where winds originating in the northern and southern hemispheres come together. This leads to a failure of the summer rains and causes drought in the low latitudes, such as the Mayan lowlands.[39] Supporting evidence for that had come from analyses, published in the 1990s, of ice cores from two Andean glaciers. Both the Quelccaya glacier in Peru and the Sajama ice cap in Bolivia indicated drought conditions towards the end of the ninth century within the Andes – at just the same time as the Mayan Classic collapse.[40]

Such circumstantial evidence would never persuade the drought-sceptics: evidence was needed from the Yucatán peninsula itself. Gill found that was also readily available. In 1996 Jason Curtis from the University of Florida and his colleagues had analysed a sediment core taken from Lake Punta Laguna in the Yucatán.[41] The core contained calcium carbonate precipitated from the shells of aquatic organisms. By measuring the changing ratio of two isotopes of oxygen within that precipitate, ^{18}O and ^{16}O, Curtis and his team could record fluctuations in the extent of water evaporation from the lake throughout the Mayan period, and hence identify periods of aridity. Their results showed that there had been a major drought between AD 536 and 590 – coinciding with Gill's 'Hiatus collapse'. This was confirmed by the analysis of sediments from a second lake, Lake Chichancanab, which also provided evidence for a severe drought at AD 863, coinciding with the Classic collapse.

Richardson Gill published his theory in 2000 but new data continued to accumulate. In 2003 Geral Haug from the University of Potsdam and his colleagues published an analysis of marine sediment cores taken from the north-west Caribbean and the Cariaco Basin.[42] The sediments were made up from the silt coming from the outflow of the Orinoco and other Venezuelan rivers into the oceans. The silt is gently deposited rather than being washed away by strong currents and so builds up in annual layers that can be counted back through time. By recording the chemical signature of each layer, notably the frequency of titanium, Haug and his colleagues could estimate the extent of rainfall that had occurred within that year throughout the river catchments of northern South America.

Haug's team found three periods of severe drought centred on AD 810, 860 and 910 – precisely the dates of the Classic collapse – along with drought between AD 150 and 200 coinciding with the Preclassic collapse. Moreover, they also found evidence for a fourth drought at AD 760, which coincides with the abandonment of the Mayan centre of Naachtun. Having the ability to count annual rings enabled Haug's team to determine how long each drought had lasted: the 810 for nine years, the 860 for three years and the 910 for six years.

Yet another source of data was published in 2007, that from a stalagmite in Belize. Stalagmites are rather like lake and marine sediments in that they have an annual layer of deposition, in their case calcite, which can be chemically analysed. Stalagmites have an advantage because the specific layers of calcite can be dated by a method known as uranium series, although as with all dating techniques there is some degree of uncertainty surrounding the precise dates. Another advantage of stalagmites is that one may be fortunate in finding them in the close vicinity of an archaeological site and hence be able to reconstruct its immediate environment.

That was the good fortune of James Webster of the United States Environmental Protection Agency, Atlanta, and his colleagues.[43] They found stalagmites within a cave known as Macal Chasm in the central lowlands, just 60 kilometres from Tikal. This had involved a descent down a 40-metre-deep and 5-metre-diameter shaft into a large chamber – a descent into the Maya underworld. The stalagmite they selected to study was just under a metre in length and later shown to have begun forming 3,300 years ago and hence covered the whole duration of the Mayan period. Webster and his colleagues could date specific rings of calcite and use them to estimate the extent of precipitation during the year in which they formed. That was by measuring their colour, luminescence, and frequency of ^{18}O and ^{13}C within their chemical composition.

The results showed that the most intense period of drought throughout the whole 3,300-year-long period was between AD 700 and 1135, with 893–922 being the most severe – coinciding with the Classic collapse. There was also a marked drought at AD 141, which when allowance is made for dating uncertainty is sufficiently close to the Preclassic collapse to be significant. Another drought was recorded at AD 517, right in the middle of Gill's Hiatus collapse. What is of particular significance is that the Macal Chasm stalagmite had continued forming into the historic period. Webster's analysis of its colour and chemical composition indicated a severe drought in 1472. Again allowing for dating uncertainty, this is sufficiently close to a known occurrence of drought between 1441 and 1471 that had resulted in widespread famine. As such, this validates the analytical methods and provides us with confidence in the identification of droughts in the second, sixth and ninth centuries coincident with the successive abandonment of Mayan royal centres.

With the accumulation of data from tree rings, lake sediments, marine sediments and stalagmites, Gill's theory that the Mayan collapse had been caused by drought becomes compelling. The idea is still resisted in some quarters, notably the archaeology establishment. When Gerald Haug's findings from marine sediments were published in 2003, Professor Norman Hammond of Boston University remained unconvinced. Referring to the northern Yucatán centre of Chichén Itzá, he asked: 'Why did the latest and greatest fluorescence of the Mayan series occur in the area that we know to be the driest?' Similarly Professor Jeremey Sabloff responded by noting that 'The Maya thrived for 1,500 years before these droughts, so it's clearly not climate alone that brought down the southern cities of the Yucatán peninsula.'[44]

Well, it is never climate alone. We should envisage these droughts as acting as a trigger to the collapse of a political system that was inherently unsustainable. Insufficient rainfall to fill the reservoirs and *aguadas* would have released many of those other factors that have been cited as the cause of the Mayan collapse – failure of yields from inadequate irrigation, disease from being forced to drink polluted water, warfare by Holy Lords in an attempt to reassert their authority.

Death of the water lily monster

It was the loss of that authority that Lisa Lucero believes was the most critical factor. She characterises the Mayan collapse as one of political disintegration rather than population demise. If rains failed to come and the reservoirs remained empty, why should the people continue to pay tribute to the Holy Lords? Put quite simply, having secured their power by claiming intimacy with supernatural forces, the Holy Lords now lost their power when the rain refused to fall. At best they may have lost the loyalty of any subjects they once possessed; at worst they may have been punished.

There may be evidence for the latter. At the archaeological site of Colha located south-east of Calakmul, a 'skull-pit' was found dating to the end of the Classic period and containing the remains of 30 individuals whose lives came to a violent end, ten each of men, women and children.[45] We know that they had been of high social status because their teeth had been filed into points and inlaid with precious stones. Cut marks show that the skulls had been de-fleshed and there were traces of other post-mortem mutilation. Whether these members of the elite had been sacrificed or self-sacrificed in a desperate attempt to appease the gods and make it rain remains unknown.

Another so-called 'royal massacre' appears to have happened at the site of Cancuén at around AD 800.[46] Here a king, queen and twelve members of their retinue appear to have been slaughtered. The king was buried in a pit with full ceremonial dress and an identifying necklace stating he was Kan Maax, Holy Lord of Cancuén. The other burials were accompanied by fine clothes, spears, jewellery, jade objects, jaguar fang necklaces and seashells. Most tellingly, these bodies had been discarded into a cistern – perhaps symbolising the Holy Lord's failure to maintain it with water.

Whether or not the people murdered or sacrificed their Holy Lords, the drought conditions removed any incentive for them to attend the royal centres. They migrated to other regions or became permanently dispersed in small but more sustainable communities within the rainforest while the centres fell into decay.

In this scenario, we should expect there to have been not a single all-encompassing collapse but a mosaic of cultural change. Drought would have been of varying intensity across the landscape, its impact mediated by local ecological factors and the capacity of people to respond.

The Preclassic drought appears to have had a severe impact on El Mirador; its reservoirs filled with silt and remained so, its people being unable to respond to the wetter conditions that returned after AD 250. Conflict would not have been an unexpected consequence: the reservoir at Tamarindito had been built in the Late Classic period during a period of regional warfare, possibly to create more defensible water supply.[47]

Patrice Bonnafoux believes that the Preclassic drought left a long memory, one that explains the marked prevalence of water imagery in the iconography of the Early Classic period between AD 250 and 560. Water-related themes are also likely to have been dominant within the rituals of that period, which in turn would reflect a genuine concern of the Holy Lords to maintain a supply of clean water, knowing that that this was the crux of their power base.[48] Bonnafoux suggests that the fading presence of water images in the iconography of the Later Classic period reflects the fading memory of the Preclassic drought during the more auspicious times. But just as such cultural complacency had gained hold, along came the Classic-period droughts.

The three phases of the Classic collapse, as identified by Gill, were closely linked to the geology of the Yucatán. The western lowlands, where rainfall was the primary source of water, collapsed first around AD 810. Then the southern lowlands, where freshwater lagoons and rivers stored some surface water for a while, until AD 860. Finally, the central lowlands and northern Yucatán, where the water table was shallower and water could have been obtained from sunken lakes and deep *cenotes*, but even this collapsed around AD 910. There were exceptions. Altun Ha in the southern lowlands relied on surface reservoirs for its water and was wiped out by AD 900, whereas Lamanai, only 40 kilometres away, continued into the Postclassic, albeit with a smaller population: Lamanai was located near to a stable source of freshwater.

The Postclassic fluorescence in the northern lowlands may reflect the continuation of water supplies in that region from the multitude of *cenotes*. Certainly, Chichén Itzá became a major centre towards the end of the Classic period and maintained a population into the Postclassic. There is some evidence of a new cultural influence from central Mexico, there may have even been an invasion and takeover. But during a period of population crisis and of mass migration, in all directions, this could have been part of a natural transfer of cultural influences. Chichén Itzá was one of the last great centres of the Maya but it was surrounded by a vacuum and its power gradually faded. By around 1200, the nearby settlement of Mayapán, supported by several coastal trading communities, became the political and cultural centre of the region.

Beauty of the beast

I find the Maya is the most paradoxical of all ancient civilisations. On the one hand it was an absolute beast: swathes of rainforest were destroyed to construct mountains of stone; warfare was endemic, human sacrifice pervasive, self-mutilation revered. Competition between Holy Lords created an upward spiral of consumption, a system that was bound to ultimately fail, drought or no drought. On the other hand, via their monstrous acts, the Maya invented writing and mathematics and created some of the most remarkable art and architecture of any human culture. And most of all in their favour, the Holy Lords chose one of the most delicate of flowers as their royal symbol, the water lily, and frequently depicted it with such beauty – perhaps that can excuse the odd bit of rampant environmental degradation and gratuitous violence.

11

WATER POETRY IN THE SACRED VALLEY

Hydraulic engineering by the Incas, AD 1200–1572

The dawn-time anticipation at the entrance to Machu Picchu was palpable. Hundreds of people were queuing at the early hour, the majority having travelled thousands of miles to complete their archaeological pilgrimage at sunrise. Buses were arriving every few minutes, having driven round a multitude of hairpin bends from the town of Aguas Calientes several kilometres below, where the hordes of visitors had stayed overnight. Others emerged along tiny paths through the dense tropical foliage, after climbing the steep hill on foot. When the gates opened and the queue began to move, I had a momentary fear of entering some form of Inca-Disney Land. But once through the turnstiles the crowds soon dispersed; some people ran up the steps to gain their first sight of Machu Picchu, others puffed their way up the climb. One by one, the groups and lone travellers settled quietly on the side of the mountain waiting for the first rays of sunlight to shine through the ragged top of Mount Yanantin on to the Inca ruins. Everyone was silent. There was a wonderful moment of shared peace and stillness, everyone feeling privileged to be witnessing this scene, a moment of communion with the Inca past.

Machu Picchu is the icon of lost civilisations, Inca and otherwise (Photograph 41). Constructed in the early 15th century AD, it provided a retreat for the Inca king; it was the Balmoral of its day.[1] Perched on a precipitous mountain ridge, 73 kilometres northwest of the Inca capital at Cusco, it was never found by the Spanish conquistadors. Having escaped the conquistadors' love of wanton destruction, the buildings of Machu Picchu remain wonderfully preserved. When brought to world attention by Hiram Bingham in 1913, his photographs splashed across the pages of *National Geographic*, Machu Picchu was immediately recognised as an architectural masterpiece. It took almost another hundred years, however, to fully appreciate that Machu Picchu was also a masterpiece of hydraulic engineering.

Viewed from the gatehouse, the golden stone buildings of Machu Picchu capture the warmth of the sun and seem to merge into the contours of the high mountain ridges above the Urubamba River. The perfectly proportioned houses and temples are linked by long staircases and wide terraces, giving a sense of freedom of movement and calm. In all directions, the mountains provide a spectacular backdrop with the snow-capped peaks of the Andes glinting in the distance. The sound of water is ever-present, emanating partly from the series of spring water fountains in the citadel but also from the rushing river below.

The Inca achievement

At the beginning of the 16th century, the Inca Empire extended 14,000 kilometres along the coastal region of the Pacific Ocean from modern-day Ecuador and Colombia in the north, through Peru, Bolivia, Chile and Argentina to the south.[2] It encompassed an estimated ten million people. This vast area included the Atacama Desert to the west, the driest place on earth, and the towering Andean Mountains with their steep snow-clad peaks and deep gorges to the east.

Despite such challenging terrain, the Incas had been expanding their empire for more than 300 years before the Spanish Conquest in 1532. Their military, political and cultural achievements ultimately derived from their management of water. The Incas possessed an expertise in hydraulic engineering that enabled them to construct irrigation systems and flood-resistant terraces on steep mountainsides, securing their food supply. The control of water was at the root of their town planning, with the removal of unwanted water by drainage systems being as important as ensuring the clean-water supply for drinking and irrigation. But their water management extended beyond an exercise in hydraulic engineering: the Incas also understood and exploited the aesthetics of water. While the Angkor kings did so primarily through large expanses of still, silent water in moats and barays that reflected the heavens above, the Inca excelled in fountains and waterfalls, combining the sights and sounds of flowing water.

The Inca achievement was impressive but short-lived at just over 300 years. The reason for their demise is simple: military conquest by the Spanish. Had that not occurred, it seems likely that many of their irrigation systems would continue to function today; indeed many of them still do – with the help of a little restoration – while Machu Picchu remains standing despite centuries of rainfall that would have undermined the foundations had there not been such an elaborate drainage plan.

We must start by asking where the Incas came from and how they acquired their expertise in hydraulic engineering?

Irrigation and civilisation before the Incas

Some would argue that the Incas' greatest achievement was their ability to assimilate the knowledge of other cultures.[3] They were superlative plagiarists, and never more so than in their successful water management. Consequently we need to start by looking at the pre-Inca cultures of South America. That is quite challenging because there is a long and complicated sequence of these cultures, many with outstanding achievements that cannot be fully credited within this book. They most definitely should not be seen as mere precursors to the Incas.

People were living in South America from at least 13,000 years ago, having spread into the continent from the north, originally crossing the Bering Straits from Siberia into the Americas. Quebrada Jaguay on the southern coast of Peru is one of the earliest known settlements at between 11,000 and 9000 BC, showing that people were seafaring, fishing for shoals of drum fish and collecting clams before the end of the last ice age (the Pleistocene).[4]

How far people were penetrating inland at this early date remains unknown, but one site several thousand years later and located in the foothills of the Andes of northern Peru is of absolute importance for our concerns regarding the origin of water management. Excavations in 2005 in the Zuña Valley, 60 kilometres east of the Pacific coast (Figure 11.1), by Professor Tom Dillehay from Vanderbilt University and his colleagues found the earliest known evidence for irrigation in the whole of the Americas, North and South.[5] A sequence of four superimposed irrigation canals between one and four kilometres in length were discovered, the earliest of which dates to at least 3400 BC and most likely 4500 BC. This is more than 3,000 years earlier than the canals in the Tucson Basin that may have inspired the Hohokam (Chapter 9). The canals channelled water from small rivers on to agricultural plots, most likely growing peanuts, manioc, squash and beans, possibly before these had become fully domesticated strains. Dillehay describes the canals as forming an 'artificially wet agro-eco-system' within a community that was primarily supported by hunting and gathering. While they would have required substantial coopera-tion between households to construct and maintain, Dillehay sees no evidence for a centralised leadership having been required.

This discovery is of particular importance because there has been a long debate as to whether the first 'civilisation' in South America devel-oped on the coast because of abundant marine resources or inland dependent upon irrigation-based farming. The former theory devel-oped following the discovery and interpretation of the site of Aspero, located on the Peruvian coast at the mouth of the Supe Valley.[6] This is a complex of ceremonial buildings, plazas, terraces and large middens covering 32 hectares. It lacks any signs of agriculture and pottery – the latter appearing in Peru at 1800 BC. When Aspero was explored by the archaeologist Michael Moseley in the 1970s he dated it to between 3000 and 2400 BC and proposed that the economy was based entirely on the extraordinarily rich coastal resources, notably shoals of sar-dines, anchovies and shell fish. This led Moseley to propose his theory of the 'maritime foundations of Andean civilisation'.[7]

In the mid-1990s, however, the Peruvian archaeologist Ruth Shady Solís began excavating a substantially larger site known as Caral, located 23 kilometres inland within the desert environment of the Supe Valley (Figure 11.1).[8] This was an urban complex covering 65 hec-tares with six monumental platform mounds – pyramids – numerous smaller mounds, two sunken circular plazas, an array of residential architecture and other buildings – all the architectural signs of a strat-ified society with a social elite. Caral was dated to between 2627 and 1977 BC, overlapping with the Early Dynastic period of the Sumerian civilisation of Mesopotamia (Chapter 3). To sustain this size and complexity of settlement within the desert-like environment of Peru, substantial levels of irrigation would have been required.

Although Solís did not find and date canals at the site, she suggested that a nearby contemporary canal most likely follows the route of a Caral-period canal and pointed to the rich collection of plant remains at the site that are indicative of irrigation-based agriculture. These included squash, beans and cotton. The last is especially important because this was cultivated to provide fibres for making fishing nets used to harvest the abundant fish shoals on the coast. Indeed, the only evidence for animal protein at both Aspero and Caral are marine resources, principally anchovies, sardines and clams. One intriguing idea is that irrigation was principally developed in order to cultivate cotton to allow large-scale fishing, which in turn provided the economic base for the social elite and their monumental architecture. The archaeologist Jonathan Haas and his colleagues have pointed to a potential wider significance of cotton, suggesting that control over its production allowed an elite to prosper from cloth production, using this for clothing, bags and adornment, as well as the supply of fibre for fishing nets.[9]

Following the dating of Caral, numerous other sites with similar pyramids and plazas have been similarly dated within the Supe Valley and adjacent valleys of northern Peru; these have been grouped together as the Norte Chico civilisation, flourishing between 3000 and 1800 BC.[10] Whether this civilisation originated on the coast, supported by the rich marine resources, or inland where cotton could be cultivated on a sufficiently large scale, remains a matter of major debate. As such, Dillehay's discovery of the earliest-known irrigation canals at an inland rather than coast setting is of some interest.

Quite why the Norte Chico culture went into decline leading to the abandonment of Aspero, Caral and other sites by 1800 BC remains unclear. The irrigation technology may have been adopted by cultures elsewhere which then grew to eclipse the Norte Chico civilisation, or successive droughts and earthquakes may have made irrigation impossible within the valleys of northern Peru.[11] By 1500 BC, however, the next great Peruvian cultures had emerged: the Chavín, who were farming in the arid coastal belt, and the Tihuanaco culture, which thrived in the less dry but cold upland areas. Both were engaged in sophisticated water management.[12]

The Chavíns occupied and had influence over a vast area of what is now the central Peruvian coast and inland desert areas, reaching their peak between 900 and 200 BC.[13] They created a sophisticated irrigation system, diverting water from rivers to their fields and building drainage systems to deal with excessive rain on their settlements in the upland areas.

In the more mountainous areas, where precipitation was not such a problem, the people of the Tiahuanaco culture dug drainage canals to create raised fields and enhance the farming land around Lake Titicaca, which became the most developed between 300 BC and AD 300.[14] By the first century AD the Tiahuanaco were also building terraced fields to maximise the available land for farming and gain the greatest advantage from the warming effects of the sun. Both the drainage canals and terraces raised the ambient temperature of the fields, reduced the danger of frost and lengthened the growing season. The sun heated the water during the day and insulated the crops during the cold high-altitude nights. The canals between the fields were multi-purpose, containing edible fish to supplement the diet and producing canal sludge for fertiliser. In areas more distant from the lake, the Tihuanaco dug 'cochas', sunken fields to access the moist soil conditions just above the ground water table. This also created a micro-climate that protected crops from the frosts.

The Nasca culture had developed by the turn of the first century AD in the arid coastal areas of southern Peru.[15] This produced an elaborate irrigation network that included canals and 'puquios', underground aqueducts that tapped into sub-surface water.[16] These were used to irrigate fields and fill reservoirs, sustaining a large population and their little-understood but remarkable cultural achievements, notably the Nasca lines.

At the turn of the first millennium, the Moche culture began to flourish in the coastal areas of what is now northern Peru, reaching its peak around AD 550. This culture was characterised by delicate ceramics, spectacular gold work and monumental architecture. It was also a highly stratified society, practising human sacrifice. The Moche built vast irrigation canals and diverted rivers, engaging in the hydraulic engineering of their whole landscape.

Two further cultures should also be noted. The Wari occupied the central and coastal area of Peru between AD 500 and 1000 and were the direct precursors of the Inca. They developed irrigation systems, terraces and a road network that the Incas were later to extend across their empire. The Chimu who rose to prominence on the north coast of Peru around AD 900 inherited much of the canal system and reservoirs of the Moche. They tapped into ground water on the higher land by building 'walk-in wells' and their own version of sunken fields called 'hoyas', excavated to access sub-surface water and to create their own micro-climate.

In the years up to the beginning of the 13th century, most of these early cultures came to an end; the people either dispersed or were absorbed into the Inca Empire. Their legacy to the Incas was expertise and ambition in irrigation and drainage – a capacity to transform the hydrology of entire landscapes. This had been developed ever since the time of the Chavíns, and potentially before, and was now inherited by the Incas. Inherited and extended to a wonderful extreme, both in a practical engineering sense and by extolling the aesthetics of water management, creating fountains and flowing channels throughout their monumental buildings and towns.

To understand why the Incas were the ultimate beneficiaries of the Chavíns, Tihuanaco, Nasca, Chimu, Moche and Wari cultures, we need to consider the history of climate change.

Cultures, climate and catastrophe

The climate and terrain of South America have not changed substantially during the last four millennia and are integral to the course of its human history. No doubt it will always continue to be so. In addition to the hyper-aridity of the coastal zone and the extreme cold in the Andes, environmental challenges are recurrent in the form of the El Niño and La Niña phenomena.[17]

El Niño is the warming of the Pacific coastal surface waters off South America causing changes in temperature and wind direction. This brings about a sequence of rain and river deluges followed by scorching droughts. The La Niña phenomenon is a similar sequence of events caused by the cooling of ocean temperatures in the equatorial Pacific. Data from ice cores from the Quelccaya ice cap in Peru[18] and from sediment cores from Lake Titicaca[19] indicate that an El Niño event in AD 565 initiated 30 years of heavy rains and floods, devastating Moche and Nasca settlements. The Nascas compounded this catastrophe by their own deforestation. They were literally swept away by water, leaving their canals high and dry above the desert floor. Similarly, excavations of Moche sites have revealed evidence of floods and erosion due to excessive rain. The most intriguing archaeological find has been of bodies decapitated, de-fleshed and encased in river mud. These appear to have been the losers of ceremonial battles that actually took place in heavy rain, presumably because of its symbolic significance.[20]

Excessive rain was followed by 30 years of drought, evidenced by huge sand dunes that had swept into the Moche buildings and monumental structures, and buried their towns. Unlike the Nascas, the Moche survived these natural disasters by dispersing into smaller settlements, creating hill forts that were heavily fortified. Indeed, rather than environmental disaster, it was internal unrest and civil war that appears to have caused the demise of the Moche, although warfare is frequently the consequence of resource stress caused by either drought or flood.

Over the next few hundred years the climate stabilised and enjoyed above-average precipitation. But around AD 950 another period of increasingly dry weather began. The Wari Empire had survived the sixth-century floods and drought because of their terraced farming methods. The high-altitude Tihuanaco near Lake Titicaca had similarly survived, although they were now much diminished in numbers. However, both the Wari and the Tihuanaco suffered from the new decline in rainfall and occasional fierce droughts that began in AD 950 and lasted until 1300. The Wari were forced to abandon their towns such as Pikillacta in the Cusco valley, blocking up the doorways with stones with an apparent intention to return that was never realised.

As from around AD 1150 the climate began to ameliorate – higher temperatures and more rainfall – encouraging the farmers to settle higher in the valleys. They used the knowledge and experience of previous generations and cultures to create agricultural terraces and irrigation systems. These were the first of the Incas who set about creating settlements that were able to cope with droughts and floods in equal measure, taking hydraulic engineering to a completely new level. With their food supply secured, the Inca population grew rapidly, and soon an empire emerged.

Who were the Incas?

The name 'Inca' comes from the ancient Quechua word meaning 'ruler'. Their empire straddled several El Niño events, droughts, flooding, earthquakes and even a volcanic eruption. But their hydraulic engineering, among other achievements, enabled the Incas to provide security and stability for their people. A valley high in the Andean Mountains became the heartland of their empire – the so-called 'Sacred Valley' along which runs the Urubamba River. Cusco was chosen to be their capital, Cusco meaning 'navel' in Quechua, the centre of the Inca world (Figure 11.1).[21]

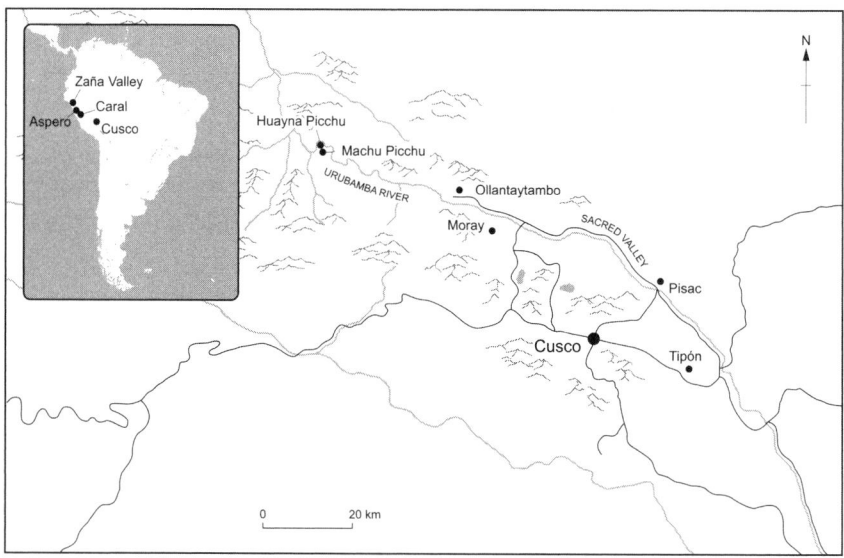

11.1 Pre-Inca and Inca sites of the Sacred Valley and Machu Picchu referred to in Chapter 11

Inca engineers created expertly constructed terraces for agriculture, complex irrigation systems and aqueducts for extraordinary water control. The hillsides of the Andes are still covered in fully functioning Inca terraces, while the Sacred Valley remains the most productive agricultural region in Peru. By terracing the hillsides, the Incas not only created suitable flat land for farming but also minimised the risks of landslides, protecting the soil from deluges of water cascading down the hillside in heavy rain. Although the sun rarely reached the lower valleys due to the shadows from the mountains, the terraces on the hillsides were able to capture the sun's warmth for the greatest extent of time and hence became especially productive.

The Incas effectively eliminated the risk of drought by diverting and controlling rivers. At Machu Picchu they were only able to site the dramatic settlement between the peaks of two mountain tops by redirecting a natural spring to provide drinking and domestic water. At Ollantaytambo, a hilltop settlement in the Sacred Valley, they directed water from a distant glacier in underground canals over a distance of several kilometres to the settlement. At Moray, also in the Sacred Valley they dug out and exaggerated natural depressions in the ground to create descending rings of perfectly concentric circlulan terraces. These provided a succession of micro-climates as one gets deeper, ideal for growing a wide variety of crops.

The Inca empire

With their supply of food and water secured, the Incas had the capacity to build a network of roads as well as ceremonial buildings and to become outstanding craftsmen in many materials, notably textiles and gold. They were also able to muster armies, extending their power over neighbouring populations. Their philosophy of government seems to have been coercion rather than conflict whenever possible. They often chose to move difficult inhabitants from outlying areas into the Inca strongholds, repopulating with loyal citizens from the centre of the empire.

Storehouses were built along the Inca trails which the local people were instructed to fill with food, acting as reserves for times of shortage and a supply of food for travelling troops. Because of the steep terrain, the trails stretched up and down hillsides with steps, tunnels and bridges. Rope bridges made of spun grass stretched from stone abutments across wide and fast-moving rivers. There were regular security posts dotted along the routes; in certain places wooden plank bridges were stretched across ravines to be removed at times of emergency. The Inca had no need of wheeled vehicles as the landscape would have been impassable. In fact the Inca never invented the wheel nor iron tools, relying on llamas to carry heavy loads and stone artefacts.

Curiously for such an advanced empire there was neither a written language nor any form of hieroglyphics. The Incas invented what seems to us like a computer code made from coloured and knotted cords known as 'quipus', looking something like a fringe. How this system worked and what information could have been transmitted remains largely unknown but the modern Quechua still weave information and messages into their cloth using traditional patterns and symbols. Information and messages were rapidly sent along the trails with a system of relay runners. Architectural plans were communicated by clay and stone maquettes, while stonemasons were most likely to have used their own body parts as units of measurement.

The Inca had a complex belief system that involved the worship of multiple gods: Inti, the sun; Pachamama, the earth; Mama Cocha, sea goddess; Mama Quilla, moon goddess; Apu Illapu, the raingiver, to name just a few. They had an elaborate calendar of rituals, including the offering of animal and child sacrifices, a practice that was deeply embedded within the cultures of the Andean region. The Inca belief in the afterlife extended to mummification and burial with accompanying ceramic jars of food and clothing.

Deceased Inca emperors were not buried. They were dressed, fed, entertained and consulted for advice at ceremonies. They were considered as powerful 'huacas', religious symbols that also included carved stones and natural phenomena such as mountains and rivers imbued with meaning. After the fall of the Inca Empire, the remaining Inca people tried to conceal the mummified remains of their ancient emperors. The majority were eventually found and confiscated by the Spanish conquistadors. They were exhibited in Lima, where only Europeans could see them disintegrating in the damp climate and then buried without ceremony.

The rule of the Inca Empire usually passed through the generations from father to son. When an Inca king died he still ruled his land so the new king had to build a new city palace and country home. In the mid-1400s, however, there was an uprising by a powerful community to the south called the Cancha. The emperor, Inca Viracocha, fled in panic and the Inca were nearly defeated. His son, Inca Yupanqui, rallied the Inca armies and defeated the Cancha. He then deposed his father and took the name Inca Pachacuti or Pachacutec, meaning 'he who shakes the earth' or 'cataclysm' (Photograph 42).

Pachacuti's reign marked the beginning of a series of invasions to the north and the south, vastly extending the Inca empire. It was Pachacuti who established Cusco as the Inca capital and commissioned temples, fortresses and palaces in the distinctive Inca style, including the 'lost city' of Machu Picchu. Pachacuti claimed to be the descendant of the Inca sun god, Inti, turning himself into a cult figure, reputedly demanding new clothes every day and gold plates to eat off.

After his death Pachacuti's military success was continued by his son, Topa Inca Yupanqui and in the next generation by Huayna Capac. By this time the Inca Empire had reached its greatest extent but was becoming difficult to manage. Huayna Capac and his son both died in 1527 of a disease that was likely to have been smallpox. This was spreading rapidly among the indigenous population having been inadvertently introduced by Christopher Columbus and the Spanish conquistadors. In the confusion about who was to succeed to the Inca throne a brutal civil war developed between the two other sons of Huayna Capac. This weakened the Inca army and although the successful brother, Atahuallpa, ruled for several years, he was captured by the Spanish leader, Francisco Pizarro. Atahuallpa was imprisoned for eight months and then killed in 1533.

The Inca Empire continued to resist Spanish influence, but the population became devastated by disease. The Spanish were brutal, destroying Inca cities, melting their gold works and enslaving the population. Eventually the Christian Spanish forbade the worship of 'false gods', and prohibited all expression of native religion and culture. By 1572 the remaining Inca strongholds had been subdued and most of the Inca cities had been looted and destroyed. But not Machu Picchu.

Bingham's rediscovery of Machu Picchu

Machu Picchu, famous to travellers around the world as the 'lost city' of the Incas, is now a UNESCO World Heritage site visited by thousands of tourists each day. It perches high above the Urubamba valley to the west of Cusco as a magnificent reminder of the brilliance of town planning at the height of the Inca Empire. It was primarily the citadel's geographical position that saved it from the Spanish and left it to become overgrown and forgotten. But the devoted Inca population also played a role, keeping its location secret until their dying day – most likely an unpleasant death from smallpox.

The rediscovery of Machu Picchu has long been attributed to Hiram Bingham III in 1911. He was the 35-year-old son of a once wealthy family of missionaries from Hawaii, educated in Connecticut, and then taught Latin American history at Yale University.[22] Bingham was driven by a need to achieve fame and fortune and was soon to describe himself as an 'explorer'. He married into a wealthy family and became increasingly interested in travelling, particularly in South America – perhaps partly an escape from a challenging home life after having fathered seven sons!

After several short visits to Peru, he organised an expedition in 1911 with the aim of being the first to climb Mount Coropuna and also to map the 73rd meridian in the process. His team of seven academics and like-minded explorers was accompanied by a far larger group of local sherpas and mules to carry the supplies: folding beds, tents, boxes of food and other 'essential' luxuries:

to be sure … some of the younger men may feel that their reputations as explorers are likely to be damaged if it is known that strawberry jam, sweet chocolate and pickles are frequently found on the menu.

During his journey to find the best way through the mountainous terrain, Hiram Bingham came upon a local farmer called Melchor Arteaga, who pointed out the mountain of Machu Picchu, meaning 'old mountain' in Quechua, and assured him that there were ruins on the summit. Almost as a distraction from his real purpose, and leaving his colleagues to relax in the camp, Hiram set off for a strenuous afternoon's hike. For the last part of the ascent, Hiram was guided by an eleven-year-old Quechua boy, Pablito Alvarez, who led him to the ruins, among which two families were living and farming. Thirty years later Bingham describes the moment of discovery in romantic terms:

Presently we found ourselves in the midst of a tropical forest, beneath the shade of whose trees we could make out a maze of ancient walls, the ruins of buildings made of blocks of granite, some of which were beautifully fitted together in the most refined style of Inca architecture.

In actual fact, it seems that Bingham had a very quick look around, found the names of three Cusco explorers written in charcoal on one of the stones and descended fairly rapidly. It was only several months later, at the end of the expedition, that he sent two of his colleagues back to Machu Picchu to clear vegetation and record the ruins.

When back at Yale, Bingham sought to use the outcome of his expedition to secure international acclaim. His first attempt was based on the discovery of bones that he initially believed to be of ancient human ancestors but which turned out to be of recent date. Next he poured his energies into an account of his ascent of Mount Coropuna but since this turned out to be lower than he had hoped there was little public or media interest. It was only later, and after further investigations at Machu Picchu in 1912, that Hiram Bingham began to appreciate the potential of what he had found. In 1913 he persuaded *National Geographic* to devote a whole issue of their magazine to his discoveries which he claimed were 2,000 years old and the ruins of an ancient Inca capital. Both assertions were incorrect.

Although Hiram Bingham is widely attributed with having discovered Machu Picchu, in reality the ancient city had merely been forgotten rather than lost. Indeed, local people had continued to use its agricultural terraces and the ruins for shelter ever since the Inca abandonment. In addition to the three Cusco explorers of 1901, a German man is rumoured to have systematically looted Machu Picchu for objects of gold and relics in the 1890s, using the cover of a saw mill in the valley below. This may explain why only one piece of gold, a bracelet, has been found during archaeological excavations at Machu Picchu. It was discovered at the base of a drainage wall as if it had been purposely buried there as a totem, fortunately ensuring that it was missed by the looters.

None of this, however, should detract from the fact that Hiram Bingham was the first to recognise the significance of Machu Picchu and that he devoted many years to the meticulous recording of the site and the study of the Inca. Machu Picchu undoubtedly propelled him to fame and fortune, gained him a professorship and provided a rich source of academic and creative material to which he returned at various times throughout his life. But Bingham did not rest on his laurels: he joined the Air Service in the First World War, later writing a book entitled *An Explorer in the Air Service*. He then joined the Republican Party, entered politics and became a Senator. It was only later on that he set about writing an account of his life in the army and then finally of his discovery of Machu Picchu, *The Lost City of the Incas*, which was first published in 1952.

Whatever the merits of his original motivation, Hiram Bingham's enthusiastic and meticulous work at Machu Picchu exposed the site to the world. It has been the subject of academic study ever since and a spectacular tourist destination. Also the subject of controversy. Forty thousand artefacts that Bingham excavated from Machu Picchu, including bronze and silver mirrors, ceramic vessels, pendants and bone pins, are still at Yale University where they were shipped for 'safe keeping' after his excavations. Negotiations continue, 100 years after their removal, to facilitate the return of a mere ten per cent of the objects.

The Central Plaza as a great drain

Machu Picchu occupies a spectacular position 2,438 metres high on a mountain ridge stretching between two geological faults near the tops of the mountains of Machu Picchu itself and Huayna Picchu.[23] The citadel is strategically located and readily defended. The Urubamba River loops round on three sides, 450 metres below, while access to the routes from Cusco was protected by rope and tree trunk bridges over deep precipices.

Although it is situated on a high ridge, Machu Picchu is 1,000 metres lower than Cusco and consequently much warmer, making it a beautiful, secure and pleasant place for Emperor Pachacuti to spend his leisure time. The climate in the 15th century would have been much the same as today with a dry winter season from May to August and a wet summer season from October to March; the temperature was mild with no frosts. Although the site could only have been chosen carefully, it presented numerous complex engineering challenges for the Inca builders. Considerable structural alteration of the rocky valley between the peaks was required before residential building could even begin.

The central plaza of the finished citadel presents itself as a large flat open grassy area the size of several football pitches. This would have been ideal for markets, meetings, and as an arena for spectacles, surrounded by terraces, staircases, buildings and platforms. In fact its underlying purpose is far more practical because the plaza provides a drainage basin, constructed from layers of large stones and gravel, secretly collecting all the water draining from the surrounding structures and guiding this to the Urubamba River below. Fertile soil was brought from nearby mountains or the flood plain in the valley to bury the stones and gravel, creating a perfectly grassed plaza. Despite the annual deluges of water the plaza has drained from the site for 600 years, there has been no subsidence; there is nothing on the surface to indicate that the plaza is anything other than merely an open space.

Beyond the plaza there is a complex of agricultural terraces around the sides of the two mountains, some no more than a few metres wide (Photograph 43). These had been intensively cultivated with no need for irrigation because of the reliable rainfall; the soil has remained in place for more than half a millennium. Hiram Bingham's first impression in 1911 that 'the Incas were good engineers' was something of an understatement.

The 172 buildings of Machu Picchu were skilfully integrated into the Inca-made terraces and linked by over 100 stone-cut flights of stairs. The layout was carefully organised with an agricultural zone and one for the residences of the Inca nobility and temples. The buildings used the classic Inca style of polished drystone walling known as 'ashlar', a technique in which stone blocks were cut to fit together tightly with no mortar. This made them more resistant to earthquakes as the walls could move and then resettle without causing damage. Another Inca feature designed to limit the effect of earthquakes is the trapezoid shape of the windows and doors. Even the walls sloped inwards to create greater stability and L-shaped stone blocks were used at the outer corners to tie each structure together. As a result of the high-quality building methods, the temples, sanctuaries and even the smaller houses have remained intact for nearly 600 years with very little damage, apart from the loss of their thatched roofs.

All of the buildings were rectangular and conformed to Inca principles of symmetry and order, except for the so-called 'Temple of the Sun' which is circular and built around a natural outcrop of stone, possibly of ritual significance. At the winter solstice the sun enters the temple's central window and shines directly on the stone floor. Fountains of spring water cascaded past the temple areas and were accessible to all for drinking water. The whole citadel was served by a network of drainage canals and ducts, keeping the city clean and dry in the heaviest of rains.

Machu Picchu had usually housed around 300 people, a community which maintained the citadel in preparation for the arrival of Pachacuti and his large retinue. Their skeletal remains have been found in cemeteries, caves and niches in the surrounding hills.[24] Building work was still underway when the citadel fell into disuse around 1540; large megaliths were abandoned with their rolling stones still in place beneath.

As one walks around the site today, it is impossible not to be drawn to the landscape beyond the citadel, a landscape that would have been imbued with meaning for the Inca. One is encircled by panoramic views of mountain peaks, their slopes covered with tropical forest, while the air is filled with the sound of the rushing Urubamba River in the gorge below. Such an engagement with the natural world was the deliberate intention of the Inca: they had shaped standing stones within the citadel to precisely mimic the shapes of the mountain tops behind and created their fountains to do the same for the river.

The fountains of Machu Picchu

Our understanding of Machu Picchu as an exemplary feat of hydraulic engineering is thanks to the work of Kenneth R. Wright. He first visited Machu Picchu in 1974 with his wife, Ruth, returning in 1994 to begin an intensive study of how the Inca hydraulic system worked. He had established his own highly successful consultancy firm, Wright Water Engineers, and had become increasingly interested in paleohydrology: the study of water systems created by ancient cultures. Wright confesses that 'what began as a modern-day engineer's curiosity about prehistoric water resources management quickly became a fixation'. In 1994 Wright and a team which included his wife and the Peruvian anthropologist Alfredo Valencia Zegarra set about the study of the water systems at Machu Picchu and other Inca sites. Ruth Wright wrote a seminal guide book for Machu Picchu with meticulous descriptions of each building indispensable for any visitor. In 2008 they were both elected honorary professors at Peru's Universidad Nacional de San Antonio Abad del Cusco.

Although Machu Picchu was well served by sufficient rainfall to sustain an agricultural system, its spectacular and defensive position also benefited from an abundant perennial spring at a geological fault on the side of the mountain of Machu Picchu itself. The Inca engineers built a highly effective collecting system at the spring and directed the water to the citadel by way of a 749-metre channel (Photograph 44). This narrow aqueduct (10–12 centimetres across and 12–13 centimetres deep) was built from tightly fitting stones and sealed with clay to minimise seepage to less than 10 per cent. The aqueduct ran along a substantial stone terrace, past a large area of agricultural terraces, to the highest point of the urban settlement. The gradient is often only 1 per cent and the water leads to the first fountain at the Inca ruler's residence, which is situated at the highest possible place to receive water by gravity flow from the spring.

The water then flows through the area of Machu Picchu that is designated as royal or religious. It feeds sixteen fountains, positioned adjacent to a long staircase, and numbered by Ruth Wright from top to bottom. Each fountain is uniquely designed but is typically positioned within a roofless chamber that provides room to arrange containers for collecting water while also allowing some privacy, perhaps for washing or performing rites (Photograph 45). The water flow is cleverly designed so that it produces a steady stream that could more easily fill the narrow opening of an 'aryballo' – the Inca water container. There is usually at least one trapezoid niche built into the wall of each fountain, frequently still used by the Quechua to position offerings. The water system was designed to function with only a small flow of water but the Inca engineers provided places where excess water could overflow into drains.

The fountain system at Machu Picchu provided domestic and drinking water. Several of the fountains seem to have been designed for ceremonial use, such as fountain number 3, which has four niches for ceremonial objects and is positioned near the Temple of the Sun.

This is the only fountain that the water can bypass via an underground channel from fountain 2 to fountain 4. All appear to have been for public use except for the last one, fountain 16, which has a significantly higher wall. This last fountain is only accessible from the building complex known as the Temple of the Condor, so-called due to the central room, which has a carved condor head and ruff in its floor with giant wings of bedrock towering above it.

The sixteen fountains at Machu Picchu are again flowing with water thanks to the work of Kenneth Wright – and a few pieces of plastic pipe as repairs. The sight and sound of flowing water goes some way to re-create the atmosphere of Ancient Machu Picchu and restore a sense of Inca power by the control of water in a challenging environment. In 1998 a fountain was cleaned and restored lower down the slopes. When water suddenly appeared, as if by magic, the Quechua workmen gathered for a traditional Inca thanksgiving prayer (here translated):

I call to the spirits of the gods of Machu Picchu, Putucusi, Intipunka and Mandor. Here Pachamama-Pacha earth, beautiful mother, do not let the fountains go dry; every year, water must flow forth so that we can drink.

The Inca Water Garden at Tipón

Machu Picchu has become the best known Inca site because of its highly dramatic position and expertly crafted buildings, but there are many other spectacular sites, all having impressive hydraulic engineering. Tipón, located 18 kilometres east of Cusco, equals Machu Picchu in the ingenuity and technical skill of its water works and has rightfully become known as the Inca Water Garden. Here, too, water still flows along the Inca-built channels.[25]

Perched in a narrow valley high above the main road out of Cusco, Tipón is not an easy place to reach, even today. My taxi wound its way through small villages and along precipitous roads, eventually arriving at a small car park with limited visitor facilities. As you follow a small path upwards the first sign of human habitation comes in the form of a beautiful Inca wall and a vertical fall of water appearing as if from nowhere like a sculpture. Climb a little higher, and the view suddenly opens out on to a wide valley of terraces, with water descending all around in a delightful series of vertical waterfalls, carefully built into the retaining walls of the terraces. The gentle sound of cascading water pervades the air and yet the stillness is undisturbed by a multitude of narrow canals flowing with water unseen along the terraces.

The main area at Tipón is a series of 13 wide terraces, covering a distance of 400 metres and descending 50 metres down the hillside, providing ample fertile agricultural land for the Inca settlement (Figure 11.2, Photographs 46 and 47). Tipón had several centuries of occupation before being developed into an impressive imperial Inca estate at around 1400. It was built for Inca Viracocha, the father of Inca Pachacuti, and is defended on three sides by rivers and cliffs and on the other by Mount Tipón, over which there is a high wall originally built by the Wari.

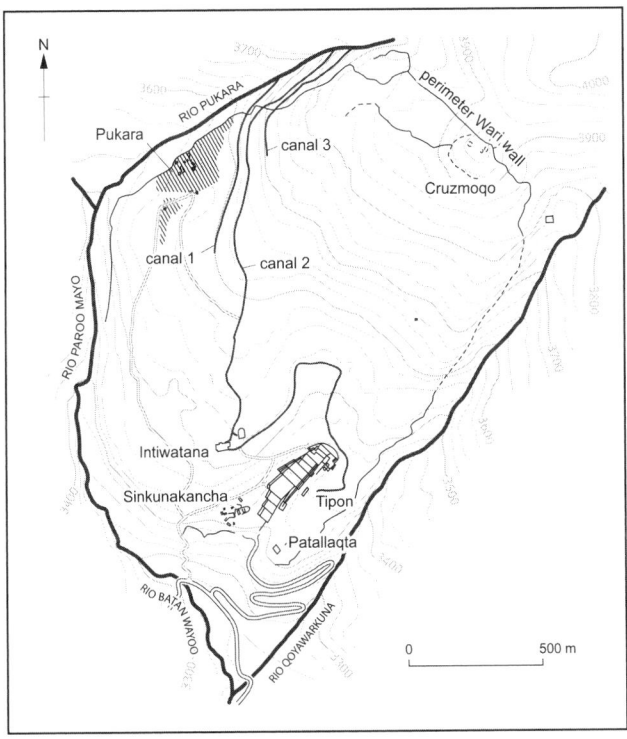

11.2 Terraces and canals at Tipón (after Wright 2006)

Tipón was cleverly designed to celebrate and maximise the potential of water in a breathtaking series of canals, aqueducts, fountains, vertical water drops and underground conduits. The invigorating sound of moving water must have provided a musical accompaniment to any activities at the site – Kenneth Wright has described it as 'hydraulic poetry'. But this was poetry with a purpose. The controlled rush of flowing and falling water was a demonstration of the power of the Inca rulers over both land and water.

Hydraulic poetry

Tipón is located at an elevation of almost 4,000 metres on the south face of a steep incline. It is subject to hard frosts, low temperatures and hardly any rain from May to August. The location was protected from drought by a perennial spring which issues on to the third of the thirteen terraces (counting down from the top) and provided water for drinking and irrigation. The Inca engineers supplemented this supply by redirecting water from the River Pukara, one kilometre away to the north-west. This flowed along an aqueduct that passed over valleys on stone-built bridges before descending 400 metres to irrigate the top two terraces at Tipón and then supplement the spring water on the main terraces.

The Tipón spring was clearly one of the main attractions of the site when the Incas began to develop the area as an Imperial estate. However, there was originally very little available land for farming in the valley. In order to overcome this problem the ambitious Inca engineers set about completely redesigning the landscape. First they filled in a deep ravine in order to create the thirteen main terraces. To do this they built a substructure of retaining walls and foundations for drainage before transporting soil from the top of an adjacent hill, which in turn created further flat land for agriculture. The expertise employed in this task was such that even after 600 years there has been minimal erosion and the terraces were farmed until recently.

At an elevation of 3,800 metres, the spring collected water from a catchment of over 153 acres. The Inca engineers sought to optimise the amount of water by building at least eight stone conduits into the hillside to channel the underground water towards the main spring.

The sub-surface flow was then directed into a head-work which concentrated the water into a single stone-built channel issuing on to the third terrace (down from the top, Photograph 48). As if to celebrate the magical arrival of the water, the Inca crafted an intricate design whereby the flow was divided into two narrow channels from which the water fell into a single basin; from here, four more channels took the water over another edge into a second basin; a third waterfall took the water into the irrigation network for the lower terraces.

The water flowed by gravity from one terrace to the next in a series of highly attractive and functional drops – waterfalls. The inset channel at the base of each fall was expertly designed to minimise the amount of splash, while the sight and sound of falling water must have been impressive. Just above the eighth terrace there was another ornate double waterfall generally referred to as the Ceremonial Fountain. Other waterfalls/fountains were carefully designed to allow for the filling of an aryballo.

Access between the terraces was mostly by 'flying stairs', steps that protruded from the terrace walls in an attractive symmetrical arrangement. In addition to the superb water irrigation, the terraces were arranged on a south-facing hillside where the retaining walls of brown andesite rock served to absorb the sun's heat and ward off overnight frosts.

The community of Tipón

Tipón had buildings for a royal residence located close to the spring, providing access to the purest water, a farming community known as Patallaqta below the terraces and a military garrison known as Sinkunakancha on a promontory of the hillside. With regard to water use, the most intriguing buildings are known as the Intiwatana located above and to the north-west of the terraces.

The Intiwatana appears to have been the focal point of religious or ceremonial activity. It had a horseshoe-shaped plaza and a pyramid-type structure of jagged rocks arranged on to protruding bedrock. The aqueduct from Pukara brought water through the middle of this complex, its flow having been artfully incorporated into its design. Regrettably, the aqueduct is now dry but the water had once entered a throne room via a waterfall and then continued in an open channel through other rooms of the complex. Two other stone thrones, adjacent to the pyramid, had long, shallow grooves carved into stone blocks at their base, suggesting the use of liquid, most probably water, for ritual activities – perhaps for quenching the thirst of mummified bodies seated on the thrones. More generally, drainage ducts throughout the complex indicate that considerable quantities of water had been an integral part of the ceremonial activities – the sound of it flowing would have been pervasive.

Finally, high above the terraces and perched on the top of Mount Tipón is the Cruzmoqo, an Inca 'huaca' or revered site. This inaccessible and spiritual location has a panoramic view over the Tipón community and consists of two volcanic outcrops of rock joined by a terrace wall. The summit has an enigmatic metre-long depression carved into the stone. This is often full of water and would have provided a mini reservoir for the Inca watchmen who may have been stationed there. Alternatively, the stone basin could have been used for ceremonial purposes. This atmospheric mountain top is also scattered with numerous petroglyphs in the form of spirals and arrows which date back 5,000 years, implying that this had been a special place for many people before the Incas.

Past and present in the Sacred Valley

My visit to Peru took me to several Inca settlements other than Tipón and Machu Picchu. I went to Pisac, located 30 kilometres from Cusco in the Sacred Valley. The modern town and cultivated fields are in the valley bottom, but the Incas chose to situate themselves high on a mountainside. Its slopes were intensively terraced with especially fine Inca domestic architecture and temples constructed close to the summit. This provided an ideal vantage point to look north-west far along the Sacred Valley, for both the Incas of 600 years ago and for me when resting after the climb (Photograph 49). The opposite direction looks across a deep gorge to another mountainside, which was peppered with what looked like animal burrows but were Inca burials.

I followed the Sacred Valley for another 20 kilometres to reach Ollantaytambo, the starting point of the Inca Trail to Machu Picchu. Here there was another magnificent Inca settlement, which had been one of the last places of Inca resistance to the conquistadors. A steep staircase of sixteen terraces led me to a temple on the summit with huge stone monoliths that had been transported from across the valley. From there I could see snow-capped peaks of the distant high Andes. It was from these glaciers that water had been carried via aqueducts to Ollantaytambo and Pisac, both of which were laced with canals, irrigation channels and fountains.

I had been impressed by all of the Inca settlements I visited during my time in Peru. They had been remarkable for not only their hydraulic engineering but also their standing architecture and the stories I learned about the rise and fall of the Inca Empire. The Inca ideology, craftworks, quipu system and achievements of Pachacuti were all worthy of study for their own sake, to quench our thirst for knowledge of the past.

But the sight of the snow-capped mountains from Ollantaytambo reminded me that the past may also provide us with lessons for the future: present-day Peru is facing a water crisis, caused partly by the retreat of those same glaciers that had once quenched the Inca thirst.

12

AN UNQUENCHED THIRST

For water and for knowledge of the past

Quelccaya glacier, located in southern Peru, is the world's largest tropical ice cap. It provided us with the evidence for climate change during the sixth century AD that is likely to have been fatal for the Moche and Nasca cultures of Peru.[1] Such evidence was derived from cores cut from deep within its ice which required expert and meticulous scientific analysis. Quelccaya's evidence about present-day climate change is more forthcoming: one can simply watch it melt. The glacier is retreating at 200 feet per year, ten times the rate of its retreat in the 1960s, a consequence of an average temperature increase of 0.15 degrees Celsius per decade since the 1950s.[2] Other glaciers in Peru are retreating even more quickly, leading scientists to predict that all glaciers below 5,500 metres will have disappeared completely by 2015.

The consequences for water supply are severe. The glaciers have always acted as natural water towers, accumulating snowfall during the cold season and then releasing it as run-off into rivers during the warm dry months between May and September. The Incas depended upon such melt waters, either by tapping a source close to the glacier itself, as with the aqueduct to Ollantaytambo, or by drawing from rivers that were fed by melt water, such as the River Pukara at Tipón.

Modern-day Peruvians are also dependent on such melt water – for their drinking water, for irrigation and to power their hydroelectric plants. Lima's ten million people depend on water from the Rimac, Chillón and Lurín rivers. The melt water is the only source for these rivers during the dry season. With the steady and potentially increasing rate of glacier retreat such water will soon entirely disappear. Quite how Lima will cope remains unclear; one plan was to drive a tunnel through the Andes to bring water directly from the mountains into reservoirs. That was shelved as being too expensive for the government ($98m). Another is to build desalination plants; these are also expensive, requiring substantial private investment. Their running costs are likely to make the water produced prohibitively expensive for use as drinking water.

Shortage of clean drinking water is one problem; the scarcity of water for irrigation is another. This is also dependent upon the glacier melt waters; their predicted loss threatens the agricultural-export industry that is critical to Peru's economy.

Peru's looming water crisis comes not only from the retreat of the glaciers. Rainfall has also been steadily decreasing, causing many springs to become dry, while the water supplies within Lima are thought unlikely to cope with two consecutive years of drought. Moreover, in 2007 Lima's state-owned water distribution utility (SEDAPAL) was reported as having 43 per cent of its water unaccounted for, two thirds of that being lost to leakage and the remainder consumed but not billed.

We can add to this catalogue of challenges the extensive pollution of water sources by heavy metals from mining operations: 75 per cent of Peru's rivers are contaminated by heavy metals. This further threatens the agricultural sector because countries will refuse to receive Peru's exports if they fear that the irrigation water has been contaminated – Japan refused vegetable imports from Peru in 2007. Such water pollution has led to many conflicts between local communities and the mining companies, some of which have been violent; local communities have been turned against each other.

Peru's water crisis threatens its food security, economy and human health; it is the cause of social unrest. One's initial reaction is to contrast the inefficiencies of the leaking water system in Lima with the meticulous attention the Incas paid to preserving every drop of water within their skilfully made aqueducts at Machu Picchu, Tipón and elsewhere; equally one is drawn to contrast the reverence that the Incas appear to have placed on the purity of water, with the cavalier attitude towards pollution of the modern-day water suppliers.

But we must be cautious. When we visit Machu Picchu and Tipón we are visiting the residences of kings and nobles; however water-challenged modern day Peru might become, I suspect that high-quality water will still flow freely within the governmental palace in Lima. Indeed, a visitor to the city would have the impression that water is abundant: in 2007 Lima opened its 'Magical Water Circuit' tourist attraction – a park with thirteen fountains, one of which is the highest in the world (at 240 feet). My guidebook describes the park as 'showcasing how water can be artistically manipulated through form, movement and light'.

Partial and biased

This takes us to one of the most obvious weaknesses with an archaeological study of ancient water management – the evidence is partial and biased. It most frequently tells us about the upper classes and only rarely about the day-to-day life for the population as a whole. Did the Inca farmers, craftsmen and soldiers have a secure water supply? The Minoan queen of Knossos may have had her own flush toilet, but did anyone else? In light of the extravagance of the fountains in water-stressed Lima today, one is tempted to think that the Inca water displays may also be masking a reality of widespread water stress for the majority of the population – and maybe the same applies to those in Nabataean Petra. As such, we must be cautious with any conclusions about the past and even more so regarding its lessons for the present.

We must certainly be careful about identifying when and where water management techniques were first invented. The Neolithic record from the Levant appears to suggest that water management did not appear until towards the end of the Neolithic period. But despite more than 100 years of archaeological research, it has only been during the last two decades that the wells from Cyprus, Sha'ar Hagolan and Atlit-Yam have been found, and less than a decade since Sumio Fujii made his discovery of a Neolithic 'barrage' in the Jafr Basin. Who knows what else might be discovered in the near future? We currently have Choga Mami (*c.* 6000-4500 BC), Las Capas (*c.* 1250 BC) and the Zuña Valley (*c.* 4500 BC) as the earliest examples of canal building in Mesopotamia, the south-west United States and South America respectively. How long will this remain the case? I suspect that the status of Choga Mami will soon be challenged as a new era of archaeological research is now set to begin within Iraq. A dilemma we face is that the earliest examples of water management are unlikely to have left any archaeological trace at all – temporary dams of clay and channels cut through silt that were washed away soon after they had completed their work.

Nevertheless, archaeologists must continue to search for the earliest evidence and write histories of origins and inventions – it is part of their academic mission to do so. Indeed, while the evidence about water management in ancient civilisations may be partial and biased, its recovery and interpretation have been one of the great achievements of archaeology. Exciting new discoveries continue to be made, including the source of Rome's Aqua Traiana, discovered in January 2010 below a pig pasture 40 kilometres north-west of the city,[3] and the ceramic-lined Maya reservoir at Uxul, announced in August 2010.[4]

Such evidence not only identifies water management as a key driver of human history but often reveals skills in hydraulic engineering that match those seen in the art and architecture of the ancient world. This enhances our understanding of the past and enriches our visits to archaeological sites, whether as tourists or researchers. So if you wish to appreciate the Nabataean achievement when visiting Petra, ensure that you pay attention to the discreetly hidden water channels and drains as well as the ostentatious rock-cut tombs. When in Rome, leave the city and go to the Parco degli Acquedotti and if in Arizona search out the Hohokam equivalent at the Park of the Canals in Phoenix. Similarly, when in Constantinople, don't just visit the Hagia Sophia but also enjoy the Basilica Cistern and Aetius reservoir; if you really wish to appreciate the Byzantine achievement take a trip to the Kurşunlugerme.

The anonymous hydraulic engineers who designed and built such structures are unquestionably the heroes of this book. So too are the many thousands of people who laboured to build the aqueducts, dams and reservoirs. The death toll in doing so must have been substantial, something that we can only appreciate from historical accounts such as that of the Hoover Dam. Just as we must acknowledge past heroes, we must also recognise the accomplishments of those of the present day: the archaeologists who have found, surveyed and excavated the evidence for ancient water management. They have often worked in enormously challenging environments for extended periods of time. For these heroes we can add some names.

While it is always questionable to pick out certain individuals over others, my personal champions are Robert Adams in Mesopotamia, John Oleson at Humayma and Bernard Philippe Groslier at Angkor. Not far behind, I would add Eberhard Zangger's geomorphological study of the Argive Plain, Sumio Fujii's excavations in the arid Jafr Basin and Ray Matheny's survey at Edzná. Hiram Bingham receives the acclaim for having (re)discovered Machu Picchu, but my vote goes to Kenneth Wright, who exposed its hydraulic engineering, as well as that of Tipón. And no one can fail to recognise the truly remarkable contribution of Joseph Needham to our understanding of water management in Ancient China, along with his precursor, Sima Qian.

China is arguably the region in greatest need of further survey and excavation to understand the role of water management in the ancient world and how this arose within its Neolithic communities. But there remains a great deal to learn in all regions: indeed, a key conclusion of my study is not how much we know about water management in the past but how little we know, and the need for extensive programmes of archaeological fieldwork using both traditional foot-slogging methods and new approaches. Remote sensing, including imagery derived from the space shuttle missions, has already made a contribution to revealing the canal systems of Mesopotamia and Angkor. This has a great deal more to offer, while the ongoing refinement of chronological sequences by the further application of radio-carbon dating remains a priority. Such dating is essential to relate cultural developments to those of environmental change, especially as derived from tree rings that can reconstruct the history of floods and droughts in such detail, as we have seen with the Hohokam and at Angkor.

Archaeological evidence does not speak for itself. It requires interpretation and then repeated reinterpretation as new evidence is recovered and our concerns evolve. We have seen the nature of those debates, such as whether the Sumerians caused their own downfall by excessive irrigation without adequate drainage, the function of the kouloures on Crete, and whether the Angkor kings were concerned with irrigation, flood control or creating their own image of heaven on earth. We have touched on other debates, such as whether irrigation in pre-Inca Peru originated in coastal regions or within the foothills of the Andes and whether the twists and turns of the siq at Ba'ja had once provided a sequence of naturally made water cisterns.

Such debates are often enlivened because we lack the aid of documentary evidence. We have no written records to help us understand the kouloures on Crete, the barays of Angkor or innumerable other examples of hydraulic engineering in the ancient world. As we have seen, however, written records rarely, if ever, provide any easy answers. The cuneiform inscriptions from Mesopotamia are open to quite contradictory interpretations while many documents were written as deliberate works of propaganda. Perhaps the one exception is the *De aqua ductu* written by Julius Sextus Frontinus in Rome in AD 97. How would we have been able to interpret the aqueducts of Rome without his treatise? Imagine how we might reinterpret the canals of Mesopotamia, Angkor, Edzná and elsewhere if we had an equivalent account for their construction and use?

Archaeological debates will always continue. Gradually, however, we are moving towards an understanding of the role of water management in the ancient world, and perhaps gaining some lessons for the present. Some lessons come without even looking for them: knowledge about the ancient world inevitably brings the present into sharper focus.

Ingenuity and ambition

The simplest lesson is that water management was both critical and pervasive throughout the ancient world, from Mesopotamia to the Maya and from the Andes to Angkor. This conclusion might, of course, reflect my own bias, having selected case studies in which I had some prior indication that water management was significant. But the most obvious two studies that I excluded – Ancient Egypt and the Indus Valley civilisation – are also the two that are most well known for having relied on irrigation. Evidence is readily available to demonstrate the essential role of water management for the Aztecs, the Assyrians and for a multitude of other cultures in the ancient world.[5]

How could it possibly be otherwise? Urbanism, a defining feature of ancient civilisation, involves large numbers of people living in close proximity and hence there will always be a substantial demand for water. But it is not simply the presence of water management that has been so striking: the sheer scale of hydraulic engineering undertaken in the ancient world is simply astounding. I did not attempt any formal comparison, but the construction of the 551-kilometre aqueduct from Vize to Constantinople in the fifth century AD, the 1,600-kilometre Grand Canal in China in the seventh century AD, the 16 square kilometres of the West Baray built for Suryavarman I of Angkor in the 11th century AD, along with the water garden at Tipón built for Inca Viracocha in the 13th century AD, must surely be the equal of such twenty-first-century projects as the Three Gorges Dam or the Red Sea–Dead Sea canal in the amount of resources they required in proportion to the scale of the economy.

The ambition within the ancient world to entirely redesign nature to meet the human desire for water is astonishing, as are the technical ingenuity and ability of past communities to do so, whether part of a civilisation or not. Recall the Mycenaean city of Tiryns and its surrounding fields which were constantly threatened by flood water that would wash away buildings and deposit gravel and silt. It may have been 1500 BC and a Bronze Age culture, but this didn't deter the construction of a dam and outlet channel to completely divert the course of the offending river. A millennium or so later, Polycrates, the tyrant of Tigani on the island of Samos, found there was a mountain between his town and the Agiades spring. Not a problem – it may only have been 530 BC but Eupalinos surveyed and then excavated a tunnel through the mountain. A similar problem existed 250 years later at Dujiangyan, China, where a mountainside blocked the desired flow of the River Min towards the Chengdu Plain. Li Bing used fire and water to crack the rock and then excavated a channel for the water to flow through – it may have taken eleven years but the Chengdu Plain has since been watered for more than 2,000 years. Petra was chosen to be the Nabataean capital, despite an annual rainfall of less than 10 millimetres. The need to provide water for 30,000 people was no obstacle: every spring was tapped and every drop of rainfall run-off was captured, allowing for fountains, cascades and even a swimming pool in the desert. Despite heavy rain during the wet season and precipitous terrain, the Inca were sufficiently audacious to select a narrow shoulder of mountainside below the peak of Machu Picchu to build a citadel.

Water and power

The scale of such hydraulic engineering implies a considerable degree of centralised planning and control. The reservoirs, dams and canal systems had required vast quantities of building material, labour and time to complete. One of the most striking phrases has come from Joseph Needham, who described how the great canals and dykes in ancient China had been built by a 'million men with teaspoons'. That phrase would seem applicable for all of the ancient civilisations we have considered.

It does appear, however, that contra the 1957 arguments from Karl Wittfogel no such centralised control was necessary for the initial development of irrigation systems and for the management of those within non-state societies – those without kings or emperors.[6] On the alluvial plain of the Tigris–Euphrates, in the Salt–Gila river basins, in the foothills of the Andes and on the coastal wetlands of Mesoamerica, relatively small communities appear to have worked cooperatively in the planning, construction and maintenance of irrigation systems without any overarching authority. Strikingly similar developments happened around the world in ignorance of each other and at quite different times throughout the last 4,500 years. Similar developments most likely happened in many other regions, notably in the Yellow River valley of China; my guess is that they also occurred in locations we were unable to visit within this book, such as Sub-Saharan Africa and Australia.

In some circumstances, therefore, it appears to have been the canal systems that enabled particular communities, families or individuals to develop a power base, rather than that power base having been a requisite for the initial development of the canals. Some were then able to exploit the opportunity the canals provided to generate food surpluses, enhance trade and control the flow of water along the system. So we have a bootlegging process: hydraulic-engineering projects enabled certain families/individuals to secure a power base. With their new authority, they were able to control labour and mate-rials to extend the scale of the hydraulic engineering and so on, until we reach the Uruk period in Mesopotamia at 3900 BC and the Classic Hohokam in the Salt–Gila rivers basin at AD 900.

Power lay in controlling access to water. The most explicit case we considered was that of the Mayan Holy Lords of the royal centres, whose power depended upon the belief that the ritual they enacted secured the rainfall. During the dry season they permitted those who pledged their allegiance access to water stored within their reservoirs. With regard to the Hohokam, a plausible interpretation for the platform mounds of the Classic period is that they provided views across the irrigation system and hence enabled whoever occupied the mound to control the flow of water. Pueblo Grande was located at the head of the irrigation system and became by far the largest known so far, and by implication the most powerful, Hohokam settlement. The power of warlords and emperors in ancient China depended not only on the food surpluses produced by irrigation but on canals for transporting grain to their armies and for moving the soldiers themselves.

The history of aqueduct construction in Ancient Rome shows a complicated relationship between water and power. Taxation and/or warfare provided the funds to build new aqueducts, which became a visual demonstration of power themselves, especially when raised upon such extravagant arches across the countryside. The water provided not only helped quench the need for drinking water and that for farming and industry, but also for bathing, fountains and even a naumachia – the artificial lake built by Augustus in 2 BC for the re-enactment of naval battles. As such the aqueducts served the needs of the resident population of Rome while providing ostentatious displays of water availability to impress visitors to the city.

In many cases, control of the water supply was legitimised by ideology – those with power claiming to have divine authority to rule. In light of the apparent ritualistic discarding of human and animal carcasses within the wells of 8500 BC on Cyprus, the earliest wells currently known, I suspect that this nexus between water, power and ritual had been present from earliest times.

Power was also secured via water by using canals to facilitate trade, which has always been, and still remains, a driver for social change. Indeed, some would argue that in Southern Mesopotamia during the third millennium BC it was trade rather than the generation of an agricultural surplus by irrigation that was the ultimate catalyst for its 'take-off'. This not only facilitated the movement of raw materials and food supplies, but also the acquisition of prestige items and exchange of ideas and information between communities and city-states. A similar argument can be made for the impact of the Grand Canal in China during the seventh century AD.

Water was also a weapon. At least that is what we are told in various written documents – finding conclusive archaeological evidence is problematic to say the least. We have a cuneiform plaque from Mesopotamia describing how Abi-eshuh, king of Babylonia, redirected the flow of the Tigris to inundate fields and impede the movement of his enemy forces. Siloam's Tunnel was built in Jerusalem in 700 BC to protect the city's water supply when under the threat of siege. Both Herodotus and Thucydides describe the use of water in Ancient Greek warfare. From Ancient Rome we have the records for the Ostrogoths blocking the aqueducts to Rome in AD 537, while those taking water to Constantinople were always under threat, with the long aqueduct from Vize being cut in AD 626. For Ancient China we have the account of how the Zhengguo canal had been proposed in the third century BC as a means of exhausting the resources of the Qin and so preventing them from undertaking a military campaign, and the construction of the 'Magic Canal' to supply armies that had gone south. We should not doubt at least the gist of these accounts: water has remained a weapon of warfare through history and into the present day.[7]

Water always on the mind

Throughout the ancient world water appears to have always been on the mind, not only of those seeking to use it to secure and maintain their power base, but of every single member of society. They needed water to quench their thirst, water their crops and livestock, to use within their craft activities and for construction – remember the amount of water needed for a mud-lined hut. Those farming the land, whether in the Jordan Valley in the eighth millennium BC or the Peruvian highland in the 14th century AD would have worried about the rains: when will they arrive? How long will they last? Will they cause a flood or be insufficient to fill the cisterns?

Concern about water provides a mental unity to humankind: it is something that we share with the ancient Maya, Hohokam and Chinese. Those who built the canals in the Yellow River Valley of China, the Salt River Basin of Arizona, in the rainforests around

Edzná and Angkor, and on the Tigris–Euphrates alluvial plain did so thousands of years apart, with no knowledge of each other and within completely different cultures. But they all shared similar ideas, plans and physical labours; they addressed the same questions about gradients and where to place head-gates; they found the same solutions imposed by the common properties of water and then engaged in the same fights against the accumulation of silt and protection against floods.

We have seen that water is pervasive within the ideologies and mythologies of the ancient world; stories about floods are common, as are those about the creation of water itself. It is difficult to generalise about their function. In some cases, such ideologies serve to legitimise the power-base of the rulers, as was argued for the Mayan Holy Lords. In others, however, it may do the opposite and promote equality of access to water and thwart the emergence of hierarchies – as may have been the case prior to the Classic Hohokam and the Neolithic in the Levant, which appear to be complex, water-thirsty societies but without a social elite. Vernon Scarborough and Lisa Lucero have argued that within semi-tropical environments ideologies that consecrated areas of the landscape as sacred functioned to protect its water supplies from becoming over-exploited.[8]

Water-related motifs are frequent in the iconography of the ancient world: in my travels I was most struck by that within the wall paintings of 1800 BC at Knossos, the imagery and designs of the Mayan pottery of the 3rd century AD with its water monsters, and the iconography depicted within the 'churning sea of milk' bas-relief on the wall of Angkor Wat dating to the 12th century AD and jade relief carvings of canal building in the Forbidden Palace, Beijing. I cling to the idea that the wavy lines incised on stone plaques and the mud walls of the Neolithic site from 9500 BC I am excavating in Wadi Faynan, southern Jordan, are representations of water – they would be one of the earliest in the world. Regrettably, I will never know.

The sight of water was valued in the past, whether the vast standing moats and barays at Angkor silently reflecting the trees and the sky above – the heavens – or the gushing torrents of water in the fountains of Machu Picchu and Tipón. The Inca sites reveal how the sound of water was also of immense importance, causing Kenneth Wright to invoke the phrase 'hydraulic poetry'. The sound of water flowing within the sealed channels down the siq at Petra in the third century BC must have been enthralling for the thirsty visitors to the city, unaware they were about to enter a watery paradise. We can still enjoy Themistius' oration in the fourth century AD, extolling the water arriving via the aqueduct of Valens into Constantinople as 'Thracian nymphs' even though water no longer flows.

To the idea, the sight and the sound of water, we must also add its touch and feel. Whether relaxing in a Roman bath, dipping one's fingers into an Inca fountain, wading knee-deep through the water when cleaning out a Hohokam canal, or swimming within the corner of the West Baray – as I saw people still doing on my recent visit – the physical sensation of engaging with water was pervasive throughout the ancient world.

Hope or despair?

Does this record of water management and water experience in the ancient world provide us with hope or despair in our ability to address the challenge of the twenty-first-century water crisis? One might feasibly argue neither, that the scale of the problems we face today and the manner in which we use water are just too different to be remotely comparable to anything in the ancient world.[9] One might argue that the way water was managed changed in the late 19th century when the power of flowing water was coupled with electrical generators to create hydroelectricity.

Although little discussed within this book, water has long been a source of mechanical power. We briefly touched upon Archimedes' screw, the mills within the basement of the Baths of Caracalla and those along the newly made water channels across the Chengdu Plain used for hulling and grinding rice, for spinning and weaving. This usage of water increased significantly in the early stages of the industrial revolution as water was used to drive various machines such as Arkwright's spinning frame, and especially when converted into steam in the early 18th century – just a few centuries after the collapse of the Angkor, Inca and Hohokam regimes.[10] But the generation of hydroelectricity became a key driver for the great dams of the 20th and 21st centuries, a motivation for their construction quite different to those of the ancient world.

Feeling gloomy

One might argue that this merely adds a further dimension to water management rather than making any qualitative change: the massive water projects of the 20th century such as the Hoover, Aswan and Three Gorges dams were designed not just to generate hydroelectricity but to simultaneously address multiple issues including flood control, navigation and irrigation. As such, it would not be unreasonable to conclude that our knowledge of the past is not only relevant for the present but also provides a reason for despair rather than hope for the future. By definition none of the ancient civilisations have survived into the modern world. The Sumerians were the cause of their own downfall by over-irrigating with insufficient drainage; the Hohokam, Maya, the pre-Inca cultures of the Andes and the Angkorians all succumbed to climate change – the impacts of either droughts or floods or a combination of both in too quick succession; the Incas and the Nabataeans were overwhelmed by the expansion of more powerful empires, the former partly being conquered by disease; the reasons for the demise of the Ancient Greek and Chinese cultures, and the fall of the Western and the Eastern Roman empires will long be debated. So however large were the investments in water management and however ingenious the technology, the ancient civilisations were ultimately unsustainable and provide us with a bleak outlook: if the twenty-first-century version of smallpox doesn't get us then climate change certainly will.

The impact of climate change is all too evident in the past and a stark warning for the future. Many present-day coastal-living populations are already suffering from rising sea levels often causing saltwater to infiltrate their freshwater supplies;[11] their ultimate fate can be seen at Atlit-Yam, Israel, with its wells from the ninth millennium BC now entirely submerged. However ingenious had been the water collection and storage system at the Bronze Age site of Jawa in the Levant, this was unable to sustain any resident population once rainfall had dipped below a critical threshold. Perhaps the starkest message from the ancient world may simply be about the need to abandon persistently drought-stricken regions which climate change will make even worse, such as the Horn of Africa. But whereas the people of Jawa had been able to find unoccupied land in which to build a new community, where could the millions of people in the Horn of Africa today possibly go?

One cannot help but draw comparisons between the increasing frequency of floods and droughts now being experienced in so many parts of the world and the past experience of those who had lived in the Mayan lowlands during the ninth century, in the Salt River Basin during the 14th century and at Angkor during the 15th century. Their economies and societies were destroyed by a succession of extreme events and the same is happening today: quite literally because on this day that I am writing (17 December 2011) flash floods on the island of Mindanao in the southern Philippines have killed more than 500 people, washed away entire villages, made 10,000 people homeless and done untold damage to the island's agricultural economy.[12]

Despair about our future prospects might also come from learning that throughout history our forebears have ravaged their environments, often contributing to their own social and economic demise. While drought was certainly a cause of the Mayan collapse, deforestation for fuel and building materials had made their environment especially sensitive to climate change, reducing its resilience to fluctuations in rainfall. Deforestation is likely to have been a major reason for the demise of the Nasca in Peru and the Neolithic cultures in the Levant, the loss of vegetation changing the dynamics of their hydrological systems and causing extensive soil erosion. Deforestation destabilised the loess soils of northern China, reducing agricultural productivity while escalating the accumulation of silt in the Yellow River. The desire of political leaders to sustain their power base by ever more extravagant works of water management irrespective of the environmental consequences pervades the ancient world. It continues today: the desire of individuals, governments, cultures and nations for power makes it impossible for humankind to find a balance between environmental sustainability and economic growth.

Finally, in my gloomy mood, we should simply note not just the existence but often the severity of the water crisis now facing each of those regions where we saw such technical ingenuity and achievement in managing the ancient water supply. I have already described this for Peru and the same is true for Jordan, where the Nabataeans had been hydraulic masters of the desert.[13] In August 2007 the Greek government announced a 'state of emergency' because of water shortages throughout the mainland and islands where the Mycenaeans and Minoans had built aqueducts and collected run-off so effectively. The islands are now suffering because of tourists with their excessive water needs.[14] Both Italy and Turkey, where the Romans undertook such elaborate aqueduct construction as exemplified in Rome and Istanbul, have had their own water crises reported.[15] The dramatic economic and demographic growth of China has put such enormous pressure on its water supply that its future must be in doubt, whether or not there is an added impact from climate change.[16] A report in the *Independent* newspaper in March 2008 described Cambodia as facing a water crisis, with the waters of the Tonle Sap lake having become severely polluted just a few kilometres south of where such elaborate water management had been achieved by the Angkor kings.[17] With regard to the Americas, my introductory chapter noted the 2011 drought in Texas and Arizona where the Hohokam had built their irrigation systems; prior to this in December 2010 an issue of the local newspaper, *The Tucson Citizen*, had declared that 'water is hands down the most important issue in Arizona today'.[18] Excessive pumping of ground water around Phoenix is causing a dramatic fall of the water table resulting in land subsidence.[19] Finally, in Mexico, where the Maya had sustained large populations in their green desert by an array of hydraulic-engineering techniques, the water crisis is focused on Mexico City. Excessive pumping of ground water, leaking pipes and failing rainfall caused the city's authorities to simply turn off the taps in April 2009, causing crowds to gather in the street chanting: 'Water, water, water'[20] – just as they may have once done outside the acropolis of Tikal as the Holy Lord cowered inside.

Reasons to be cheerful

We can, of course, turn all of the above round and use the same evidence as reason to hope for the future. The ancient civilisations and cultures may have disappeared but they were no mere fly-by-nights. We may count the Hohokam's irrigation as a failed experiment in desert adaptation, leaving their descendants of the 14th century AD with a lifestyle little different from that of their fore-bears of the first century AD. But let us not forget that this 'experiment' lasted for more than a millennium, enduring many episodes of climatic turmoil. So too did the Mayan, Khmer and Mesopotamian civilisations. These all had to adapt and change to survive, just as we need to do today.

Although the ancient civilisations may have disappeared in name their hydraulic engineering has often continued, in some cases with a little help from minor restoration. At Dujiangyan I saw water flowing through the irrigation system towards the Chengdu plain just as it has done every day since 256 BC; in Rome I sat by the Trevi fountain, it being fed by a restored Aqua Virgo; in southern Jordan I saw Bedouin still making use of reservoirs built by the Nabataeans; and in Cambodia people were swimming in the West Baray still full of water. When in Phoenix I saw the Arizona Canal flowing along the same course as a Hohokom canal, and in Peru water was flowing along Inca aqueducts, not only at Machu Picchu and Tipón, but also at Pisac and Ollantaytambo.

Should we use the technical ingenuity and skill of our forebears as a sign of hope that our current problems will be resolved by new inventions? That is certainly the belief of many people today, placing their faith in new methods to generate electricity, desalinate water and genetically engineer crops to require less water. For them, new technology to sustain economic growth is the answer. Certainly the past testifies to human creativity and this should give us cause for optimism. The application of remote sensing and digital technology to manage water flow within the Yellow River appears to be solving some of the long-term problems of that river. Maybe one day this will be recognised as an equivalent breakthrough to Li Bing's engineering of the River Min.

The ancient legacy is not only one of inventions and physical structures, some continuing to function while others are in ruins: we have also inherited the ancient attitudes to water, or at least some of them. The early texts, notably Homer and the Old Testament, emphasised the role of water for cleansing the body both physically and spiritually. This is surely what many of the stone basins, water vases and secluded fountains found throughout the ancient world would once have been used for. Spiritual cleansing with water continues today, this being a rite in all major religions, whether as Christian baptism or washing prior to Muslim prayer. Bathing also takes on a ritualistic nature for many people and the modern-day spa experience is the direct descendant of that of the Roman baths, which itself derived from the Greeks.

A third trait inherited from the past is the ostentatious display of water by city fountains. While this is closely tied to the political control of water, a means of legitimising that authority, my suspicion is that our attraction to such fountains goes back far earlier than the ancient world. It originated, I guess, sometime in our deep prehistory when an attraction to water became embedded within our DNA.

The final reason to think that our glasses of water should be considered as half full rather than half empty is that we do have knowledge about the ancient world to guide us in the present and future: understanding the past enables us to see the present more clearly.

Reminders, rather than lessons

Although one must be cautious about drawing any lessons from the past, there are certainly reminders of appropriate ways of behaving that we should have already learned from elsewhere. I've previously noted how the fate of past societies demonstrates the reality of climate change and warns us against either ignoring or denying scientists' projections for the future and their potential impact on communities around the world.

Control your own water supply – or at least hold those who control your water supply accountable is another warning from the ancient world, this one about the potent relationship between water and power. While the motives of long-dead Holy Lords, Nabataean and Angkor kings, Roman, Chinese and Inca emperors are even harder to evaluate than those of our politicians and business leaders today, it is quite evident that throughout history the water supply was manipulated by those with authority for their own aggrandisement. This continues to be the case throughout the world, whether by multinational companies or governments – elected and otherwise. As Maude Barlow described in her 2007 book *Blue Covenant*, there is an escalating battle for water rights: on one side there are transnational water and food corporations, most first world governments and international institutions, including the World Bank and International Monetary Fund, who treat water as a commodity; on the other side there is an ever-growing global water justice movement, consisting of environmentalists, human rights activists and thousands of grass roots communities fighting for control of their local water supply. Our knowledge of the ancient world makes the latter's case compelling and provides succour to their cause.

Cut down your trees at your peril is a persistent message from the past. We are all informed about the impacts of deforestation on the build-up of greenhouse gases and the destabilisation of soil resulting in massive mudflows. We have several demonstrations of the consequences of deforestation from throughout the ancient world, but another does no harm.

Cities are thirsty beasts is a reminder about the water-dilemma of urbanisation. Within this book we visited a few of the great cities of the ancient world: Rome, Constantinople, Chang'an, Angkor, Tikal. Cities, especially those with cosmopolitan populations, were the centres of economic, artistic and intellectual innovation: civilisations cannot exist without them. But cities are thirsty; they need reservoirs, aqueducts, canals and drains for both supplying water and removing the waste. Urbanisation continues apace today, the planet now having more people living within cities than without. So we shouldn't be at all surprised about the global water crisis and must heed a warning from the ancient world about the consequences of continued urban growth for necessary investment in water management if such cities are to be sustained.

Value local knowledge about the water supply and its use in farming. This is the know-how possessed by farmers who have worked the land for generations and is unlikely to be written down. A good case can be made that the salinisation of Mesopotamian soils was a consequence of the loss of the fallowing system, arising from demands for farmers to generate short-term yields at the expense of long-term sustainability, overriding their traditional means of land management. I suspect that the demise of the Classic period Hohokam, the decision not to rebuild the irrigation system after the floods of 14th century, can also be attributed, in part at least, to the loss of local know-how and levels of community cooperation that had once been maintained through the ballcourt system.

As with my other 'reminders', ensuring that local knowledge is valued and maintained is easy to say but often hard to implement, especially in large, state-controlled farming systems. I learned of the complexities when I visited the Gezira irrigation system in Sudan in March 2010. This is an extensive complex of canals and ditches, amounting to a length of 4,300 kilometres, that distributes water from the Nile to tenant farmers. It was originally designed and built by the British in the 1920s and has primarily supported cotton, although a diversity of crops was being grown during my visit. A 1990 World Bank Report into Gezira's low productivity had concluded that the centralised control of the scheme involving a large irrigation bureaucracy had restricted the capacity of tenants to decide what crops to grow, where and how – an appreciation that local know-how was not being utilised.[21] But attempts to move to self-governance have reduced rather than enhanced the tenants' sense of ownership.[22] Moreover, my impression from conversations was that as farmers have been allowed to do as they wish, crops are no longer being grown in rotation and organised to reduce pest damage. It may be the case that those decades of rigid state control had eroded local know-how and cooperation to such an extent that it cannot now be recovered.[23]

Don't waste water is hardly a reminder that we need from the past, especially when our water bills are escalating and we have a constant flow of media images about a drought-stricken world. The ancient world can remind us of what can be achieved when sufficient attention is paid to capturing, conserving and recycling water. While both the Minoan palaces and Mayan courtyards captured run-off water, the Nabataeans were perhaps the masters of saving every drop, especially evident both within Petra and in their desert settlements of the Negev. The Nabataeans also remind us about the value of ensuring there is redundancy in the system – ensuring a back-up exists should one supply of water fail. Their achievements can certainly send a reminder to the present-day leaders in Jordan: Amman is described as a water-stressed city but suffers from an appallingly inefficient and wasteful water supply system,[24] of the type that the Nabataeans would never have tolerated. We have seen how the Nabataeans, Greeks, Romans and Incas all designed systems to recycle water, from drinking to washing to use for irrigation and animals. One can't help but feel that the further restoration of their attitudes to water would be as valuable as has been the restoration of their water systems.

A final reminder is perhaps the most important and is a message that that has been the most unheeded throughout much of the ancient world and into modern times. It is that which comes from Yu the Great, and which was heeded by Li Bing when constructing Dujiangyan, the most successful of all the hydraulic engineering projects of the ancient world: *Work with nature, not against it.*

Unquenched thirst

Ever since the Neolithic, the world has had an unquenchable thirst for water. Meeting that need was a key driver of social, economic and political change within the ancient world, one that played a fundamental role in both the rise and then fall of ancient civilisations. That unquenchable thirst continues today, perhaps more desperate than it has ever been before.

It remains contentious whether knowledge of water management and hydraulic engineering in the ancient world can play any significant role in enabling us to tackle the twenty-first-century water crisis in a more effective manner. I hope it can, but I am an unerring optimist. What such knowledge certainly can help with, however, is the quenching of another type of human thirst: that for knowledge about the human past.

I've had the good fortune of travelling to many parts of the world to research this book, visiting some of the most intriguing archaeological sites on the planet, ranging from the magnificent citadel of Machu Picchu to empty ditches in the Sonoran Desert that had once been Hohokam canals. I've had the privilege of sitting in libraries reading the academic reports of excavations and environmental studies, marvelling at the exploits and achievements of archaeologists. I am not sure whether it is with regret or pleasure that I now must inform you that the more I have seen and read, the thirstier I have become; the more that I have learned about the past the more I need to know. I now feel compelled to explore those civilisations not covered within this book, to visit many more archaeological sites around the world and continue with my own excavations in Wadi Faynan. While I hope you have learned something about the past from the pages of my book, I also hope that your thirst has remained equally unquenched.

NOTES

1. *Thirst*

1. This is the Hoover Dam bypass, otherwise known as the Mike O'Callaghan–Pat Tillman Memorial Bridge, which is itself a remarkable construction, having the widest concrete arch in the western hemisphere and being the second-highest bridge in the United States. It opened on 19 October 2010.

2. Cited in Hiltzik (2010, III).

3. For a full, and quite brilliant, account of the building and impact of the dam see Hiltzik (2010).

4. Solomon (2010) provides an outstanding account of the economic, political and environmental impact of the Aswan and Three Gorges dams within a comprehensive study of the role of water in human history from the ancient civilisations to the present day.

5. The most outstanding introduction to ancient civilisations is provided by Trigger (2003).

6. UN Water (2007); see also Clarke and King (2004) for statistics and predictions about current and future access to water, and Roddick (2004) for a challenging set of short essays about the current water crisis. For more in-depth academic studies see the essays within the biennial reports on freshwater resources in Gleick (2007, 2009a). Solomon (2010, Chapter 14) provides an overview of the current global water crisis, and how this is set to become more severe.

7. See http://www.azcentral.com/news/articles/2011/09/25/20110925arizona-water-drought-may-deepen.html, accessed 16 November 2012.

8. Such figures are obviously problematic. These have been taken from the relevant BBC News webpages.

9. This is the claim within http://www.nation.com.pk/pakistan-news-newspaper-daily-english-online/Politics/19-Aug-2010/Heavily-funded-FFC-fails-to-deliver/.

10. For the Red–Dead Sea canal see Lipchin (2006).

11. For the North–South Water Transport Project in China, and other monumental water management projects in China see Watts (2010, Chapter 3).

12. For a summary of the man-made river project in the Libyan desert see Pearce (2006).

13. Hiltzik (2010).

14. Ibid. (2010).

15. Figures for the Three Gorges Dam project are taken from Li Jinlong and Yi Chang (2005). See Gleick (2009c) for an assessment of the project.

16. Pearce (2006).

17. For the rarity of 'water wars' between national states see Barnaby (2009). For disputes between multi-nationals and local populations about access to water see Barlow (2007), while Gleick (2006a) considers water and terrorism.

18. Solomon (2010, Chapter 15) provides an excellent account of the disputes about water in the Middle East and North Africa. For a more detailed account of 'The Middle East Water Question' up until the end of the 20th century see Allan (2002).

19. Wittfogel (1957).

20. Scarborough (2003). He proposed that the study of water management provides an opportunity for understanding resource use and control in past societies, arguing that water management typically entails a series of decisions rooted within ritual and religious belief.

21. For a concise summary of water management in Ancient Egypt see El-Gohary (2012), while Solomon (2010, 26–37) provides a more discursive account of how water management related to the history of Ancient Egypt.

22. Solomon (2010, 38) makes this comparison between Egypt and Mesopotamia.

23. See Scarborough (2003) for summary studies of the Aztecs and Sinhala; Geertz (1980) provided a classic study of nineteenth-century Bali. Angelakis et al. (2012) provide several case studies of ancient water management for regions not covered in this volume, including Ancient Iran, pre-Columbian societies in Ancient Peru and for the historical periods on Cyprus and in Barcelona, Spain. Scarborough and Lucero (2010) review the relationship between water and cooperation in the semi-tropics, providing summaries of water management in West Africa, Bali and Amazonia. Wilkinson and Rayne (2010) provide an outstanding review of hydraulic landscapes in northern Mesopotamia (northern Iraq, Syria and southern Turkey) that covers the Assyrian, Parthian, Sasanian, Roman/Byzantine and Early Islamic periods.

24. For qanats see Forbes (1956) and Beekman et al. (1999), and for noria see Solomon (2010, 34).

2. The water revolution

1. Caran et al. (1996) describe a well at San Marcos Necoxtla, Mexico, that

is 10 metres wide and 5 metres deep and filled with cultural debris dated between 9,863 and 5,950 years ago.

2. For descriptions and analysis of chimpanzees using leaves as tools for transporting water see Matsuzawa et al. (2006).

3. See Fleagle (2010) for a study of what is called 'Out of Africa 1', the initial hominin colonisation of Eurasia.

4. 'Ubeidiya has large quantities of stone and animal bone fragments, and some questionable fragments of hominin remains; see the acount within Fleagle (2010).

5. Shea (1999) describes the formation of the so-called 'living floors' of 'Ubeidiya.

6. Ohalo II has been described in numerous publications, including: Nadel and Hershkovitz (1991), who report on its subsistence base; Nadel et al. (1995) on the site chronology; and Nadel and Werker (1999) on the brush-wood huts.

7. See Stringer (2011) for a recent interpretation of both archaeological and genetic evidence for *Homo sapiens* dispersal.

8. The extent and nature of the language and symbolic capacities of Neanderthals and modern humans are subjects of significant debate amongst archaeologists. See Mithen (2005) for one interpretation of the evidence.

9. For climate change in the Levantine region see Brayshaw et al. (2011).

10. Bar-Yosef (1998) and Bar-Yosef and Valla (1991) provide a review of the Natufian culture in the Levant.

11. For a description of 'Ain Mallaha see Valla et al. (1999). Mithen (2003, Chapter 4) provides an interpretation of the Early and Later Natufian.

12. See Brayshaw et al. (2011) and Robinson et al. (2011).

13. See Kenyon (1957) for a fantastic account of her work at Jericho; Kenyon and Holland (1981) provide the academic account of the stratigraphy of the tell.

14. Key works by Childe in which the Neolithic of Europe became defined include Childe (1936) and Childe (1957).

15. For a review of the PPNA and PPNB, see Kuijt and Goring-Morris (2002).

16. There are a great number of theories as to how domesticated plants and animals arose, and farming began; see Mithen (2003) and Barker (2006).

17. I have been using 'years ago', whereas archaeologists often use 'BP', standing for 'Before Present', which for academic purposes was set as 1950. Consequently to convert a BP to a BC date, one must subtract 1950 years. These can only be undertaken in calibrated radiocarbon dates which have taken into account the impact of changing carbon dioxide levels in the atmosphere.

18. Por (2004) describes the hydrological context for 'Ain el-Sultan.

19. For a review of the Neolithic of the Levant with an extensive bibliography regarding sites such as Gilgal, Jericho and Netiv Hagdud see Kuijt and Goring-Morris (2002). For Netiv Hagdud see Bar-Yosef and Gopher (eds, 1997) and for Gilgal see Bar-Yosef et al. (2010).

20. Por (2004) describes the hydrological context for Netiv Hagdud.

21. For Zahrat edh-Dhra' see Edwards et al. (2002); for Dhra' see Finlayson et al. (2003); and for WF16 see Finlayson and Mithen (eds, 2007) and Mithen et al. (2010).

22. Brayshaw et al. (2011) provide state-of-the-art computer simulations for the early Holocene climate of the Levant.

23. For Göbekli Tepe see Schmidt (2002) and for a site with similar art known as Jerf el Ahmar see Stordeur et al. (1997). Watkins (2010) provides an interpretation of the monumental architecture at these sites, although see also Mithen et al. (2011) and Finlayson et al. (2011).

24. For architecture at Jerf el Ahmar see Stordeur et al. (1997).

25. Kuijt and Goring-Morris (2002) provide a review of the PPNB covering architecture, economy, art and religion. See also Mithen (2003, Chapters 9 and 10).

26. Kirkbride (1968) provides an interim report and Byrd (2005) a full description of the architecture at Beidha.

27. The following draws on Rambeau et al. (2011).

28. The following draws on Gebel (2004).

29. For Ghuwayr, see Simmons and Najjar (1996; 1998).

30. Bar-Yosef (1996).

31. The following text regarding Wadi Abu Tulayha draws on Fujii (2007a, 2007b, 2008).

32. Peltenberg et al. (2000) described the early wells on Cyprus.

33. See Galili and Nir (1993) and Galili et al. (1993) for a description and interpretation of the wells at Atlit-Yam.

34. Garfinkel et al. (2006) provide an outstanding description of the well at Sha'ar Hagolan and its excavation.

35. Kuijt et al. (2007) describe the Pottery Neolithic landscape modification at Dhra'.

36. For excavations and interpretation at Jawa see Helms (1981) and Helms (1989).

37. Whitehead et al. (2008).

38. For a review of the Early Bronze Age in the Levant see Philip (2008) and of water management techniques see Lovell within Finlayson et al. (2011b).

39. Rast and Schaub (1974) describe the cistern at Bab edh-Dhra, other examples of which are found throughout the region. An example at Horvat Tittora is described by Rast and Schaub (2003).

40. For Tell Handaquq see Mabry et al. (1996).

41. Bienert (2004) describes the tunnels at Khirbet Zeraqoun, located 60 metres below ground and accessed by three shafts which are thought to have been used for construction.

42. The irrigation systems and a review of on-going research at Tell Deir Allah are described by Van der Kooij (2007). Kaptijn (2010) provides an excellent consideration of Iron Age irrigation in the Zerqa region of Jordan, drawing similarities between that of the twentieth century, the Mamluk period (AD 1250–1516) and the Iron Age, examining the link between irrigation systems and power relations.

43. The following text draws on http://relijournal.com/religion/water-meaning-and-importance-in-the-old-testament/ accessed on 5 December 2011.

44. For Siloam's tunnel see Frumkin and Shimron (2005).

45. The biblical references are: 'As for the other events of Hezekiah's reign, all his achievements and how he made the pool and the tunnel by which he brought water into the city, are they not written in the book of the annals of the kings of Judah?' (2 Kings 20:20); 'When Hezekiah saw that Sennacherib had come and that he intended to wage war against Jerusalem, he consulted with his officials and military staff about blocking off the water from the springs outside the city, and they helped him. They gathered a large group of people who blocked all the springs and the stream that flowed through the land. "Why should the kings of Assyria come and find plenty of water?" they said' (2 Chronicles 32:2–4); 'It was Hezekiah who blocked the upper outlet of the Gihon Spring and channelled the water down to the west side of the City of David. He succeeded in everything he undertook' (2 Chronicles 32:30).

3. 'The black fields became white / the broad plain was choked with salt'

1 There are several excellent introductions to Sumerian civilisation. This chapter draws on three key works: Postgate (1992), Pollock (1999) and Algaze (2008). Tamburrino (2009) provides a review of water technology in Mesopotamia.

2. This comes from the Atra-Hasis epic about the creation and early history of man, cited in Tamburrino (2010, 33).

3. Algaze (2008, 5)

4. Wilkinson (2000) reviews regional surveys in Mesopotamia and refers to the 1960s and 1970s as the 'golden age' of archaeological survey.

5. For use of imagery from the space shuttle for Mesopotamian archaeology see Hritz and Wilkinson (2006).

6. For Choga Mami see Oates (1968; 1969), and for its agricultural base see Helbaek (1972).

7. Algaze (2008, 88) notes that in the fourth millennium fleeces required plucking by hand because unlike modern sheep populations the fourth-millennium sheep still moulted their coats annually at the end of the spring and their soft woolly undercoats were still covered by a layer of bristly kemp hairs.

8. Adams (1981, 11) and also cited in Algaze (2008, 92) who emphasises the significance of the textile industry.

9. These figures for the size of Mesopotamian settlements are taken from Algaze (2008).

10. Postgate (1992).

11. Quoted in Solomon (2010, 46).

12. Postgate (1992, 45).

13. For Adams's archaeological achievements see Adams (1981) and Adams and Nissen (1972).

14. The following text is drawn from Gibson (1974); those wishing to follow this line of interpretation should consult Fernea (1970).

15. Postgate (1992, 178).

16. Tamburrino (2009) describes the role of water in Sumerian mythology.

17. These figures are taken from Algaze (2008), who provides extensive discussion of productivity and trade.

18. Algaze (2008, 100).

19. Such is the description provided by Pollock (1999).

20. See discussion of use of fallow in Gibson (1974).

21. Russell cited in Jacobsen (1958, 67).

22 Jacobsen and Adams (1958).

23. Powell (1985).

24. Artzy and Hillel (1988).

25. Gibson (1974).

4. 'Water is the best thing of all' – Pindar of Thebes 476 BC

1. Clarke (1903, 598).

2. Clarke (1903, 598).

3. Clarke (1903, 597).

4. For overviews of Minoan Crete see Castleden (1992), Dickinson (1994) and Warren (1989).

5. Warren (1989, 69).

6. Dickinson (1994, 24–5).

7. For extensive academic studies of Knossos see Cadogan et al. (2004) and Evely et al. (2007).

8. The following text draws on Angelakis et al. (2006) and Mays (2010), who review Minoan aqueducts and other forms of hydraulic engineering. See also Koutsoyiannis and Angelakis (2007) and Koutsoyiannis et al. (2008).

9. Angelakis et al. (2012) provide a technical account of the design and energy efficiency of the Minoan terracotta pipes.

10. Long-distance aqueducts using either terracotta pipes or open channels of terracotta or stone also supplied the Minoan sites of Tylissos and Malia, bringing water from mountain springs (Angelakis et al. 2012).

11. Angelakis et al. (2012).

12. Mays (2010) describes the water-collecting system at Phaistos.

13. The following draws on Angelakis et al. (2012).

14. Cadogan (2007).

15. Cadogan (2007, 108).

16. Cadogan (2007, 107).

17. For an overview of the Mycenaeans see Taylour (1983).

18. Homer, *The Illiad*, 1.313.

19. See Gutgluede (1987) for the significance of bathing in *The Odyssey*.

20. Mays (2010) reviews the hydraulic technology of the Mycenaeans.

21 Balcer (1974, 148).

22. The following text draws on Knauss (1991).

23. The following text draws on Balcer (1974), Zangger (1994) and Maroukian et al. (2004).

24. Crouch (1993) provides a seminal study of water management in Ancient Greek cities, which I draw upon for the description of the water supply to Athens. Zarkadoulas et al. (2012) also provide an excellent account of urban water management in Ancient Greece, with interesting views about the relationship between democracy and the absence of major hydraulic engineering projects in Athens.

25. A detailed account of the water supply of Classical Athens is provided by Chiotis and Chioti (2012).

26. The Great Drain is described in Chiotis and Chioti (2012).

27. Crouch (1993) describes the water management system at Corinth.

28. The following text draws on Goodfield and Toulmin (1965), Apostol (2004) and Mays (2010).

29. Kienast (1995).

30. Apostol (2004).

31. Cited in Crouch (1993, 49).

32. For Archimedes' screw and other Greek and Roman mechanical water-lifting devices see Oleson (1984).

33. For the various inventions of Ctesibius see Rihil (1999) and Tuplin and Rihil (2002).

34. See Crouch (1984) for an excellent study of water management within a Hellenistic city, that of Morgantina, Sicily.

5. A watery paradise in Petra

1. My account of the Nabataeans in this chapter primarily draws on articles within Politis (ed., 2007), Bienert and Häser (eds, 2004) and Bedal (2004).

2. Diodorus, *Historical Library*, 19.94.7. Bedal (2004, 6) explains that Diodorus of Sicily wrote in the first century BC, basing his account on the fourth-century BC eyewitness account of Hieronymous of Cardia.

3. Oleson (2001, 2007a) provides a comprehensive overview of Nabataean water management techniques.

4. Strabo, *Geography*, XVI.4.26.

5. The following draws on Oleson (1995, 2001, 2007a, 2007b).

6. Evenari (1982, 9).

7. It has not been formally demonstrated that all of these walls, terraces and mounds are Nabataean in date. Evenari cautiously refers to 'ancient farmers' rather than making any distinction between what might belong to Bronze Age, Nabataean, Roman or Byzantine periods.

8. Evenari (1982, 194)

9. Evenari et al. (1982, 415).

10. Everani et al. (1982, 414–15).

11. I primarily draw on Oleson (2007a). See Oleson (2007b) for a summary of his long-term research at Humayma.

12. Population estimates based on hydrological modelling are also provided in Foote et al. (2011). These broadly agree with those provided by Oleson (2007a, b).

13. The following draws on Ortloff (2005).

14. Akasheh (2004) describes how the growth of tourism in Wadi Musa village is causing a serious imbalance in the Wadi Musa watershed. As more roads and concrete buildings are constructed, there is a loss of precious soil in the watershed, a soil cover that is necessary for soaking up a large proportion of the rain water. The threat of major flash floods is thereby increased as well.

15. Bedal (2004, 95).

16. Bedal (2004). Bellwald (1999) believes that the camels are carrying goods, although he acknowledges the depicted goods are difficult to see:

'photographs taken at night in a sharp sidelight show the upper dromedary of the upper group has a basket-like saddle for goods ... while the lower one carries a cone-shaped load covered by a blanket adorned with long tassels'.

17. Oleson (2007a).

18. Akasheh (2004).

19. Bedal (2004).

20. See the description by Joukowsky (2004).

21. The following draws on Bedal (2004) and Bedal and Schryver (2007).

22. Bedal and Schryver (2007) describe charred plants remains recovered by flotation of deposits from excavation of the garden terrace area. As well as the species I refer to these include a wide range of weed species, legumes and cereals. All such remains had been charred, and they may reflect the economic activity within Petra more generally rather than the specific plants being grown within the garden.

23. Strabo, *Geography*, XVI.4.21.

6. Building rivers and taking baths

1. Haut and Viviers (2012) provide a particularly impressive review and set of photographs for Roman hydraulic engineering in the Middle East. De Feo et al. (2012) provide a wider-ranging review, while Martini and Drusiani (2012) focus on Rome itself.

2. See Fabre, Fiches and Paillet (1991) for a study of the aqueduct of Nîmes and the Pont du Gard.

3. See DeLaine (1997) for a study of the building costs of the Baths of Caracalla.

4. Water flowing over the ground from fountains was not necessarily a waste because this washed away filth. Solomon (2010, 88) states that Rome's 'provision of copious amounts of fresh public water washed away so much filth and disease as to constitute an urban sanitary breakthrough unsurpassed until the nineteenth century's great sanitary awakening in the industrialised West'.

5. Hodge (1992, 11).

6. Bono and Boni (1996, 126).

7. Nielsen (1990, 2).

8. Nielsen (1990) provides a comprehensive account of the spread of bath-houses throughout the colonies and provinces and notes how they were always one of the first establishments to be built.

9. Hodge, (1992, 1).

10. Cited in Veyne (1997, 199).

11. Seneca, *Epistles*, 56, 1–2. The bath in question was at Baiae in southern Italy.

12. Nielsen (1990, 6–13) provides a detailed review of the forerunners of the Roman baths, looking both towards Greece and more locally within Italy itself, in terms of washing in hot water within lavatrinae and swimming in cold water, in rivers and pools. Hodge (1992) provides a more general review of the predecessors of Rome with regard to hydraulic engineering in general.

13. Nielsen (1990, 33) describes the specific architectural changes to the Stabian baths arising from the construction of the aqueduct feed. She describes as well how the introduction of the 'tubulation' also resulted in significant innovation and improvements to the baths. This was a wall and sometimes vault heating system which meant that chambers could be kept considerably hotter and more comfortable to be in, although it meant that there was also a greater demand for a cold rinse.

14. See Nielsen (1990, 115–48) for a detailed and comprehensive review of 'the bathing institution'.

15. Nielsen (1990, 151).

16. Hodge (1992).

17. De Feo et al. (2012).

18. The text that follows primarily draws on Ashby (1935) and Evans (1997).

19. Martini and Drusiani (2012) describe the *Cloaca Maxima* and other sewers of ancient Rome.

20. Frontinus' treatise *De aqua ductu* has been subject to intense discussion, beginning with Rodolfo Lanciani's commentary in 1881. In my text that follows, I have primarily drawn on references to Frontinus contained within Evans (1997), which includes a full translation.

21. This was the Great Fire of Rome of AD 64 that, according to Tacitus, *Annals*, XV.40, burned for five and a half days.

22. Frontinus, *De aqua ductu*, I.16, translated in Hodge (1992, 1).

23. Martini and Drusiani (2012, 462–3) describe the fountain in Piazzale degli Eroi, which displays water arriving from the Peschiera aqueduct, constructed between 1949 and 1908.

24. Martini and Drusiani (2012).

25. Hodge (1992, 93–125) provides a detailed account of how aqueducts were constructed, with the various requirements for cleaning and maintenance.

26. Marcus Vipsanis Agrippa (63–12 BC) was a great Roman statesman and general, the son-in-law of the Emperor Augustus and the father-in-law of the Emperor Tiberius. His work in Rome included promoting public festivals and displays of art as well as the overhaul of the water supply system

27. Nielsen (1990).

28. Bono and Boni (1996, 132).

29. The following text primarily draws on Piranomonte (1998).

30. For a detailed and comprehensive analysis of the construction of the Baths of Caracalla, and all of the costs involved see DeLaine (1997, 193). She writes that 'The Baths of Caracalla could have supported an average minimum work-force over the four years of the main construction period, of 7,200 men directly involved in producing material and construction, plus 1,800 men and pairs of oxen for transport in the immediate vicinity of Rome, rising to 10,400 and 3,200 respectively during peak periods. The overall figure is highest in 213 when a work-force of 13,100 was required. This is a conservative estimate which does not include the pavilions and porticoes of the outer precinct, and does not allow for the building of the aqueduct and the access road. In addition, on average some 200 men were needed to produce the brick, and the same to produce the lime in various country districts within the 100-km radius of the city, rising to 300 and 240 men respectively at peak periods, while a further number which cannot be calculated was required to produce the form-work timbers, scaffolding poles, ropes and baskets. More men would have been required for the transport of all of these to Rome. An average figure of 500 men should be about right, rising at times to nearer 700.'

31. The following text draws on Bono et al. (2001) and Crow et al. (2008).

32. Crow et al. (2008) note that Hadrian had donated an aqueduct to Nicaea in AD 123.

33. Çeçen (1996).

34. Crow et al. (2008, 1).

35. Themistius, *Oratio*, 11.151a–2b, cited by Crow et al. (2008, 9).

36. Themistius, *Oratio*, 13.167c–168c, cited by Crow et al. (2008, 9–10).

37. See Crow et al. (2008, 16).

38. For a detailed description of the Kurşunlugerme see Crow et al. (2008, 57–61 for the water course, 93–7 for the architecture and 157–76 for the decoration).

39. For further details see Crow et al. (2008, 17–20).

40 For a full description see Ciniç (2003).

41. Theophanes, *Chronicles*, cited in Crow et al. (2008, 19–20).

7. A million men with teaspoons

1. This claim is attributed to various writers by Du and Koenig (2012, 169). Solomon (2010, 107) states that 'for well over a millennium, China was the human civilization's leader in harnassing water as energy to do useful work.'

2. Cited in Gillet and Mowbray (2008).

3. For example, Du and Koenig (2012) for water supply to ancient cities and Kidder et al. (2012) for the catastrophic flooding of a Han-period settlement.

4. For an excellent biography of the brilliant but rather eccentric Joseph Needham see Winchester (2008).

5. Needham (1971).

6. Liu and Xu (2007).

7. This comparison between the Han and Rome was made by Kidder et al. (2012, 30). Solomon (2010, 105) notes that not only were their periods of greatest wealth, power and influence contemporaneous, but that their empires were of a comparable size, they both flourished at the edges of the civilised world and the proximate causes of their demise were barbarian attacks on their northern frontiers.

8. The following draws on Kidder et al. (2012).

9. As stated in February 2012 by Hu Siyi, the Vice-Minister of China's Water Resource Ministry; see http://www.china.org.cn/environment/2012-02/16/content_24653422.htm (accessed 22 April 2012).

10. Reported by a representative from China's Ministry of Housing and Urban–Rural Development at a meeting in Singapore in July 2011; see http://news.xinhuanet.com/english2010/china/2011-07/07/c_13969749.htm (accessed 22 April 2012).

11. Du and Koenig (2012, 187).

12. The following account of water management in Chang'an draws on Du and Koening (2012).

13. Solomon (2010, 97).

14. Solomon (2010, 112).

15. Solomon (2010, 112)

16. Solomon (2010, 107)

17. For the ecological challenge caused by the Three Gorges Dam see Gleick (2009c) and Watts (2010, Chapter 3).

18. *China Daily*, (16 June 2002). See Guangqian Wang et al. (2007).

19. Cited from a report in the *Guardian* (29 June 2011).

20. Cited in http://www.waterworld.com/index/display/article-display/0785922114/articles/waterworld/world-regions/far-east_se_asia/2010/03/Lee-Kuan-Yew-Water-Prize-awarded-to-Yellow-River-Conservancy-Commission.html.

8. *The hydraulic city*

1. For the Khmer civilisation this chapter draws on the excellent reviews by Coe (2003) and Higham (2001). Coe provides an anthropological approach as far as is possible with the existing evidence, while Higham's work is a more

conventional study of historical developments with accounts of the kings and their accomplishments. Freeman and Jacques (2003) provide an outstanding study of the art and architecture of Angkor, with detailed and beautifully illustrated descriptions of each monument. Unless otherwise stated, the text within this chapter draws on these three works.

2. See Poole and Briggs (2005) for an outstanding narrative and photographic study of Tonle Sap.

3. Bernard Philippe Groslier (1926–86) was one of the key figures in the archaeological research of south-west Asia during the twentieth century. See http://www.efeo.fr/biographies/notices/groslier.htm for a summary biography. He had many publications, the most significant for this study being Groslier (1952; 1956; 1960) and Groslier (1979), which was his most explicit statement of Angkor as the 'hydraulic city'.

4. See Mouhot (2001). My account of the discovery of Angkor also draws on Coe (2003).

5. Cited in Higham (2001, 60).

6. Cited in Higham (2001, 65).

7. Zhou Daguan's 1296 account of his visit was translated into French in 1319, which in turn has been translated into English: Zhou (1993).

8. Goloubew (1936; 1941).

9. Groslier (1966, 75).

10. These, and other figures relating to the irrigation and agricultural capacity of Angkor are drawn from Acker (1998).

11. Van Liere (1980).

12. Stott (1992, 55).

13. Acker (1998).

14. Acker (1998, 31).

15. Groslier (1974, 112).

16. Siribhadra and Moore, cited in Acker (1998, 35–6).

17. Fletcher et al. (2008) provides a succinct summary of the Greater Angkor Project (GAP) and is a key source for the following text.

18. Evans et al. (2007). See also Fletcher et al. (2004).

19. Fletcher et al. (2004, 137).

20. Fletcher et al. (2008).

21. The following draws on Buckley et al. (2010).

22. Fletcher et al. (2007).

9. Almost a civilisation

1 Rogge et al. (2002) describe excavations of Hohokam canals at Sky Harbor airport.

2 For an overview of the Hohokam see the edited volumes by Crown and Judge (1991) and Fish and Fish (2007). Unless otherwise stated this chapter draws on material from these volumes.

3. See Bahr (2007) for a consideration of the relationship between the Akimel O'odham and the Hohokam.

4. See Crown (1991) for the history of research on the Hohokam.

5. See Taylor (within Doelle 2009) for the history of protection of Casa Grande..

6. Andrews and Bostwick (2000) describe the Pueblo Grande site in the context of the Hohokam culture.

7. Masse (1991) describes subsistence in the Sonoran Desert.

8. See Mithen (2003, chapters 26 and 27) for Palaeo-Indians and extinction of the mega-fauna.

9. Wallace (2007) considers Hohokam origins, while Doyel (1991, 227–8), Wilcox (1991, 273–4) and McGuire and Villalpando (2007) consider the relationship between Mesoamerica and the Hohokam.

10. Mabry (ed. 2008) describes Las Capas and the early canals in the Tucson Basin.

11. Description and discussion of the Hohokam canal and irrigation system pervades much of the literature on the Hohokam. Especially useful are Gregory (1991), Masse (1991), Wilcox (1991), Andrews and Bostwick (2000), and Doyel (2007).

12. For Hohokam settlement patterns see Masse (1991), Gregory (1991) and Wilcox (1991).

13. Elson (2007) describes the architecture of Hohokam ballcourts.

14. Whittlesey (2007) describes the iconography on Hohokam ceramics and how these may relate to their belief system. See Bayman (2007) for a description of other Hohokam crafts.

15 Whittlesey (2007, 71).

16. Haury (1976).

17. Wilcox (1991). See also Ravesloot (2007) for changing views of Snaketown.

18. The following draws on Gregory (1991) and Masse (1991).

19. For Hohokam trade and its social implications see Doyel (1991).

20. The social changes associated with the Classic period are described and discussed in Wilcox (1991), Masse (1991) and Gregory (1991). The Hohokam culture has been divided into numerous additional phases, as described in Crown (1991).

21. Doelle (2009) edits an edition of *Archaeology Southwest* from the Centre for Desert Archaeology with several articles describing various aspects of Casa Grande.

22. Masse (1991, 219).

10. Life and death of the water lily monster

1 Matheny (1976). I should also acknowledge the work of Adams (1980), who began fieldwork in the Maya region in 1958. I am nyself cautious about this comparison because Groslier can justifiably be considered to have made a more profound academic contribution to our understanding of South-East Asian archaeology than Matheny has to the Maya.

2. For biographical information about Ray Matheny see Jackson (2001).

3. Matheny's long-term research at Edzná is published in Matheny et al. (1983). Accounts of the site can be found in standard text books regarding the Mayan civilisation, and a summary of its hydraulic engineering in Mays and Gorokhovich (2010).

4. The following text draws on several general works concerning Mayan civilisation, notably Coe (1999), Demarest (2004), Lucero (2006), Webster (2002).

5. For Kaminalijuyú see Demarest (2004, 72–86).

6. Norman Hammond excavated Cuello – see Hammond (2000) for a summary.

7. I am grateful to Norman Hammond for drawing my attention to this particular relief. It could, of course, be interpreted in other ways and the role of sacrifice, if any, within Maya ball games remains unclear.

8. Lucero (2006, 160).

9. There is an extensive literature on the collapse of the Ancient Maya. This text primarily draws on Gill (2000), Gill et al. (2007) and Webster (2002).

10. Lucero (2006, 190) describes how day-to-day life at Saturday Creek and other minor settlements that had access to water had continuity across the so-called 'Mayan collapse': 'Political shifts occurring elsewhere in the southern Maya lowlands had little impact on a community that was not much involved in Classic Maya politics to begin with.'

11. Back (1995) describes the geology of the Yucatán and its implications for water management.

12. Cited by Gill (2000, 255).

13. Cited by Back (1995, 241).

14. The mobility of the Ancient Maya has been stressed by Lucero (2006).

15. Saturday Creek is described by Lucero (2006).

16. Demarest (2004).

17. Scarborough (1998, 139).

18. The following text draws on Scarborough (2003, 108–13), Scarborough and Gallopin (1991) and Lucero (2000).

19. Scarborough (1998; 2003).

20. Scarborough (1998; 2003), Scarborough and Gallopin (1991).

21. The following text regarding water management at Edzná draws on Matheny (1976) and Matheny et al. (1983).

22. The following text draws on Folan (1992), Folan et al. (1995), Gunn et al. (2002) and Gunn, Foss et al. (2002).

23. Davis-Salazaar (2006).

24. Davis-Salazaar (2006).

25. Dunning et al. (1999).

26. Beach et al. (2009); see also Fedick et al. (2000).

27. Dunning et al. (1997).

28. Beach and Dunning (1997).

29. For example, see McAnany (1990) and Johnston (2004).

30. The Uxul discoveries were reported at http://www.sciencedaily.com/releases/2010/08/100826083803.htm (accessed on 23 December 2011).

31. Bonnafoux (2011).

32. Gill (2000, 264) and Lucero (2006, 160) stress the significance of water lilies.

33. See Guillermo (2007) for a description of the sacred *cenote* and a study of the human remains.

34. Bonnafoux (2011) provides a review of the iconography of water in Mayan art and related cultures of Mesoamerica.

35. Scarborough (2003, 126).

36. Lucero (2006).

37. Scarborough (1998) emphasises the significance of mirror images and surface tensions in the performance of Mayan ritual.

38. Karlén (1984)

39. Gill et al. (2007).

40. Thompson et al. (1985; 1986; 1998); Shimada et al. (1991).

41. Curtis et al. (1996); see also Hodell et al. (1995).

42. Haug (2003); Gill et al. (2007).

43. Webster et al. (2007).

44. Hammond and Sabloff were quoted in *New Scientist*, 13 March 2003, within an article entitled 'Intense droughts blamed for the Mayan collapse' by Gaia Vince.

45. Massey (1986).

46. Reported in http://news.nationalgeographic.com/news/2005/11/1117_051117_maya_massacre.html.

47. Beach and Dunning (1997).

48. See Scarborough (1998) and Bonnafoux (2011) for the role of water in Mayan ritual.

11. Water poetry in the Sacred Valley

1. See Berger et al. (1998) for radiocarbon dating of Machu Picchu.

2. There is a multitude of books that provide detailed studies and more general reviews of the Incas. Unless otherwise stated, this book draws on Bruhns (1994) and Moseley (2001). Morris and Hagen (2011) is a particularly impressive recent publication.

3. For reviews of pre-Inca Southern American archaeology see Bruhns (1994), Burger (1992) and Moseley (2001).

4. Sandweiss et al. (1998). For debates about the earliest settlement in the Americas see Mithen (2005, Chapters 23–29).

5. The following draws on Dillehay et al. (2005). Whether or not the field-work was undertaken in 2005 remains unclear as this is not referred to in the publication.

6. For Apsero I draw on descriptions in Solís et al. (2001) and Haas et al. (2004).

7. Moseley (1975).

8. Solís et al. (2001).

9. Haas et al. (2005).

10. Haas et al (2004). I should note that there has been a substantial academic dispute between Solís and Haas regarding the primacy of their research and ideas.

11. The case of periodic environmental change caused by El Niño and earth-quake damage to settlements between 3800 and 1600 BC has been made by Sandweiss et al. (2009).

12. For studies on pre-Inca canal systems in Peru see Ertsen (2010), Ortloff et al. (1985) and Nordt et al. (2004).

13. Burger (1992).

14. For the farming systems of the Tiahuanaco see McAndrews et al. (1997) and Kolata (1986).

15. For the Nasca culture see Silverman and Proulx (2002).

16. Schreiber and Rojas (1995) describe the puquios of the Nasca.

17. Fagan (1999) provides an excellent study of the impact of the El Niño phenomenon on ancient civilisations.

18. Thompson et al. (1985; 1986).

19. Barker et al. (2001).

20. Verano (2000), accessed at http://www.scielo.cl/scielo.php?pid=S0717-73562000000100011&script=sci_arttext (24 December 2011), describes these finds while Taggart (2010) provides a wider perspective on Moche sacrifice.

21. Salazar and Salazar (2004) provides a beautifully illustrated study of the

Inca sites within the Sacred Valley, combining good archaeological descriptions with highly speculative interpretations.

22. Bingham (1952).

23. The following text draws on Wright et al. (1997, 1999) and Wright and Valencia Zegarra (2001). See also Wright (2008).

24. For a recent analysis of skeletal remains at Machu Picchu see Turner et al. (2009).

25. The following text draws on Wright (2006).

12. *An unquenched thirst*

1. For the reconstruction of climate change from the Quelccaya ice cap, see Thompson et al. (1985).

2. The following text regarding glacier retreat and water insecurity in Peru draws on Lubovich (2007).

3. http://news.discovery.com/history/roman-aqueduct-emperor.html (accessed 2 January 2012).

4. http://www.livescience.com/11171-ancient-mayan-reservoirs-discovered-city-ruins.html (accessed 2 January 2012).

5. For example, see Arco and Abrams (2006), Scarborough (2003), Solomon (2010), Wilkinson and Rayne (2010), and Scarborough and Lucero (2010).

6. This reinforces the conclusion drawn by Scarborough (2003).

7. See Gleick (2006) for a review of water and terrorism.

8. Scarborough and Lucero (2010).

9. See Clarke and King (2004) for a global snapshot of how water is currently used within the world.

10. See Solomon (2010) for an outstanding account of what he describes as 'water and the making of the modern industrial society'.

11. For the intrusion of saltwater into freshwater aquifers see Barlow and Reichard (2009) and Tiruneh and Motz (2004).

12. As reported on the BBC news website 17 December 2011, http://www.bbc.co.uk/news/world-asia-pacific-16229394 (accessed 2 January 2012).

13. For the water situation in Jordan see chapters in Mithen and Black (2011).

14. Reported in http://www.nytimes.com/2007/08/03/world/europe/03iht-dry.4.6976449.html (accessed 2 January 2012).

15. http://www.wsws.org/articles/2007/aug2007/anka-a22.shtml and http://news.bbc.co.uk/1/hi/world/europe/2137280.stm, accessed 2 January 2012.

16. For an assessment of China's water challenge see Gleick (2009b), Solomon (2010) and Watts (2010).

17. http://www.independent.co.uk/news/world/asia/a-poisoned-paradise-water-water-everywhere-798416.html (accessed 2 January 2012).

18. http://tucsoncitizen.com/three-sonorans/2010/12/27/arizona-water-crisis-begins-in-2011/ (accessed 2 January 2012).

19. For the impact of excessive pumping of ground water in the Phoenix region and throughout the US see Alley and Reilly (1999) and US Geological Fact Sheet 103-03 (2003).

20. http://www.time.com/time/world/article/0,8599,1890623,00.html, (accessed 2 January 2012).

21. Plusquellec (1990).

22. This is drawn from the abstract of Mathot's 2011 MSc Thesis for Wageningen University concerning the management reforms of the Gezira irrigation scheme, which was accessed on 28 April 2012 at http://www.iwe. wur.nl/NR/rdonlyres/76AE196F-E01A-473E-A538-F2F682329772/146787/ IW1197AbstractThesisKoenMathotvertrouwelijk.pdf.

23. This was, in fact, the understanding gained by Beth Reed, who joined me on a visit to Gezira. I take responsibility if this is an inaccurate representation of the situation.

24. For water supply issues in Amman see Darmame and Potter (2011).

BIBLIOGRAPHY

Acker, R. 1998. New geographical tests of the hydraulic thesis at Angkor. *South East Asia Research* 6, 5–47

Adams, R. E. W. 1980. Swamps, canals, and the locations of Ancient Maya cities. *Antiquity* LIV, 206–214

Adams, R. McC. 1981. *Heartland of Cities*. Chicago: University of Chicago Press

Adams, R. McC. and Nissen, H. J. 1972. *The Uruk Countryside*. Chicago: University of Chicago Press

Akashel, T. S. 2004. Nabataean and modern watershed management around the Siq and Wadi Musa in Petra. In *Men of Dikes and Canals: The Archaeology of Water in the Middle East* (eds. H.-D. Bienert and J. Häser), pp. 108–20. Orient-Archäologie Band 13

Algaze, G. 2008. *Ancient Mesopotamia at the Dawn of Civilization: The Evolution of an Urban Landscape*. Chicago: University of Chicago Press

Allan, J. A. 2002. *The Middle East Water Question: Hydropolitics and the Global Economy*. London: I. B. Tauris

Alley, W. A. and Reilly, T. E. 1999. Sustainability of ground water resources. *US Geological Survey Circular* 1186

Andrews, J. P. and Bostwick, T. W. 2000. *Desert Farmers at the River's Edge: The Hohokam and Pueblo Grande*. Phoenix: Pueblo Grande Museum and Archaeological Park

Angelakis, A. N., Dialynas, E. G. and Despotakis, V. 2012. Evolution of water supply technologies through the centuries in Crete, Greece. In *Evolution of Water Supply through the Millennia* (eds. A. N. Angelakis, L. W. Mays, D. Koutsoyiannis and N. Mamassis), pp. 226–58. London: IWA Publishing

Angelakis, A. N., Mays, L. W., Koutsoyiannis, D. and Mamassis, N. 2012 (eds.). *Evolution of Water Supply through the Millennia*. London: IWA Publishing

Angelakis, A. N., Savvakis, Y. and Charalampakis, G. 2006. Minoan aqueducts: a pioneering technology. *IWA 1st International Symposium on Water and Wastewater Technologies in Ancient Civilisations*. Iraklio, Greece, National Agricultural Research Foundation

Apostol, T. M. 2004. The Tunnel of Samos. *Engineering and Science* 1, 30–40

Arco, L. J. and Abrams, E. M. 2006. An essay on energetics: the construction of the Aztec chinampa system. *Antiquity* 80, 906–18

Artzy, M. and Hillel, D. 1988. A defense of the theory of progressive salinisation in Ancient Southern Mesopotamia. *Geoarchaeology* 3, 235–8

Ashby, T. 1935. *The Aqueducts of Ancient Rome*. Oxford: Clarendon Press

Back, W. 1995. Water management by early people in the Yucatán, Mexico. *Environmental Geology* 25, 239–42

Bahr, D. M. 2007. O'odham traditions about the Hohokam. In *The Hohokam Millennium* (eds. S. K. Fish and P. R. Fish), pp. 123–30. Santa Fe: School for Advanced Research Press

Balcer, J. 1974. The Mycenaean dam at Tiryns. *American Journal of Archaeology*, Vol. 78, 141–9

Barker, G. 2006. *The Agricultural Revolution in Prehistory: Why Did Foragers Become Farmers?* Oxford: Oxford University Press

Barker, G., Adams, R., Creighton, O. et al. 2007a. Chalcolithic (*c*.5000–3600 cal.BC) and Bronze Age (*c*.3600–1200 cal.BC) settlement in Wadi Faynan: metallurgy and social complexity. In *Archaeology and Desertification: The Wadi Faynan Landscape Survey, Southern Jordan* (eds. G. Barker, D. Gilbertson and D. Mattingly). Oxford: Council for British Archaeology in the Levant/Oxbow Books

Barker, G., Gilbertson, D. and Mattingley, D. (eds.). 2007b. *Archaeology and Desertification: The Wadi Faynan Landscape Survey, Southern Jordan.* Oxford: Council for British Archaeology in the Levant/Oxbow Books

Barker, P. A., Seltzer, G. O., Fritz, S. C., Dunbar, R. B., Grove, M. J., Tapla, P. M., Cross, S. L., Rowe, H. D. and Broda, S. P. 2001. The history of South American tropical precipitation for the past 25,000 years. *Science* 291, 640–43

Barlow, M. 2007. *Blue Covenant: The Global Water Crisis and the Coming Battle for Rights to Water*. New York and London: New Press

Barlow, P. M. and Reichard, E. G. 2009. Saltwater intrusion in the coastal regions of North America. *Hydrogeology Journal* 18, 247–60

Bar-Yosef, O. 1996. The walls of Jericho: an alternative explanation. *Current Anthropology* 27, 150–62

Bar-Yosef, O. 1998. The Natufian culture in the Levant: threshold to the origins of agriculture. *Evolutionary Anthropology* 6, 159–72

Bar-Yosef, O. and Gopher, A. (eds.). 1997. *An Early Neolithic Village in the Jordan Valley. Part I: The Archaeology of Netiv Hagdud*. Cambridge, Mass.: Peabody Museum of Archaeology and Ethnology, Harvard University

Bar-Yosef, O., Gopher, A. and Goring-Morris, N. 2010. *Gilgal: Early Neolithic Occupations in the Lower Jordan Valley. The Excavations of Tamar Noy*. Oxford: Oxbow Books

Bar-Yosef, O. and Valla, F. R. 1991 (eds.). *The Natufian Culture in the Levant*. Ann Arbor, Mich.: International Monographs in Prehistory

Barnaby, W. 2009. Do nations go to war over water? *Nature* 458, 282–3

Bayman, J. M. 2007. Artisans and their crafts in Hohokam society. In *The Hohokam Millennium* (eds. S. K. Fish and P. R. Fish), pp. 75–82. Santa Fe: School for Advanced Research Press

Beach, T. and Dunning, N. 1997. An Ancient Maya reservoir and dam at Tamarindito, El Petén, Guatemala. *Latin American Antiquity* 8, 20–29

Beach, T., Luzzadder-Beach, S., Dunning, N., Jones, J., Lohse, J., Guderjan, T., Bozarth, S., Millspaugh, S. and Bhattacharya, T. 2009. A review of human and natural changes in Maya lowland wetlands over the Holocene. *Quaternary Sciences Reviews* 28, 1710–24

Bedal, L.-A. 2004. *The Petra Pool-Complex: A Hellenistic Paradeisos in the Nabataean Capital*. Piscataway: Georgias Press

Bedal, L.-A. and Schryver, J. G. 2007. Nabataean landscape and power: evidence from the Petra garden and pool complex. In *Crossing Jordan: North American Contributions to the Archaeology of Jordan* (eds. T. E. Levy, P. M. M. Daviau and R. W. Youker), pp. 375–83. London: Equinox Publishing Ltd

Beekman, C. S., Weigand, P. C. and Pint, J., 1999. Old World irrigation technology in a New World context: qanats in Spanish Colonial Mexico. *Antiquity* 73, 440–46

Bellwald, U. 1999. Streets and hydraulics: the Petra National Trust Siq Project 1996–1999: the archaeological remains. In *Men of Dikes and Canals: The Archaeology of Water in the Middle East* (eds. H.-D. Bienert and J. Häser), pp. 73–94. Orient-Archäologie Band 13

Berger, R., Chohfi, R., Valencia Zegarra, A., Yepez, W. and Carrasco, O. 1988. Radiocarbon dating Machu Picchu, Peru. *Antiquity* 62, 707–10

Bienert, H. 2004. The underground tunnel system in Wadi ash-Shellalah, northern Jordan. In *Men of Dikes and Canals: The Archaeology of Water in the Middle East* (eds. H.-D. Bienert and J. Häser). Orient-Archäologie Band 13

Bienert, H.-D. and Häser, J. (eds) 2004. *Men of Dikes and Canals: The Archaeology of Water in the Middle East*. Orient-Archäologie Band 13

Bingham, H. 1952. *Lost City of the Incas*. London: Phoenix House

Bonnafoux, P. 2011. Waters, droughts, and Early Classic Maya worldviews. In *Ecology, Power, and Religion in Maya Landscapes*, (eds. C. Isendahl and B. L. Person), pp. 31–48. *Acta Mesoamerica*, Vol. 23

Bono, P. and Boni, C. 1996. Water supply of Rome in antiquity and today. *Environmental Geology* 27, 126–34

Bono, P., Crow, J. and Bayliss, R. 2001. The water supply of Constantinople: archaeology and hydrogeology of an Early Medieval city. *Environmental Geology* 40, 1325–33

Brayshaw, D., Black, E., Hoskins, B. and Slingo, J. 2011. Past climates of the Middle East. In *Water, Life and Civilisation: 20,000 Years of Climate, Environment and Society in the Jordan Valley* (eds. S. Mithen and E. Black), pp. 25–50. Cambridge: Cambridge University Press/UNESCO

Bruhns, K., 1994, *Ancient South America*. Cambridge: Cambridge University Press

Buckley, B. M., Anchukaitis, K. J., Penny, D., Fletcher, R., Cook, E. R., Sano, M., le Canh Lam, Wichienkeeo, A., Minh, T. T. and Hong, T. M. 2010. Climate as a contributing factor in the demise of Angkor, Cambodia. *Proceedings of the National Academy of Sciences* 107, 6748–52

Burger, R. L. 1992. *Chavín and the Origins of Andean Civilization*. New York: Thames & Hudson

Byrd, B. F. 2005. *Early Village Life at Beidha, Jordan: Neolithic Spatial Organization and Vernacular Architecture*. Oxford: Oxford University Press

Cadogan, G. 2007. Water management in Minoan Crete, Greece: the two cisterns of one Middle Bronze Age settlement. *Water Science and Technology: Water Supply* 17, 103–11

Cadogan, G., Hatzaki, E. and Vasilakis, A. (eds.). 2004. *Knossos: Palace, City, State. Proceedings of the Conference in Herakleion Organised by the British School at Athens and the 23rd Ephorate of Prehistoric and Classical Antiquities of Herakleion, in November 2000, for the Centenary of Sir Arthur Evans's Excavations at Knossos*, London: British School at Athens Studies 12

Caran, S. C., Neeby, J. A., Winsborough, B. M., Sorensen, F. R. and Valastros, S., Jr 1996. A late Palaeoindian/early archaic water well in Mexico: possible oldest water management feature in the New World. *Geoarchaeology* 11, 1–35

Castleden, R. 1992. *Minoan Life in Bronze Age Crete*. London: Routledge

Çeçen, K. 1996. *The Longest Roman Supply Line*. Istanbul: Turkiye Sınai Kalkınma Bankası

Childe, V. G. 1936 (1965). *Man Makes Himself*. 4th edn. London: Watts

—. 1957. *The Dawn of European Civilization*. 6th edn. London: Routledge and Kegan Paul,

Chiotis, E. D. and Chioti, L. E. 2012. Water supply of Athens in antiquity. In *Evolution of Water Supply through the Millennia* (eds. A. N. Angelakis, L. W. Mays, D. Koutsoyiannis and N. Mamassis), pp. 407–42. London: IWA Publishing

Ciniç, N. 2003. *Yerebatan Cistern and Other Cisterns of Istanbul.* Istanbul: Duru Basım Yayın Reklamcılık ve Gıda San, Tic Ltd

Clarke, R. and King, J. 2004. *The Atlas of Water.* Brighton: Earthscan

Clarke, T. H. M. 1903. Prehistoric sanitation in Crete. *British Medical Journal,* September 12, 1903, 597–9

Coe, M. D. 1999. *The Maya.* 6th edn. New York: Thames & Hudson

—. 2003. *Angkor and the Khmer Civilization.* London: Thames & Hudson

Crouch, D. 1984. The Hellenistic water system of Morgantina, Sicily: contributions to the history of urbanisation. *American Journal of Archaeology,* 88, 353–65

Crouch, D. P. 1993. *Water Management in Ancient Greek Cities.* Oxford: Oxford University Press

Crow, J., Bardill, J. and Bayliss, R. 2008. *The Water Supply of Byzantine Constantinople.* Society for the Promotion of Roman Studies, *Journal of Roman Studies* Monograph No. 11

Crown, P. L. 1991. The Hohokam: current views of prehistory and the regional system. In *Chaco and Hohokam: Prehistoric Regional Systems in the American Southwest* (eds. P. L. Crown and W. J. Judge), pp. 135–57. Santa Fe: School of American Research Press

Crown, P. L. and Judge, W. J. (eds.). 1991. *Chaco and Hohokam: Prehistoric Regional Systems in the American Southwest.* Santa Fe: School of American Research Press

Curtis, J., Hodell, D. and Brenner, M. 1996. Climate variability on the Yucatán Peninsula (Mexico) during the past 3,500 years and implications for the Maya cultural evolution. *Quaternary Research* 46, 37–47

Darmame, K. and Potter, R. B. 2011. Political discourses and public narratives on water supply issues in Amman. *Water, Life and Civilisation: 20,000 years of Climate, Environment and Society in the Jordan Valley,* (eds. S. J. Mithen and E. Black), pp. 455–465. Cambridge: Cambridge University Press/UNESCO

Davis-Salazaar, K. 2006. Late Classic Maya drainage and flood control at Copán, Honduras. *Ancient Mesoamerica* 17, 125–38.

De Deo, G., Laureano, P., Mays, L. W. and Angelakis, A. N. 2012. Water supply management technologies in the Ancient Greek and Roman civilisations. In *Evolution of Water Supply through the Millennia* (eds. A. N. Angelakis, L. W. Mays, D. Koutsoyiannis and N. Mamassis), pp. 351–82. London: IWA Publishing

DeLaine, J. 1997. *The Baths of Caracalla: A Study in the Design, Construction and Economics of Large Scale Building Projects in Imperial Rome.* Journal of Roman Archaeology, Supplementary Series, N. 23

Demarest, A. 2004. *Ancient Maya: The Rise and Fall of a Rainforest Civilization*. Cambridge: Cambridge University Press

Dickinson, O. P. T. K. 1994. *The Aegean Bronze Age*. Cambridge: Cambridge University Press

Doelle, W. H. 2009. Hohokam heritage: the Casa Grande community. *Archaeology Southwest* 25, No. 4

Doyel, D. E. 1991. Hohokam exchange and interaction. In *Chaco and Hohokam: Prehistoric Regional Systems in the American Southwest*, (eds. P. L. Crown and W. J. Judge), pp. 225–52. Santa Fe: School of American Research Press

—. 2007. Irrigation, production and power in Phoenix Basin Hohokam society. In *The Hohokam Millennium* (eds. S. K. Fish and P. R. Fish), pp. 83–9. Santa Fe: School for Advanced Research Press

Du, P. and Koenig, A. 2012. History of water supply in pre-modern China. In *Evolution of Water Supply through the Millennia* (eds. A. N. Angelakis, L. W. Mays, D. Koutsoyiannis and N. Mamassis), pp. 169–226. London: IWA Publishing

Dunning, N., Beach, T. and Rue, D. 1997. The paleoecology and ancient settlement of the Petexbatún Region, Guatemala. *Ancient Mesoamerica* 8, 255–66

Edwards, P. C., Meadows, J., Metzgar, M. C. and Sayei, G. 2002. Results from the first season at Zahrat adh-Dhra' 2: a new Pre-Pottery Neolithic A site on the Dead Sea plain in Jordan. *Neo-Lithics* 1/02: 11–16

El-Gohary, F. A. 2012. A historical perspective on the development of water supply in Egypt. In *Evolution of Water Supply through the Millennia* (eds. A. N. Angelakis, L. W. Mays, D. Koutsoyiannis and N. Mamassis), pp. 127–46. London: IWA Publishing

Elson, M. D. 2007. Into the earth and up to the sky. In *The Hohokam Millennium* (eds. S. K. Fish and P. R. Fish), pp. 49–55. Santa Fe: School for Advanced Research Press

Ertsen, M. W. 2010. Structuring properties of irrigation systems: understanding relations between humans and hydraulics through modeling. *Water History* 2, 165–83.

Evans, D., Pottier, C., Fletcher, R., Hemsley, S., Tapley, I., Milner, A. and Barbetti, M. 2007. A comprehensive archaeological map of the world's largest preindustrial complex at Angkor, Cambodia. *Proceedings of the National Academy of Sciences* 104, 14277–82

Evans, H. B. 1997. *Water Distribution in Ancient Rome: The Evidence of Frontinus*. Ann Arbor: University of Michigan Press

Evely, D., Hughes-Brock, H. and Momigliano, N. (eds.). 2007. *Knossos: A Labyrinth of History*. London: British School at Athens

Evenari, M., Shanan, L. and Tadmor, N. 1982. *The Negev: The Challenge of a Desert* (2nd edn). Cambridge, Mass.: Harvard University Press

Fabre, G., Fiches, J.-L. and Paillet, J.-L. 1991. Interdisciplinary Research on the Aqueduct of Nîmes and the Pont du Gard. *Journal of Roman Archaeology* 4, 63–88

Fagan, B. M. 1999. *Floods, Famines, and Emperors: El Niño and the Fate of Civilizations*. New York: Basic Books

Fedick, S. L., Morrison, B. A., Andersen, B. J., Boucher, S., Ceja Acosta, J. and Mathews, J. P. 2000. Wetland manipulation in the Yalahau region of the northern Maya lowlands. *Journal of Field Archaeology* 27, 131–52

Fernea, R. A. 1970. *Shaykh and Effendi: Changing Patterns of Authority among the El Shabana of Southern Iraq*. Cambridge, Mass.: Harvard University Press

Finlayson, B., Kuijt, I., Arpin, T., Chesson, M., Dennis, S., Goodale, N., Kadowki, S., Maher, S., Smith, S., Schurr, M. and McKay, J. 2003. Dhra' Excavation Project, 2002 Interim Report. *Levant* 25, 1–38

Finlayson, B. and Mithen, S. (eds.). 2007. *The Early Prehistory of Wadi Faynan, Southern Jordan: Archaeological Survey of Wadis Faynan, Ghuwayr and al-Bustan and Evaluation of the Pre-Pottery Neolithic A Site of WF16*. Oxford: Council for British Research in the Levant/Oxbow Books

Finlayson, B., Mithen, S. J., Najjar, M., Smith, S., Maričević, D., Pankhurst, N. and Yeomans, L. 2011. Architecture, sedentism and social complexity: communal building in Pre-Pottery Neolithic A settlements: new evidence from WF16. *Proceedings of National Academy of Sciences* doi10.1073/pnas 1017642108

Fish, P. R. and Fish, S. K. 2007. Community, territory and polity. In *The Hohokam Millennium*, (eds. S. K. Fish and P. R. Fish), pp. 39–47. Santa Fe: School for Advanced Research Press.

Fish, S. K. and Fish, P. R. (eds.). 2007. *The Hohokam Millennium*. Santa Fe: School for Advanced Research Press

Fleagle, J. 2010. *Out of Africa 1: The First Hominin Colonization of Eurasia*. Berlin: Springer Verlag

Fletcher, R., Evans, D., Tapley, I. and Milne, A. 2004. Angkor: extent, settlement pattern and ecology: preliminary analysis of an AIRSAR survey in September 2000. *Bulletin of the Indo-Pacific Prehistory Association* 24, 133–8

Fletcher, R., Penny, D., Evans, D., Pottier, C., Barbetti, M., Kummu, M., Lustig, T. and ASPARA. 2008. The water management network of Angkor, Cambodia. *Antiquity* 82, 658–70

Folan, W. J. 1992. Calakmul, Campeche: a centralised urban administrative center in the northern Petén. *World Archaeology* 24, 158–68

Folan, W. J., Marcu, J., Pincemin, S., Carrasco, M. R. D., Fletcher, L. and Lopez, A. M. 1995. Calakmul: new data from an Ancient Maya capital in Campeche, Mexico. *Latin American Antiquity* 6, 310–34

Foote, R., Wade, A., el Bastawesy, M., Oleson, J. and Mithen, S. J. 2011. A millennium of rainfall, settlement and water management at Humayma, southern Jordan, 2050–11050 BP (100 BC–AD 900). In *Water, Life and Civilisation: Climate, Environment and Society in the Jordan Valley*, (eds. S. J. Mithen and E. Black), pp. 302–33. Cambridge: Cambridge University Press/UNESCO

Forbes, R. J. 1956. Hydraulic engineering and sanitation. In *A History of Technology* (eds. C. Singer et al.), pp. 663–94. Oxford: Oxford University Press

Freeman, M. and Jacques, C. 2003. *Ancient Angkor*. Thailand: Amarin Printing and Publishing (Public) Co. Ltd

Frumkin, A. and Shimron, A. 2005. Tunnel engineering in the iron age: geo-archaeology of the Siloam Tunnel, Jerusalem. *Journal of Archaeological Science* 33, 227–37

Fujii, S. 2007a. Wadi Abu Tulayha: a preliminary report of the 2006 summer field season of the Jafr Basin Prehistoric Project, Phase 2. *Annual of the Department of Antiquities of Jordan* 51, 373–401

—. 2007b. PPNB barrage systems at Wadi Abu Tulayha and Wadi ar-Ruway-shid Ash-Sharqi: a preliminary report of the 2006 spring field season of the Jafr Basin Prehistoric Project, Phase 2. *Annual of the Department of Antiquities of Jordan* 51, 403–27

—. 2008. Wadi Abu Tulayha: A Preliminary Report of the 2007 Summer Field Season of the Jafr Basin Prehistoric Project, Phase 2. *Annual of the Department of Antiquities of Jordan* 52, 445–78

Galili, E. and Nir, Y. 1993. The submerged Pre-Pottery Neolithic water well of Atlit-Yam, northern Israel and its palaeoenvironmental implications. *The Holocene*, 3, 265–70

Galili, E., Weinstein-Evron, M., Hershkovitz, I., Gopher, A., Kislev, M., Lernau, O., Kolska-Horwitz , L. and Lernau, H. 1993. Atlit-Yam: a prehis-toric site on the sea floor off the Israeli coast. *Journal of Field Archaeology*, 20, 133–57

Garfinkel, Y. 1989. The Pre-Pottery Neolithic A site of Gesher. *(Mitekufat Haeven) Journal of the Israel Prehistoric Society* 22, 145

Garfinkel, Y., Vered, A. and Bar-Yosef, O. 2006. The domestication of water: the Neolithic well at Sha`ar Hagolan, Jordan Valley, Israel. *Antiquity* 80, 686–96

Gebel, H. G. K. 2004. The domestication of water: evidence from early Neolithic Ba'ja. In *Men of Dikes and Canals: The Archaeology of Water in the Middle East* (eds. H.-D. Bienert and J. Häser), pp. 25–35. Orient-Archäologie Band 13

Geertz, C. 1980. *Negara: The Theatre State in Nineteenth-Century Bali.* Princeton: Princeton University Press

Gibson, M. 1974. Violation of fallow and engineered disaster in Mesopotamian civilization. In *Irrigation's Impact on Society* (eds. T. E. Dowling and M. Gibson), pp. 7–19. Tucson: University of Arizona Press

Gill, R. B. 2000. *The Great Maya Droughts: Water, Life and Death.* Albuquerque: University of New Mexico Press

Gill, R. B., Mayewski, P. A., Nyberg, J., Haug, G. H. and Peterson, L. C. 2007. Drought and the Maya collapse. *Ancient Mesoamerica* 18, 283–302

Gillet, K. and Mowbray, H. 2008. *Dujiangyan: In Harmony with Nature.* Beijing: Matric International Publishing House

Gleick, P. H. 2006. Water and terrorism. In *The World's Water 2006–2007: The Biennial Report on Freshwater Resources,* (ed. P. H. Gleick), pp. 1–28. Washington DC: Island Press

—. (ed.) 2006. *The World's Water 2006–2007. The Biennial Report on Freshwater Resources.* Washington DC: Island Press

—. 2009a. *The World's Water 2008–2009. The Biennial Report on Freshwater Resources.* Washington DC: Island Press

—. 2009b. China and Water. In *The World's Water 2008–2009: The Biennial Report on Freshwater Resources,* (ed. P.H. Gleick), pp. 79–97. Washington DC: Island Press

—. 2009c. Three Gorges Dam project. In *The World's Water 2008–2009: The Biennial Report on Freshwater Resources* (ed. P. H. Gleick), pp. 139–49. Washington DC: Island Press

Goloubew, V. 1936. Reconnaissances aériennes au Camboge. *Bulletin de l'École Française d'Extrême Orient* 36, 465–78

—. 1941. L'hydraulique urbaine et agricole à l'époque des rois d'Angkor. *Bulletin Economique de l'Indochine* 1, 1–10

Goodfield, J. and Toulmin, S. 1965. How was the tunnel of Eupalinos aligned? *Isis,* 56, 46–55

Groslier, B.-P. 1956. Milieu et évolution en Asie. *Bulletin de la Société des Études Indochinoises* 27, 51–83

—. 1956. *Angkor: Hommes et Pierres.* Paris: Arthaud

—. 1960. Our knowledge of Khmer civilisation: a reappraisal. *Journal of the Siam Society* 48, 1–28

—. 1966. *Indochina.* London: Methuen and Co.

—. 1974. Agriculture et religion dans l'empire angkorien. Étude Rurales, 53–6, 95–117

—. 1979. Le cité hydraulique Angkorienne: exploitation ou surexploitation du sol? *Bulletin de l'École Française d'Extrême-Orient* 66, 161–202

Gregory, D. A. 1991. Form and variation in Hohokam settlement patterns. In *Chaco and Hohokam: Prehistoric Regional Systems in the American Southwest* (eds. P. L. Crown and W. J. Judge), pp. 159–63. Santa Fe: School of American Research Press

Guangqian Wang, Baosheng Wu and Tiejian Li 2007. Digital Yellow River model. *Journal of Hydro-Environment Research* 2007. Doi:10.1016/j.jher 2007.03.001

Guillermo, A. A. de. 2007. Sacrifice and ritual body mutilation in Postclassical Maya society: taphonomy of the human remains from Chichén Itzá's Cenote Sagrado. In *New Perspectives on Human Sacrifice and Ritual Body Treatments in Ancient Maya Society* (eds. V. Tiesler and A. Cucina), pp. 190–208. New York: Springer Verlag

Gunn, J. D., Foss, J. E., Folan, W. J., Carrasco, M. R. D. and Faust, B. B. 2002. Bajo sediments and the hydraulic system of Calakmul, Campeche, Mexico. *Ancient Mesoamerica* 13, 297–315

Gunn, J. D., Matheny, R. T. and Folan, W. J. 2002. Climate-change studies in the Maya area: a diachronic analysis. *Ancient Mesoamerica* 13, 79–84

Gutgluede, J. 1987. A detestable encounter, Odyssey VI. *The Classical Journal* 83, 97–102

Haas, J., Creamer, W. and Ruiz, A. 2004. Dating the Late Archaic occupation of the Norte Chico region in Peru. *Nature* 432, 1020–23

Haas, J., Creamer, W. and Ruiz, A. 2005. Power and the emergence of complex polities in the Peruvian preceramic. *Archaeological Papers of the American Anthropological Association* 14, 37–52

Hammond, N. 2000. The Maya lowlands: pioneer farmers to merchant princes. In *The Cambridge History of the Native Peoples of the Americas, Vol. II: Mesoamerica, part 1* (eds. R. E. W. Adams and M. J. Macleod), pp. 197–249. Cambridge: Cambridge University Press

Haug, G. H., Günther, D., Peterson, L. C., Sigman, D. M., Hughen, K. A. and Aeschlimann, B. 2003. Climate and the collapse of Maya civilization. *Science* 299, 1731–5

Haury, E. W. 1976. *The Hohokam: Desert Farmers and Craftsmen*. Tucson: University of Arizona Press

Haut, B. and Viviers, D. 2012. Water supply in the Middle East during Roman and Byzantine periods. In *Evolution of Water Supply through the Millennia* (eds. A. N. Angelakis, L. W. Mays, D. Koutsoyiannis and N. Mamassis), pp. 319–50. London: IWA Publishing

Helbaek, H. 1972. Samarran irrigation agriculture at Choga Mami in Iraq. *Iraq* 34, 35–48

Helms, S. 1981. *Jawa, Lost City of the Black Desert*. New York: Cornell University Press

—. 1989. Jawa at the beginning of the Middle Bronze Age. *Levant* 21, 141–68

Higham, C. 2001. *The Civilization of Angkor*. London: Weidenfeld & Nicolson

Hiltzik, M. 2010. *Colossus: The Turbulent, Thrilling Saga of the Building of the Hoover Dam*. New York: Free Press

Hodell, D., Curtis, J. and Brenner, M. 1995. Possible role of climate in the collapse of the Classic Maya civilization. *Nature* 375, 391–4

Hodge, A. T. 1992. *Roman Aqueducts and Water Supply*. London: Duckworth

Hritz, C. and Wilkinson, T. J. 2006. Using shuttle radar topography to map ancient water channels in Mesopotamia. *Antiquity* 80, 415–24

Jackson, L. 2001. In the jungle. *Bingham Young University Magazine*, Fall 2001. http://magazine.byu.edu/?act=view&a=729

Jacobsen, T. 1958. *Salinity and Irrigation in Antiquity. Diyala Basin Archaeological Project: Report on Essential Results*. Baghdad

Jacobsen, T. and Adams, R. A. 1958. Salt and silt in Ancient Mesopotamian agriculture. *Science* 3334, 1251–8

Johnston, K. J. 2004. Lowland Maya water management practices: the household exploitation of rural wells. *Geoarchaeology* 19, 265–92

Joukowsky, M. S. 2004. The water installations of the Petra Great Temple. In *Men of Dikes and Canals: The Archaeology of Water in the Middle East*, (eds. H.-D. Bienert and J. Häser), pp. 121–41. Orient-Archäologie Band 13

Kaptijn, E. 2010. Community and power: irrigation in the Zerqua Triangle, Jordan. *Water History* 2, 145–63

Karlén, W. 1984. Dendrochronology, mass balance and glacier front fluctuations. In *Climatic Changes on a Yearly Millennia Basis: Geological, Historical and Instrumental Records* (eds. N.-A. Morner and W. Karlén), pp. 263–71. Dordrecht: D. Reidel

Kenyon, K. 1957. *Digging up Jericho*. London: Ernest Benn

Kenyon, K. and Holland, T. 1981. *Excavations at Jericho, Volume 3: The Architecture and Stratigraphy of the Tell*. London: British School of Archaeology in Jerusalem

Kidder, T. R., Lii, H. and Li, M. 2012. Sanyangzhuang: early farming and a Han settlement preserved beneath Yellow River flood deposits. *Antiquity* 86, 30–47

Kienast, Hermann J. (1995). *Die Wasserleitung des Eupalinos auf Samos (Samos XIX.)*. Bonn: Rudolph Habelt

Kirkbride, D. 1966. Five seasons at the Pre-Pottery Neolithic village of Beidha in Jordan. *Palestine Exploration Quarterly* 98, 8–72

Kirkbride, G. 1968. Beidha: Early Neolithic village life south of the Dead Sea. *Antiquity* 42, 263–74

Knauss, J. 1991. Arkadian and Boiotian Orchomenos, centres of Mycenaean hydraulic engineering. *Irrigation and Drainage Systems* 5, 363–381

Kolata, A. L. 1986. The agricultural foundations of the Tiwanaku state: a view from the heartland. *American Antiquity* 51, 748–62

Koutsoyiannis, D. and Angelakis, A. 2007. Agricultural hydraulic works in Ancient Greece. *The Encyclopedia of Water Science.* (ed. S. W. Trimble), pp. 24–77. London: CRC Press

Koutsoyiannis, D., Zarkadoulas, N., Angelakis, A. N. and Tchobanoglous, G. 2008. Urban water management in Ancient Greece: legacies and lessons. *Journal of Water Resources, Planning and Management – ASC,* 134, 45–54

Kujit, I., Finlayson, B. and Mackay, J. 2007. Pottery Neolithic landscape modification at Dhra'. *Antiquity* 81, 106–18

Kuijt, I. and Goring-Morris, N. 2002. Foraging, farming and social complexity in the Pre-Pottery Neolithic of the south-central Levant: a review and synthesis. *Journal of World Prehistory* 16, 361–440

Li Jinlong and Yi Chang 2005. *The Magnificent Three Gorges Project*

Lindner, M. 2004. Hydraulic engineering and site planning in Nabataean–Roman southern Jordan. In *Men of Dikes and Canals: The Archaeology of Water in the Middle East,* (eds. H.-D. Bienert and J. Häser), pp. 65–9. Orient-Archäologie Band 13

Lipchin, C. 2006. *A Future for the Dead Sea Basin: Water Culture among the Israelis, Palestinians and Jordanians.* Fondazioni Eni Enrico Mattei (FEEM) Working Paper 22

Liu, L. and Xu, H. 2007. Rethinking Erlitou: legend, history and Chinese archaeology. *Antiquity* 81, 886–901

Lubovich, K. 2007. The coming crisis: water insecurity in Peru. *Foundation for Environmental Security and Sustainability Issue Brief.* September 2007

Lucero, L. J. 2006. *Water and Ritual: The Rise and Fall of the Classic Maya Rulers.* Austin: University of Texas Press

Mabry, J. (ed.) 2008. *Las Capas: Early Irrigation and Sedentism in a Southwestern Floodplain (AP28).* Archaeology Southwest, Anthropological papers No. 28

Mabry, J., Donaldson, L., Gruspier, K., Mullen, G., Palumbo, G., Rawlings, N. M. and Woodburn, M. A. 1996. Early town development and water management in the Jordan Valley: investigations at Tell el-Handaquq North. *Annual of the American Schools of Oriental Research* 53: 115–54

Maroukian, H., Gaki-Papanastassiou, K. and Piteros, Ch. 2004. Geomorphological and archaeological study of the broader area of the Mycenaean dam of Megalo Rema and Ancient Tiryns, southeastern Argive Plain, Peloponnesus. *Bulletin of the Geological Society of Greece* XXXVI, 1154–63

Martini, P. and Drusiani, R. 2012. History of the water supply of Rome as a paradigm of water services development in Italy. In *Evolution of Water Supply through the Millennia* (eds. A. N. Angelakis, L. W. Mays, D. Koutsoyiannis and N. Mamassis), pp. 442-65. London: IWA Publishing

Masse, W. B. 1991. The quest for subsistence sufficiency and civilization in the Sonoran Desert. In *Chaco and Hohokam: Prehistoric Regional Systems in the American Southwest* (eds. P. L. Crown and W. J. Judge), pp. 195–223. Santa Fe: School of American Research Press

Massey, K. M. 1986. The human skeletal remains of a skull pit at Colha, Belize. Abstract to unpublished thesis, University of Minnesota

Matheny, R. T. 1976. Maya lowland hydraulic systems. *Science* 193, 639–46

Matheny, R. T., Gurr, D. L., Forsyth, D. W. and Huak, F. R. 1983. *Investigations at Edzná, Campeche, Mexico, Vol. 1: The Hydraulic System*. New World Archaeological Foundation, Brigham Young University, Provo, Utah: Papers of the New World Archaeological Foundation 46

Matsuzawa, T., Tomanaga, M. and Tanaka, M. (eds.) 2006. *Cognitive Development in Chimpanzees*. Berlin: Springer

Mays, L. and Gorokhovich, Y. 2010. Water technology in the ancient American societies. In *Ancient Water Technologies*, (ed. L. Mays), pp. 171–200. Dordrecht: Springer

Mays, L. W. 2010. A brief history of water technology during antiquity: before the Romans. In *Ancient Water Technologies* (ed. L. Mays), pp. 1–29. Dordrecht: Springer

McAnany, P. A. 1990. Water storage in the Puuc region of the northern Maya lowlands: a key to population estimates and architectural variability. In *Precolumbian Population History in the Mayan Lowlands* (eds. T. P. Culbert and D. S. Rice), pp. 263–84 Albuquerque: University of New Mexico

McAndrews, T. L., Albarracin-Jordan, J. and Bermann, M. 1997. Regional settlement patterns in the Tiwanaku Valley of Bolivia. *Journal of Field Archaeology* 24, 67–83

McCorriston, J. and Weisberg, S. 2002. Spatial and temporal variation in Mesopotamian agricultural practices in the Khabur Basin, Syrian Jazira. *Journal of Archaeological Science* 29, 485–98

McGuire, R. and Villalpando, C. E. 2007. The Hohokam and Mesoamerica. In *The Hohokam Millennium* (eds. S. K. Fish and P. R. Fish), pp. 57–63. Santa Fe: School for Advanced Research Press

Mithen, S. 2003. *After the Ice: A Global Human History 20,000–5000 BC*. London: Weidenfeld & Nicolson

Mithen, S. 2005. *The Singing Neanderthals: The Origins of Music, Language, Body and Mind*. London: Weidenfeld & Nicolson

Mithen, S. J., Finlayson, B., Najjar, M., Jenkins, E., Smith, S., Hemsley, S.,

Maričević, D., Pankhurst, N., Yeomans, L. and Al-Amarahar, H. 2010. Excavations at the PPNA site of WF16: a report on the 2008 season. *Annual of the Department of Antiquities of Jordan* 53, 115–26

Mithen, S. J., Finlayson, B. Smith, S., Jenkins, E., Najjar, M. and Maričević, D. 2011. An 11,600-year-old communal structure from the Neolithic of southern Jordan. *Antiquity* 85, 350–64

Morris C. and von Hagen, A. 2011. *The Incas: Lords of the Four Quarters.* London: Thames & Hudson

Moseley, M. E. 1973. *The Maritime Foundations of Andean Civilization.* Menlo Park: Cummings

—. 2001. *The Incas and Their Ancestors: The Archaeology of Peru.* London: Thames & Hudson

Mouhot, H. 2001. *Travels in the Central Parts of Indo-China (Siam), Cambodia, and Laos, during the Years 1858, 1859, and 1860: Volume 1.* Boston: Adamant Media Corporation

Nadel, D., Carmi, I. and Segal, D. 1995. Radiocarbon dating of Ohalo II: archaeological and methodological implications. *Journal of Archaeological Science* 22, 811–22

Nadel, D. and Hershkowitz, I. 1991. New subsistence data and human remains from the earliest Levantine epipalaeolithic. *Current Anthropology* 32, 631–5

Nadel, D. and Werker, E. 1999. The oldest ever brush hut plant remains from Ohalo II, Jordan Valley, Israel (19,000 BP), *Antiquity* 73, 755–64

Needham, J. 1971 (with contributions from Lu Gwei-Dien and Ling Wang). *Science and Civilisation in China, Vol. 4: Physics and Physical Technology.* Cambridge: Cambridge University Press

Nielsen, I. 1990. *Thermae et Balnea: The Architecture and Cultural History of Roman Public Baths.* Aarhus: Aarhus University Press

Nordt, L., Hayashida, F., Hallmark, T. and Crawford, C. 2004. Late prehistoric soil fertility, irrigation management and agricultural production in northwest coastal Peru. *Geoarchaeology* 19, 21–46

Oates, J. 1968. Prehistoric investigations near Mandali, Iraq. *Iraq* 30, 1–20.

—. 1969. Choga Mami 1967–68: A preliminary report. *Iraq* 31, 115–52.

Oleson, J. P. 1984. *Greek and Roman Mechanical Water-Lifting Devices. The History of a Technology.* Dordrecht: D. Reidel

—. 1995. The origins and design of Nabataean water-supply systems. In *Studies in the History and Archaeology of Jordan V,* (ed. G. Bisheh), pp. 707–19. Amman: Department of Antiquities

—. 2001. Water supply in Jordan through the ages. In *The Archaeology of Water Management* (ed. B. MacDonald, R. Adams and P. Bienkowski), pp. 603–34. Sheffield: Sheffield University Press

—, J. 2007a. Nabataean water supply, irrigation and agriculture: an overview. In *The World of the Nabataeans* (ed. K. D. Politis), pp. 217–49. Stuttgart: Franz Steiner Verlag

—, J. 2007b. From Nabataean King to Abbasid Caliph: the enduring attraction of Hawara/al-Humayma, a multi-cultural site in Arabia Petraea. In *Crossing Jordan: North American Contributions to the Archaeology of Jordan* (eds. T. E. Levy, P. M. M. Daviau and R. W. Youker), pp. 447–55. London: Equinox

Ortloff, C. 2005. The water supply and distribution system of the Nabataean city of Petra (Jordan), 200 BC–AD 300. *Cambridge Archaeological Journal* 15, 93–109

Ortloff, C., Feldman R. A. and Mosely M. E. 1985. Hydraulic engineering and historical aspects of the pre-Columbian intravalley canal system of the Moche Valley, Peru. *Journal of Field Archaeology* 12, 77–98

Pearce, F. 2006. *When Rivers Run Dry: What Happens When Our Water Runs Out?* London: Eden Project Books

Peltenberg, E., Colledge, S., Croft, P., Jackson, A., McCartney, C. and Murray, M. A. 2000. Agro-pastoralist colonization of Cyprus in the 10th millennium BP: initial assessments. *Antiquity* 74, 844–53

Peterson, L. and Haug, G. 2005. Climate and the collapse of the Maya civilization. *American Scientist* 93, 322–9

Philip, G. 2008. The Early Bronze Age I–III. In *Jordan: A Reader* (ed. R. Adams), pp. 161–226. London: Equinox Publishing

Piranomonte, M. 1998. *The Baths of Caracalla*. Rome: Soprintendenza Archeologica di Roma

Plusquellec, H. 1990. *The Gezira Irrigation Scheme in Sudan: Objectives, Design and Performance*. World Bank Technical Paper No. 120, p. 106. Washington: IBRD/World Bank

Pollock, S. 1999. *Ancient Mesopotamia: The Eden That Never Was*. Cambridge: Cambridge University Press

Politis, K. D. (ed.) 2007. *The World of the Nabataeans*. Stuttgart: Franz Steiner Verlag

Poole, C. and Briggs, E. 2005. *Tonle Sap: Heart of Cambodia's Natural Heritage*. Bangkok: River Books

Por, D. F. 2004. The Levantine waterway, riparian archaeology, paleolimnology, and conversation. In *Human Paleoecology in the Levantine Corridor* (eds. I. Goren-Inbar and J. Speth), pp. 5–20. Oxford: Oxbow Books

Postgate, J. N. 1992. *Early Mesopotamia: Society and Economy at the Dawn of History.* London and New York: Routledge

Powell, M. A. 1985. Salt, seed and yields in Sumerian agriculture: a critique of the theory of progressive salinization. *Zeitschrift der Assyrologie* 75, 7–38

Rambeau, C., Finlayson, B., Smith, S., Black, S., Inglis, R. and Robinson, S. 2011. Palaeoenvironmental reconstruction at Beidha, southern Jordan (*c.* 18,000-8500 BP): implications for human occupation during the Natufian and Pre-Pottery Neolithic. In *Water, Life and Civilisation: Climate, Environment and Society in the Jordan Valley* (eds. S. J. Mithen and E. Black), pp. 245–68. Cambridge: Cambridge University Press/UNESCO

Rast, W. and Schaub, R. 1974. Survey of the southeastern plain of the Dead Sea. *Annual of the Department of Antiquities of Jordan* 19, 5–53

—. 2003. *Bab edh-Dhra': Excavations at the Town Site (1975–1981).* Winona Lake: Eisenbrauns

Ravesloot, J. C. 2007. Changing views of Snaketown in a larger landscape. In *The Hohokam Millennium* (eds. S. K. Fish and P. R. Fish), pp. 91–7. Santa Fe: School for Advanced Research Press

Rihil, T. E. 1999. *Greek Science.* Oxford: Oxford University Press

Robinson, S., Black, S., Sellwood, B. and Valdes, P. J. 2011. A review of palaeoclimates and palaeoenvironments in the Levant and Eastern Mediterranean from 25,000 to 5000 years BP: setting the environmental background for the evolution of human civilisation. In *Water, Life and Civilisation: Climate, Environment and Society in the Jordan Valley* (eds. S. J. Mithen and E. Black), pp. 71–93. Cambridge: Cambridge University Press/UNESCO

Roddick, A. 2004. *Troubled Water: Saints, Sinners, Truths and Lies about the Global Water Crisis.* Chichester: Anita Roddick Books

Rogge, A. E., Phillips, B. G. and Droz, M. S. 2002. Two Hohokam canals at Sky Harbor International airport. *Pueblo Grande Museum Anthropological Papers* No. 12

Rollefson, G. O. 1989. The aceramic Neolithic of the southern Levant: The view from 'Ain Ghazal. *Paléorient* 9, 29–38

Rollefson, G. O. and Kohler-Rollefson, I. 1989. The collapse of early Neolithic settlements in the southern Levant. In *People and Culture Change: Proceedings of the Second Symposium on Upper Palaeolithic, Mesolithic and Neolithic Populations of Europe and Mediterranean Basin* (ed. I. Hershkowitz), pp. 59–72. Oxford BAR Reports, International Series 508

Salazar, F. and Salazar, E. 2004. *Cusco and the Sacred Valley of the Incas.* Alkamari EIRL Cuesta Santa Ana 528, Cusco

Sandweiss, D. H., McInnis, H., Burger, R., Cano, A., Ojeda, B., Paredes, R., del Carman Sandweiss, M. and Glascock, M. D. 1998. Quebrada Jaguay: early South American maritime adaptations. *Science*, 281, 1830–32

Sandweiss, D. H., Solís, R. S., Mosely, M. E., Keefer, D. K. and Ortloff, C. 2009. Environmental change and economic development in coastal Peru between 5,800 and 3,600 years ago. *Proceedings of the National Academy of Sciences* www.pnas.org/cgi/doi/10.1073/pnas.0812645106

Scarborough, V. 1998. Ecology and ritual: Water management and the Maya. *Latin American Antiquity* 9, 135-50,

—. 2003. *The Flow of Power: Ancient Water Systems and Landscapes*. Santa Fe: SAR Press

Scarborough, V. L. and Gallopin, G. G. 1991. A water storage adaptation in the Maya lowlands. *Science* 251, 658-62

Scarborough, V. L. and Lucero, L. J. 2010. The non-hierarchical development of complexity in the semitropics: water and cooperation. *Water History* 2, 185-205

Schmidt, K. 2002. The 2002 excavations at Göbekli Tepe (Southeastern Turkey) - impressions from an enigmatic site, *Neo-lithics* 2/02: 8-13

Schreiber, K. J. and Rojas, J. L. 1995. The Puquios of Nasca. *Latin American Antiquity* 6, 229-54

Shea, J. J. 1999. Artefact abrasion, fluvial processes and 'living floors' from the Early Palaeolithic site of 'Ubeidiya (Jordan Valley, Israel). *Geoarchaeology* 14, 191-207

Shimada, I., Schaaf, C. B., Thompson, L. G. and Mosley-Thompson, E. 1991. Cultural impacts of severe droughts in the prehistoric Andes: application of a 1,500-year ice core precipitation record. *World Archaeology* 22, 247-70

Silverman, H. and Proulx, D. H. 2002. *The Nasca*. Malden: Blackwell

Simmons, A. and Najjar, M. 1996. Current investigations at Ghwair I, a Neolithic settlement in southern Jordan. *Neolithics* 2, 6-7

—. 1998. Al-Ghuwar I, a pre-pottery Neolithic village in Wadi Faynan, southern Jordan: a preliminary report of the 1996 and 1997/98 seasons. *Annual of the Department of Antiquities of Jordan* 42, 91-101

Smith, S., Wade, A., Black, E., Brayshaw, D., Rambeau, C. and Mithen, S. 2011. From global climate change to local impact in Wadi Faynan. In *Water, Life and Civilisation*, (eds. S. Mithen and E. Black), pp. 218-44. Cambridge: Cambridge University Press/UNESCO

Solís, R. S., Haas, J. and Creamer, W. 2001. Dating Caral, a preceramic site in the Supe Valley on the central coast of Peru. *Science* 292, 723-6

Solomon, S. 2010. *Water: The Epic Struggle for Wealth, Power and Civilization*. New York: HarperCollins

Stekelis, M. 1966. *Archaeological Excavations at 'Ubeidiya 1960-1963*. Jerusalem: Israel Academy of Sciences and Humanities

Stordeur, D., Helmer, D. and Wilcox, G. 1997. Jerf el Ahmar, un nouveau site de l'horizon PPNA sur le moyen Euphrate Syrien. *Bulletin de la Société Préhistorique Française* 94, 282-5

Stott, P. 1992. Angkor: shifting the hydraulic paradigm. In *The Gift of Water* (ed. J. Rigg), pp. 47–58. London: School of Oriental and African Studies

Stringer, C. 2011. *The Origin of Our Species*. London: Allen Lane

Taggart, C. 2010. Moche human sacrifice: the role of funerary and warrior sacrifice in Moche ritual organization. *Totem: The University of Western Ontario Journal of Anthropology* Volume 8, Article 10. http://ir.lib.uwo.ca/totem/vol18/iss1/10

Tamburrino, A. 2010. Water technology in Ancient Mesopotamia. In *Ancient Water Technologies* (ed. L. Mays), pp. 29–51. Dordrecht: Springer

Taylour, W. W. 1983. *The Mycenaeans*. London: Thames & Hudson

Thompson, L. G., Davis, M. E., Mosley-Thompson, E., Sowers, T. A., Henderson, K. A., Zagorodnov, V. S., Lim, P.-N., Mikhalenko, V. N., Campen, R. K., Bolzan, J. F., Cole-Dai, J. and Francou, B. 1998. A 25,000-year tropical climate history from Bolivian ice cores. *Science* 282, 1858–64.

Thompson, L. G., Mosley-Thompson, E., Bolzan, J. F. and Koci, B. R. 1985. A 1500-year record of tropical precipitation in ice cores from the Quelccaya ice cap, Peru. *Science* 229, 971–3

Thompson, L. G., Mosley-Thompson, E., Dansgaard, W. and Grootes, P. M. 1986. The little ice age as recorded in the stratigraphy of the tropical Quelccaya ice cap. *Science* 234, 361–4

Tiruneh, N. D. and Motz, L. J. 2004. Climate change, sea level rise, and saltwater intrusion. ASCE Conference, Bridging the Gap: Meeting the World's Water and Environmental Resources Challenges. *Proceedings of World Water and Environmental Resources Congress 2001*. Proc. doi:10.1061/40569(2001)315

Trigger, B. 2003. *Understanding Early Civilizations*. Cambridge: Cambridge University Press

Tuplin, C. S. and Rihil, T. E. (eds.) 2002. *Science and Mathematics in Ancient Greek Culture*. Oxford: Oxford University Press

Turner, B. L., Kamenov, G. D., Kingston, J. D. and Armelagos, G. J. 2009. Insights into immigration and social class at Machu Picchu, Peru based on oxygen, strontium, and lead isotope analysis. *Journal of Archaeological Science* 36, 317–32

UN Water 2007. *Coping with Water Scarcity: Challenges of the Twenty First Century. Prepared for World Water Day 2007.* http://wwww.org/wwd07/download/documents/escarcity.pdf, Retrieved December 2008

UNEP 2007. *GE04 Global Environment Outlook (Environment for Development)*. New York: United Nations Environment Programme

US Geological Survey (2003). Fact Sheet 103-03: Ground water depletion across the Nation

Valla, F. R. et al. 1999. Le Natoufien final and les nouvelles fouilles à Mallaha (Eynan), Israël 1966–1977. *Journal of the Israel Prehistoric Society* 28, 105–76

van der Kooij, G. 2007. Irrigation systems at Dayr Alla. In *Studies in the History and Archaeology of Jordan*, Vol. IX (ed. F. Al-Khraysheh), pp 133–44. Amman: Department of Antiquities

Van Liere, W. J. 1980. Traditional water management in the lower Mekong Basin. *World Archaeology* 11, 265–80

Verano, J. W. 2000. Paleonthological analysis of sacrificial victims at the Pyramid of the Moon, Moche River valley, northern Peru. Chungará (Arica) v.32 n.1 Arica ene. 2000 doi: 10.4067/S0717-73562000000100011

Veyne, P. 1997. *The Roman Empire*. Cambridge, Mass.: Belknap Press of Harvard University Press

Wade, A., Holmes, El-Bastawesy, P., Smith, S., Black, E. and Mithen, S. J. 2011. The hydrology of Wadi Faynan. In *Water, Life and Civilisation: 20,000 Years of Climate, Environment and Society in the Jordan Valley*, (eds. S. J. Mithen and E. Black), pp. 157–74. Cambridge: Cambridge University Press/UNESCO

Wallace, H. D. 2007. Hohokam beginnings. In *The Hohokam Millennium* (eds. S. K. Fish and P. R. Fish), pp. 13–21. Santa Fe: School for Advanced Research Press

Warren, P. 1989. *The Aegean Civilizations: From Ancient Crete to Mycenae*. Oxford: Phaidon

Watkins, T. 2010. New light on Neolithic revolution in south-west Asia. *Antiquity* 84, 621–34

Watts, J. 2010. *When a Billion Chinese Jump*. London: Faber & Faber

Webster, D. L. 2002. *The Fall of the Ancient Maya*. London: Thames & Hudson

Webster, J. W., Brook, G. A., Railsback, L. B., Cheng Hai, Edwards, R. L., Alexander, C. and Roeder, P. P. 2007. Stalagmite evidence from Belize indicating significant droughts at the time of the Preclassic abandonment, the Maya hiatus and the Classic Maya collapse. *Palaeogeography, Palaeoclimatology and Palaeoecology* 280, 1–17

Whitehead, P. G., Smith, S. J., Wade, A. J., Mithen S. J., Finlayson, B. L., and Sellwood, B. 2008. Modelling of hydrology and potential population levels at Bronze Age Jawa, northern Jordan: a Monte Carlo approach to cope with uncertainty. *Journal of Archaeological Science* 35, 517–29

Whittlesey, S. M. 2007. Hohokam ceramics, Hohokam beliefs. In *The Hohokam Millennium* (eds. S. K. Fish and P. R. Fish), pp. 65–74. Santa Fe: School for Advanced Research Press

Wilcox, D. E. 1991. Hohokam social complexity. In *Chaco and Hohokam: Prehistoric Regional Systems in the American Southwest* (eds. P. L. Crown and W. J. Judge), pp. 253–75. Santa Fe: School of American Research Press

Wilkinson, T. J. 2000. Regional approaches to Mesopotamian archaeology: the contribution of archaeological surveys. *Journal of Archaeological Research* 8, 219–67

Wilkinson, T. J. and Rayne, L. 2010. Hydraulic landscapes and imperial power in the Near East. *Water History* 2, 115–44

Wittfogel, K. 1957. *Oriental Despotism: A Comparative Study of Total Power.* New Haven: Yale University Press

Wright, K. R. 2006. *Tipón: Water Engineering Masterpieces of the Inca Empire.* Reston: American Society of Civil Engineers

—. 2008. A true test of sustainability: the triumph of Inca civil engineering over scarce water resources survives today. *Water Environment and Technology* 20, 79–87

Wright, K. R., Kelly, J. M. and Valencia Zegarra, A. 1997. Machu Picchu: ancient hydraulic engineering. *Journal of Hydraulic Engineering* 123, 838–43

Wright, K. R., Valencia Zegarra, A. and Lorah, W. L. 1999. Ancient Machu Picchu drainage engineering. *Journal of Irrigation and Drainage Engineering* 125. Accessed at http://www.waterhistory.org/histories/machupicchu/ 29 December 2011

Wright, R. M. and Valencia Zegarra, A. 2001. *The Machu Picchu Guidebook.* Boulder: Johnson Books

Zangger, E. 1994. Landscape changes around Tiryns. *American Journal of Archaeology* 98, 189–212

Zarkadoulas, N., Koutsoyiannis, D., Mamassis, N. and Angelakis, A. N. 2012. A brief history of urban water management in ancient Greece. In *Evolution of Water Supply through the Millennia* (eds. A. N. Angelakis, L. W. Mays, D. Koutsoyiannis and N. Mamassis), pp. 259–70. London: IWA Publishing

Zhou, Daguan. 1993. *The Customs of Cambodia.* Translated from the French by J. Gilman d'Arcy Paul (3rd edn). Bangkok: The Siam Society

INDEX